Many-Body Physics with Ultracold Gases

École de Physique des Houches
Session XCIV, 28 June to 23 July 2010

Many-Body Physics with Ultracold Gases

Edited by

Christophe Salomon, Georgy V. Shlyapnikov, and Leticia F. Cugliandolo

OXFORD
UNIVERSITY PRESS

OXFORD
UNIVERSITY PRESS

Great Clarendon Street, Oxford, OX2 6DP,
United Kingdom

Oxford University Press is a department of the University of Oxford.
It furthers the University's objective of excellence in research, scholarship,
and education by publishing worldwide. Oxford is a registered trade mark of
Oxford University Press in the UK and in certain other countries

First Edition published in 2013

Published in the United States of America by Oxford University Press
198 Madison Avenue, New York, NY 10016, United States of America

British Library Cataloguing in Publication Data
Data available

ISBN 978-0-19-966188-6

École de Physique des Houches
Service inter-universitaire commun à l'Université Joseph Fourier de Grenoble et à l'Institut National Polytechnique de Grenoble

Subventionné par l'Université Joseph Fourier de Grenoble,
le Centre National de la Recherche Scientifique,
le Commissariat à l'Énergie Atomique

Directeur:
Leticia F. Cugliandolo, Université Pierre at Marie Curie – Paris VI, France

Directeurs scientifiques de la session XCIV:
Christophe Salomon, Laboratoire Kastler Brossel, École Normale Supérieure, Paris, France
Georgy V. Shlyapnikov, LPTMS, CNRS, Université Paris-Sud XI, France
Leticia F. Cugliandolo, LPTHE, Université Paris VI, Paris, France

Previous sessions

I	1951	Quantum mechanics. Quantum field theory
II	1952	Quantum mechanics. Statistical mechanics. Nuclear physics
III	1953	Quantum mechanics. Solid state physics. Statistical mechanics. Elementary particle physics
IV	1954	Quantum mechanics. Collision theory. Nucleon–nucleon interaction. Quantum electrodynamics
V	1955	Quantum mechanics. Non-equilibrium phenomena. Nuclear reactions. Interaction of a nucleus with atomic and molecular fields
VI	1956	Quantum perturbation theory. Low temperature physics. Quantum theory of solids. Ferromagnetism
VII	1957	Scattering theory. Recent developments in field theory. Nuclear and strong interactions. Experiments in high energy physics
VIII	1958	The many body problem
IX	1959	The theory of neutral and ionized gases
X	1960	Elementary particles and dispersion relations
XI	1961	Low temperature physics
XII	1962	Geophysics; the Earth's environment
XIII	1963	Relativity groups and topology
XIV	1964	Quantum optics and electronics
XV	1965	High energy physics
XVI	1966	High energy astrophysics
XVII	1967	Many body physics
XVIII	1968	Nuclear physics
XIX	1969	Physical problems in biological systems
XX	1970	Statistical mechanics and quantum field theory
XXI	1971	Particle physics
XXII	1972	Plasma physics
XXIII	1972	Black holes
XXIV	1973	Fluid dynamics
XXV	1973	Molecular fluids
XXVI	1974	Atomic and molecular physics and the interstellar matter
XXVII	1975	Frontiers in laser spectroscopy
XXVIII	1975	Methods in field theory
XXIX	1976	Weak and electromagnetic interactions at high energy
XXX	1977	Nuclear physics with heavy ions and mesons
XXXI	1978	Ill condensed matter
XXXII	1979	Membranes and intercellular communication

XXXIII	1979	Physical cosmology
XXXIV	1980	Laser plasma interaction
XXXV	1980	Physics of defects
XXXVI	1981	Chaotic behavior of deterministic systems
XXXVII	1981	Gauge theories in high energy physics
XXXVIII	1982	New trends in atomic physics
XXXIX	1982	Recent advances in field theory and statistical mechanics
XL	1983	Relativity, groups and topology
XLI	1983	Birth and infancy of stars
XLII	1984	Cellular and molecular aspects of developmental biology
XLIII	1984	Critical phenomena, random systems, gauge theories
XLIV	1985	Architecture of fundamental interactions at short distances
XLV	1985	Signal processing
XLVI	1986	Chance and matter
XLVII	1986	Astrophysical fluid dynamics
XLVIII	1988	Liquids at interfaces
XLIX	1988	Fields, strings and critical phenomena
L	1988	Oceanographic and geophysical tomography
LI	1989	Liquids, freezing and glass transition
LII	1989	Chaos and quantum physics
LIII	1990	Fundamental systems in quantum optics
LIV	1990	Supernovae
LV	1991	Particles in the nineties
LVI	1991	Strongly interacting fermions and high T_c superconductivity
LVII	1992	Gravitation and quantizations
LVIII	1992	Progress in picture processing
LIX	1993	Computational fluid dynamics
LX	1993	Cosmology and large scale structure
LXI	1994	Mesoscopic quantum physics
LXII	1994	Fluctuating geometries in statistical mechanics and quantum field theory
LXIII	1995	Quantum fluctuations
LXIV	1995	Quantum symmetries
LXV	1996	From cell to brain
LXVI	1996	Trends in nuclear physics, 100 years later
LXVII	1997	Modeling the Earth's climate and its variability
LXVIII	1997	Probing the Standard Model of particle interactions
LXIX	1998	Topological aspects of low dimensional systems
LXX	1998	Infrared space astronomy, today and tomorrow
LXXI	1999	The primordial Universe
LXXII	1999	Coherent atomic matter waves
LXXIII	2000	Atomic clusters and nanoparticles
LXXIV	2000	New trends in turbulence
LXXV	2001	Physics of biomolecules and cells
LXXVI	2001	Unity from duality: gravity, gauge theory and strings

LXXVII	2002	Slow relaxations and nonequilibrium dynamics in condensed matter
LXXVIII	2002	Accretion discs, jets and high energy phenomena in astrophysics
LXXIX	2003	Quantum entanglement and information processing
LXXX	2003	Methods and models in neurophysics
LXXXI	2004	Nanophysics: coherence and transport
LXXXII	2004	Multiple aspects of DNA and RNA
LXXXIII	2005	Mathematical statistical physics
LXXXIV	2005	Particle physics beyond the Standard Model
LXXXV	2006	Complex systems
LXXXVI	2006	Particle physics and cosmology: the fabric of spacetime
LXXXVII	2007	String theory and the real world: from particle physics to astrophysics
LXXXVIII	2007	Dynamos
LXXXIX	2008	Exact methods in low-dimensional statistical physics and quantum computing
XC	2008	Long-range interacting systems
XCI	2009	Ultracold gases and quantum information
XCII	2009	New trends in the physics and mechanics of biological systems
XCIII	2009	Modern perspectives in lattice QCD: quantum field theory and high performance computing
XCIV	2010	Many-body physics with ultracold gases

Publishers

- Session VIII: Dunod, Wiley, Methuen
- Sessions IX and X: Herman, Wiley
- Session XI: Gordon and Breach, Presses Universitaires
- Sessions XII–XXV: Gordon and Breach
- Sessions XXVI–LXVIII: North Holland
- Session LXIX–LXXVIII: EDP Sciences, Springer
- Session LXXIX–LXXXVIII: Elsevier
- Session LXXXIX– : Oxford University Press

Preface

The summer school 'Many-Body Physics with Ultracold Gases', session XCIV of the Les Houches School of Physics, was held from 28/06/2010 to 23/07/2010 in Les Houches, France. A total of 54 students from 23 countries and 33 professors or seminar speakers from 9 countries attended the school. As usual, the school had a great scientific atmosphere, aided by the splendid weather and spectacular mountain scenery!

For more than fifteen years after the discovery of Bose–Einstein condensation (BEC) in dilute atomic gases, the field of ultracold quantum matter underwent a dramatic growth accompanied by striking developments. Because of the extremely low temperatures (in the nanokelvin range) at which these dilute gases are prepared, with densities eight orders of magnitude lower than liquid helium, they manifest collective quantum behaviour. As of now, the field of quantum gases is highly interdisciplinary, bringing together researchers from atomic physics and quantum optics, condensed matter physics, nuclear and high-energy physics, and non-linear physics. This is due to an extraordinary level and wide spectrum of scientific activity and a high potential for future developments. Emblematic examples are: creation of two-dimensional (2D) and one-dimensional (1D) quantum gases by tightly confining the motion of particles in one or two directions; observation of the Mott insulator transition for bosonic and fermionic atoms in optical lattices; the creation of spinor Bose–Einstein condensates (BECs), representing a gaseous analogue of condensed matter magnetic systems; achievement of the strongly interacting regime for fermions, which brings in analogies with neutron matter; creation of ultracold ground-state polar molecules; and observation of Anderson localization of matter waves. Applications of quantum gases include atom lasers—devices for the generation of coherent matter waves, atom interferometry, high-precision atomic clocks, quantum information, and quantum computing.

The lectures covered practically all domains of the physics of quantum gases. Three Nobel Prize laureates in Physics, C. Cohen-Tannoudji, W.D. Phillips, and W. Ketterle, taught at the school. Main courses, colloquia, and seminars illustrated the rich physics of experimental and theoretical many-body physics. This volume gathers most of the lectures given at the school.

The physics of correlated atoms in optical lattices is covered from both theoretical and experimental sides. The lectures of A. Georges and T. Giamarchi introduce the theory for bosonic and fermionic atoms, which in particular includes the Bose–Hubbard and Fermi–Hubbard models and the description of the Mott transition. Special attention is given to one-dimensional lattice bosons and fermions.

The lectures of I. Bloch provide a basic introduction to experimental aspects of atoms in optical lattices and discuss the effect of strong interactions in the Hubbard model. This is followed by a presentation of detection methods, which allow one to obtain the density and (quasi-)momentum distributions and correlations between particles in trapped lattice systems. The lectures then introduce the super-exchange spin–spin interactions and highlight the current status of the observations of magnetic phenomena. The lectures are completed by the description of novel methods providing single-site and single-atom resolution for atoms in optical lattices.

The lectures of D. Petrov are related to few-body physics with cold atoms and they show how one can obtain exact solutions for three-body and four-body problems. In particular, these lectures address the interaction between weakly bound bosonic molecules formed at the repulsive side of the Feshbach resonance in the studies of two-species Fermi gases and the remarkable collisional stability of these molecules, issues that were at the core of the studies of molecular BEC regimes in Fermi gases.

The lectures of I. Cirac cover the subject of entanglement in quantum many-body systems, which is the key issue for quantum information processing. The introductory part of the lectures highlights the entropy of entanglement and the quantum mutual information. It is then discussed how certain entanglement measures display special features under phase transitions in many-body systems and spin lattices with short-range interactions are considered. This is followed by the description of area laws for those systems. The lectures are completed by introducing an efficient way of describing many-body quantum states, which is based on projected entangled-pair states (PEPS), and present examples in 1D where PEPS reduce to the so-called matrix product states playing a central role in the renormalization procedures and algorithms.

The lectures of N. Cooper deal with rapidly rotating quantum gases: one of their goals is to establish a connection with the fractional quantum Hall effect for electrons in semiconductors. Special attention is paid to optically induced gauge fields, which mimic rotation and may provide a breakthrough in experiments on the observation of strongly correlated states with neutral atoms.

The lectures of L. Santos and the lectures of D. Jin and J. Ye belong to the domain of dipolar quantum gases, which include atoms with large magnetic moments (such as chromium and dysprosium) and electrically aligned polar molecules. The lectures of L. Santos emphasize how the long-range and anisotropic dipole–dipole interaction makes quantum degenerate regimes different from those of short-range interacting atoms. They give an overview of the studies of dipolar gases from the beginning of this domain up to now.

The lectures of D. Jin and J. Ye focus on experiments on creating polar molecules in the ground ro-vibrational state. Being aligned by an electric field, these molecules have a large dipole moment, up to a few debyes, and the properties of their ultracold clouds are dominated by a strong dipole–dipole interaction. The lectures also discuss difficulties in cooling these molecules to the quantum degenerate regime, which are to a large extent related to ultracold chemical reactions for the case of KRb molecules.

The lecture of F. Chevy covers the physics of ultracold Fermi gases, with an emphasis on the problem of superfluidity in spin $\frac{1}{2}$ fermionic systems. They show how experiments with ultracold fermions give invaluable insights into this problem and emphasize that ultracold Fermi gases may serve as 'quantum simulators' of complex condensed matter systems through measurements of the equation of state.

The lecture of D. Pekker and E. Demler focuses on the discussion of the search for ferromagnetism in ultracold Fermi gases and presents arguments to explain why it has not yet been found with a long lifetime, in particular in experiments of W. Ketterle at MIT.

The lecture of A. Aspect introduces a new domain of quantum gases, ultracold atoms in disordered potentials. It presents the issue of Anderson localization from transparent physical grounds as a destructive interference phenomenon and explains how the localization has been obtained in dilute quasi-one-dimensional clouds of bosonic atoms in random laser speckle potentials.

We close this preface by thanking the institutions that provided financial support for the school: the INTERCAN European network, the IFRAF institute, CNRS, the Les Houches theoretical physics school, and the Joseph Fourier University. Finally, special thanks goes to the administrative staff of the Les Houches Center who provided such excellent support to the participants and organizers of the school. We warmly thank Brigitte Rousset, Murielle Gardette, and Isabel Lelièvre for their professional expertise and countless help.

August 2012,

C. Salomon
G. Shlyapnikov
L.F. Cugliandolo

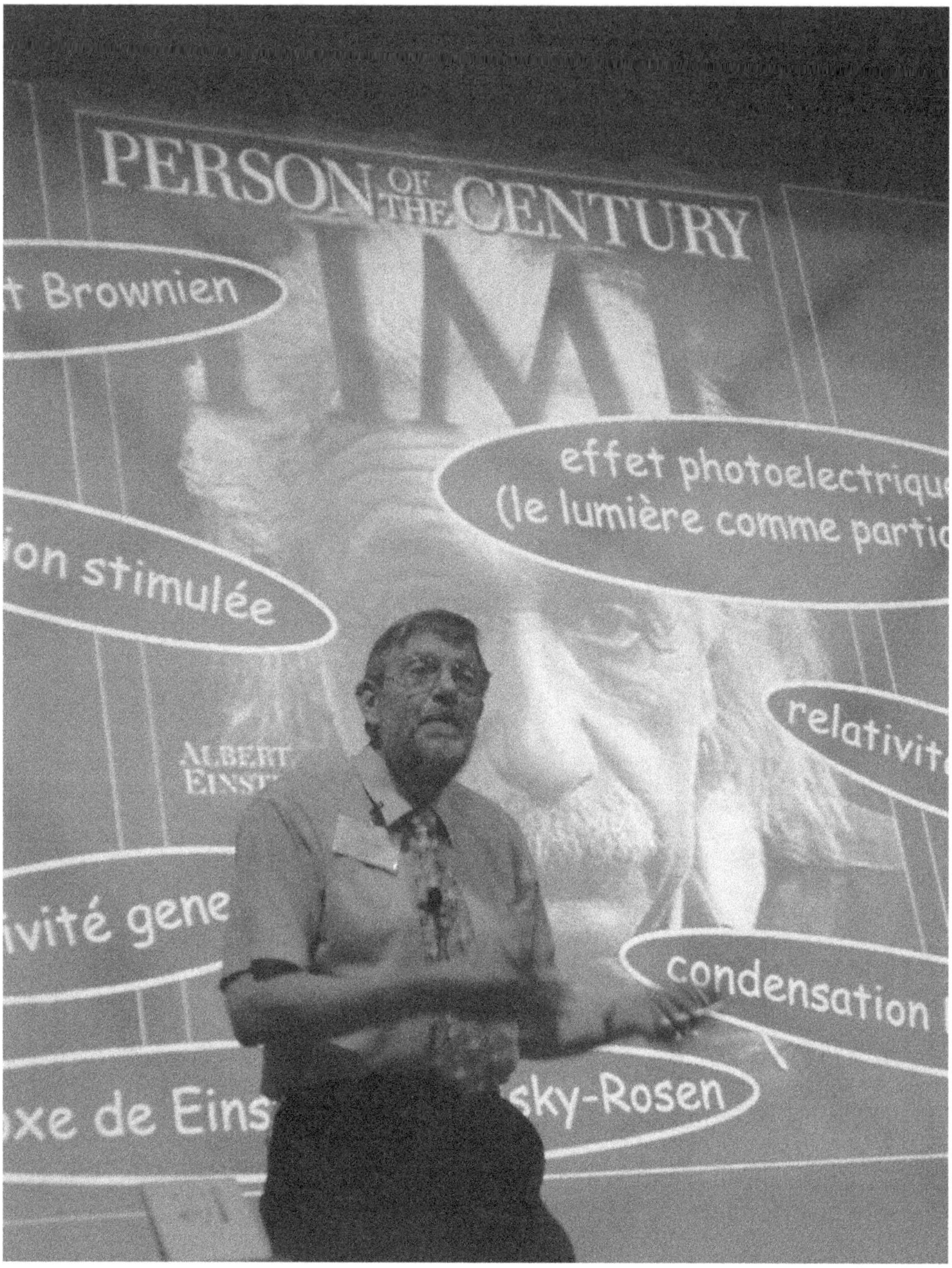

W.D. Phillips giving his public lecture at Les Houches village.

The students and teachers of the school.

Wolfgang Ketterle with mountain guide Michel Rousset on Mont Blanc summit, and with Jook Walraven at the school.

Top: The lecture room. Bottom: sunny sunset at the school. From left to right: Christophe Salomon, Antoine Georges, Immanuel Bloch, Thierry Giamarchi, Georgy Shlyapnikov and Boris Altshuler.

Contents

List of participants

Organizers

SALOMON Christophe
Laboratoire Kastler Brossel, École Normale Supérieure, 24 rue Lhomond, 75231 Paris, France

SHLYAPNIKOV Georgy
LPTMS, CNRS, Université Paris-Sud XI, Bât. 100, 91405 Orsay cedex, France

Lecturers

ALTSHULER Boris
Columbia University, Department of Physics, 824 Pupin Hall, 538 W 120 St., New York, NY 10027, USA

ASPECT Alain
Institut d'Optique, 2 Avenue Augustin Fresnel, 91127 Palaiseau, France

BLOCH Immanuel
Max-Planck-Institut für Quantenoptik, Hans-Kopferman-Strasse 1, D-85748 Garching, Germany

CAUX Jean-Sébastien
Institute for Theoretical Physics, University of Amsterdam, Science Park 904, Postbus 94485, 1090 GL, Amsterdam, NL

CHEVY Frédéric
Laboratoire Kastler Brossel, Département de Physique de l'ENS, 24 rue Lhomond, 75231 Paris cedex 05, France

CIRAC Ignacio
Max-Planck-Institut für Quantenoptik, Hans-Kopferman-Strasse 1, D-85748 Garching, Germany

COHEN-TANNOUDJI Claude
Laboratoire Kastler Brossel, Département de Physique de l'ENS, 24 rue Lhomond, 75231 Paris Cedex 05, France

COOPER Nigel R.
University of Cambridge, Cavendish Laboratory, JJ Thomson Avenue, Cambridge CB3 0HE, UK

DALIBARD Jean
Laboratoire Kastler Brossel, Département de Physique de l'ENS, 24 rue Lhomond, 75005 Paris, France

DEMLER Eugene
Harvard University, Department of Physics, Lyman 231, 15 Oxford St., Cambridge, MA 02138, USA

ESSLINGER Tilman
ETH Zurich, Institute for Quantum Electronics, HPF D4, 8093 Zurich, Switzerland

FALLANI Leornardo
LENS and Physics Department, Universitegli Studi di Firenze, Via Nello Carrara 1 50019 Sesto Fiorentino, Italy

GEORGES Antoine
Collège de France et École polytechnique, Centre de physique théorique, Ecole Polytechnique, 91128 Palaiseau cedex, France

GIAMARCHI Thierry
DPMC-MaNEP, University of Geneva, 24 Quai Ernest-Ansermet, 1211 Geneva 4, Switzerland

GREINER Markus
Harvard University, Department of Physics, Jefferson 353, 17 Oxford Street, Cambridge, MA 02138, USA

GRIMM Rudolf
University of Innsbruck, Institute for Experimental Physics, Technikerstrasse 25, 6020 Innsbruck, Austria

HART Russell
University of Innsbruck, Institute for Experimental Physics, Technikerstrasse 25, 6020 Innsbruck, Austria

HO Jason Tin-Lun
Physics Department, the Ohio State University, 191 West Woodruff Avenue, Columbus, Ohio 43210, USA

JIN Deborah S.
JILA, University of Colorado, 440 UCB, Boulder, CO 80309-0440, USA

KETTERLE Wolfgang
Massachusetts Institute of Technology, Room 26-243, 77 Massachusetts Avenue, Cambridge, MA 02139, USA

KOEHL Michael
University of Cambridge, Department of Physics, Room 834, Rutherford Building, Cavendish Laboratory, JJ Thomson Avenue, Cambridge CB3 0HE, UK

LEWENSTEIN Maciej
ICFO-Institut for Photomic Sciences, Parc Mediterrani de la Tecnologia s/n 08860 Castelldefels Barcelona, Spain

LUKIN Mikhail D.
Harvard University, Department of Physics, Lyman 231, 15 Oxford Street, Cambridge, MA 02138, USA

OBERTHALER Markus K.
University of Heidelberg, Kirchhoff Institut für Physik, Im Neuenheimer Feld 227, 69120 Heidelberg, Germany

PETROV Dmitry S.
LPTMS, University Paris-Sud, Bât. 100, 91405 Orsay Cedex, France

PHILLIPS William D.
National Institute of Standards & Technology, Physics Laboratory, Mail Stop 8424, Gaithersburg, Maryland 20899, USA

PITAEVSKII Lev
INO-CNR BEC Center & dipartimento di fisica, Universita di Trento, Via Sommarive 14, 38123, Povo, Trento, Italy

PROKOF'EV Nikolay
Hasbrouck Laboratory, Physics Department, Massachusetts University, Amherst, MA 01003, USA

REICHEL Jakob
Laboratoire Kastler Brossel de l'ENS, 24 rue Lhomond 75231 Paris Cedex 05, France

SANTOS Luis
University of Hannover, Institut für Theorische Physik, Theoretical Quantum Optics Group, Appelstrasse 2, 30167 Hannover, Germany

STRINGARI Sandro
BEC Center, Dipartimento di fisica, Universita di Trento, via Sommarive 14, 38050 Povo, Italy

TSVELIK Alexei
CMPMS Department, Bldg 510A, Brookhaven National Laboratory, Upton, NY 11973, USA

WALRAVEN Jook T.M.
Van der Waals-Zeeman Instituut, Universiteit van Amsterdam, Valckenierstraat 65/67, 1018 XE Amsterdam, NL

Participants

BAARSMA Jildou
Institute for Theoretical Physics, Utrecht University, Leuvenlaan 4, 3584 CE Utrecht, NL

BAUR Stefan
Laboratory of Atomic and Solid State Physics, Cornell University, Ithaca, NY 14853, USA

BEATTIE Scott
AMOP Group, Cavendish Laboratories, University of Cambridge, JJ Thompson Avenue, Cambridge CB3 0HE, UK

BERTAINA Gianluca
Dipartimento di Fisica, Università di Trento, and INO-CNR, BEC Center, Via Sommarive 14, 38123 Povo, Trento, Italy

BISSBORT Ulf
Institut für Theoretische Physik, Johann Wolfgang Goethe Universität Frankfurt, Max-von-Laue Strasse 1, 60438 Frankfurt am Main, Germany

BISSET Russell
University of Otago, Department of Physics, Science III Building, 730 Cumberland Street, Dunedin 9016, New Zealand

BOUDJEMÂA Abd E-Aali
Laboratoire de physique théorique et de physique des matériaux, Université Hassiba Ben Bouali de Chlef, 02000 Chlef, Algeria

CAMPBELL Andrew
School of Physics & Astronomy, University of Birmingham, Edgebaston, Birmingham B15 2TT, UK

de CARVALHO RIBEIRO Sofia Isabel
Departamento de Fisica, Universidade do Minho, Campus de Gualtar, 4710-057 Braga, Portugal

EISMANN Ulrich
Laboratoire Kastler Brossel, Département de Physique de l'École Normale Supérieure, 24 rue Lhomond 75231, Paris, Cedex 05 France

ENDRES Manuel
Max-Planck Institute of Quantum Optics, Hans-Kopferman-Strasse 1, D-85748 Garching, Germany

FENG Chao
Department of Physics, University of Queensland, St Lucia, Brisbane, Queensland 4072, Australia

GARTMAN Rafal
Institute of Physics, Nicolaus Copernicus University, Grudziadzka 5, 87-100 Torun, Poland

GISCARD Pierre-Louis
University of Oxford, Clarendon Laboratory, Parks Road, Oxford OX1 3PU, UK

GROND Julian
Institut für Physik, Fachbereich Theoretische Physik, Universität Graz, Universittsplatz 5, 8010 Graz, Austria

HOINKA Sascha
Swinburne University, CAOUS, H38, 1 Alfred Street, Mailroom, Hawthorn VIC 3122, Australia

HUANG Bo
Institut Experimentalphysik, Technikerstrasse 25/4, 6020 Innsbruck, Austria

HUANG Wujie
MIT, Room 26-259, 77 Massachusetts Avenue, Cambridge, MA 02139, USA

JACQMIN Thibaut
LCFIO, 2 avenue Augustin Fesnel, Campus Polytechnique 91127 Palaiseau cedex, France

KOZIK Evgeny
ETH Zürich, Institute for Theoretical Physics, Wolfgang-Pauli Str. 27, HIT K 31.5, 8093 Zürich, Switzerland

KRAUSER Jasper
Institut für Laser-Physik Luruper Chaussee 149, 22761 Hamburg, Germany

KULKARNI Manas
Department of Physics & Astronomy, State University of New York at Stony Brook, Stony Brook, NY-11790, USA

LOBSER Daniel
JILA, University of Colorado, 440 UCB Boulder CO 80309, USA

MACIA Adrian
Universitat politècnica de Catalunya, Campus Nord C/Jordi Girona 1-3, Edifici B4-B5 Dep. 201, 08034 Barcelona, Spain

MAKOTYN Philip
JILA, University of Colorado, 440 UCB Boulder CO 80309, USA

MALATSETXEBARRIA Eneko
Materials Physics Center CSIC-UPV/EHU and Donostia International Physics Center (DIPC), Paseo Manuel de Lardizabal 5, E-20018 Donostia, Spain

MATVEEVA Natalia
BEC Center, Physics department of Trento University, Via Sommarive 14, I-38123 Povo, Italy

MAZZA Leonardo
Max-Planck-Institut für Quantenoptik Hans-Kopfermann-Strasse 1, 85748 Garching, Germany

MEYER Nadine
School of Physics & Astronomy, University of Birmingham, Edgbaston, Birmingham B15 2TT, UK

MUNOZ FREGOSO Benjamin
The University of Chicago, 929 East 57th St, GCIS E, Chicago, IL, 60637, USA

NAVON Nir
Laboratoire Kastler-Brossel, Ecole Normale Supérieure, 24 rue Lhomond 75005 Paris, France

NONNE Héloise
LPTM-Université de Cergy-Pontoise, Site de St Martin, Pontoise, 2 Avenue Adolphe Chauvin 95302 Cergy-Pontoise, France

OZAWA Tomoki
University of Illinois at Urbana-Champaign, 1110 W. Green Street, Urbana, IL 61801, USA

PAPOULAR David
Laboratoire de Physique Théorique & Modèles Statistiques, Université Paris-Sud, Bat. 100, 15 Rue G. Clémenceau Orsay, France

PASQUIOU Benjamin
Laboratoire de Physique des Lasers, Institut Gallilée, Université Paris 13 nord 99 avenue Jean-Baptiste Clément 93430 Villetaneuse, France

PIKOVSKI Alexander
Institut für theoretische physik, Leibniz Universität Hannover, Appelstr. 2, 30167 Hannover, Germany

PIRAUD Marie
Institut d'Optique, 2 avenue Augustin Fresnel, Campus Polytechnique, 91127 Palaiseau cedex, France

PUNK Matthias
Technische Universität München, Physik Department, c/o Daniela Neufang, James-Franck-Str. 85747 Garching, Germany

ROHRINGER Wolfgang
Institute of Atomic & Subatomic Physics, Stadionallee 2, Vienna University of Technology, 1020 Vienna, Austria

RONZHEIMER Jens Philipp
Fakultät für Physik, LMU München, Schellingstrasse 4, 80799 Müchen, Germany

SADEGZADEH Kayvan
Cavendish Laboratory, JJ Thomson Avenue, Cambridge CB3 0HE, UK

SCHMIDT Richard
Technische Universität München Physik Department I James-Franck-Str. 1, 85747 Garching, Germany

SCHUSTER Tobias
Kirchhoff Institut für Physik, Im Neuenheimer Feld 227 69120 Heidelberg, Germany

STRUCK Julian
Institut für Laser-Physik, Universität Hamburg, Luruper Chaussee 149, 22761 Hamburg, Germany

SUGAWA Seiji
Quantum Optics Group, Dept. of Physics, Graduate School of Science, Kyoto University, Kitashirakawa-Oiw akecho, Sakyoku, Kyoto-shi, Kyoto 606-8502, Japan

SWANSON Mason
The Ohio State University, Physics Department, 191 W. Woodruff Avenue Columbus OH 43210, USA

TIELEMAN Olivier
Institute for Theoretical Physics, Leuvenlaan 4, 3584 CE Utrecht, NL

TOMADIN ANDREA
IQOQI Technikerstr. 21A, A-6020 Innsbruck, Austria

TRYPOGEORGOS DIMITRIS
Clarendon Laboratory, Oxford University, Parks Road, Oxford OX1 3PU, UK

VIDANOVIC IVANA
Scientific Computing Laboratory Institute of Physics Pregrevica 118, Zemun 11000 Belgrade, Serbia

VO CHRISTOPH
Max-Planck-Institut für Quantenoptik Hans-Kopfermann-Strasse 1, 85748 Garching, Germany

WU CHENG-HSUN
Massachusetts Institute of Technology, 77 Massachussetts Avenue, Bldg 26-265, Cambridge, MA, 02139, USA

YUDIN DMITRY
Max-Planck-Institut für Quantenoptik Hans-Kopfermann-Strasse 1, 85748 Garching, Germany

ZHANG SHIZHONG
Department of Physics, Chong Yuet Ming Physics Building, The University of Hong Kong, Pokfulam Road, Hong Kong

1
Strongly correlated bosons and fermions in optical lattices

Antoine GEORGES

Centre de Physique Théorique, École Polytechnique,
CNRS, 91128 Palaiseau Cedex, France;
Collège de France, 11 place Marcelin Berthelot, 75005 Paris, France;
DPMC-MaNEP, University of Geneva, 24 Quai Ernest Ansermet,
1211 Geneva 4, Switzerland

Thierry GIAMARCHI

DPMC-MaNEP, University of Geneva, 24 Quai Ernest Ansermet,
1211 Geneva 4, Switzerland

Chapter Contents

1.1 Introduction

The effect of interactions on many-particle quantum systems has proven to be one of the most fascinating problems in physics. From the fundamental physics point of view, this is a formidable challenge that combines the difficulties of quantum mechanics and statistical physics. Indeed even in very small clumps of matter there are more particles than stars in the universe; when these particles interact one is thus totally unable to solve the coupled equations by brute force. This is even more so when one deals with quantum particles, which behave as interfering waves, and must in addition obey the principles of symmetrization and antisymmetrization. As a consequence of this complexity, beautiful new physics emerges from the collective behaviour of these particles, something that could not even be guessed at simply from looking at the solutions of small numbers of coupled particles. However, even finding the proper tools to tackle such a type of problem is a Herculean task. Fortunately, some important concepts allow us to understand the main physical properties of many of these systems. However, many systems defy our understanding and we need to build new tools to tackle them. This forces us to progress in our efforts to understand these systems theoretically, either analytically or numerically.

The pressure to solve these problems goes way beyond the academic realm. Understanding how electrons behave in solids led to technological revolutions such as silicon-based electronics and the transistor, the control of spin in magnetic storage and electronics, and the fascinating applications of superconductivity. Hence, this endeavour is intimately connected with our ability to engineer and control solids, and to make devices for use in our everyday life.

Recently, a new type of physical system, cold atoms in optical lattices, has provided a marvellous laboratory to tackle the effects of strong correlations in quantum systems. These systems, made of light and neutral atoms, constitute a welcome alternative to the standard realization in solid state physics. Because in these systems interactions are short-ranged and controllable, and because optical lattices can be engineered in a flexible way and phonon modes are absent, these systems can be viewed as model realizations. In addition, they have opened the path to novel branches of physics that use the control and flexibility of these systems (mixtures of bosons and fermions; the possibility of changing the potentials rapidly; isolated quantum systems; etc.) In particular, they have enabled the realization of quantum systems in reduced dimensionality, in which quite remarkable novel physics can occur.

In this chapter, we give an introduction to the physics of interacting quantum systems, both bosonic and fermionic. We review the main concepts and tools that form cornerstones of our understanding of such systems and point out the challenges that interactions pose. These lectures cannot, of course, make any claim of completeness given the broad scope of the problem, and we encourage the reader to search the literature for more information on the subject.

The plan of the chapter is as follows. In Section 1.2, we will give an introduction to the physics of quantum particles in periodic lattices. The presentation is essentially targeted at the case of cold atomic systems. We will examine how interactions should be taken into account and define the basic models, such as the Hubbard model, that can be used to describe such interacting systems. In Section 1.3, we examine, for the case of bosons, how the combined effects of lattice and interaction can turn the system into an insulator, the so-called Mott insulator, and discuss the corresponding physics. Section 1.4 discusses what happens when the system is one-dimensional. In that case the fact that two particles cannot cross without interacting leads to novel physics effects. This section discusses this new physics and the methods needed to access it. We then move in Section 1.5 to the case of fermions. We discuss first Fermi-liquid theory and the concept of *quasi-particles*, Landau's remarkable description of the low-energy excitations of interacting fermion systems. In a nutshell, this approach implies that the effect of interactions does not qualitatively change the nature of low-energy excitations as compared with a non-interacting system, except, of course, if the interactions are strong enough to lead to a instability of the system and, for example, destroy metallic behaviour. Fermi liquid theory and the concepts behind it have been the basis of our understanding of the properties of most solids. We will then see in Section 1.6 how, like the method for bosons, the combination of a lattice and strong interactions can turn a Fermi liquid into a Mott insulator. We look in Section 1.7 at the properties of one-dimensional fermions, show how Fermi liquid theory fails because low-energy excitations are now collective modes instead of quasi-particles, and examine the corresponding physics for both the conducting and insulating phases. Finally, in Section 1.8, we draw some conclusions and give some perspectives.

1.2 Optical lattices

Before dealing with the effects of interactions let us first look at the properties of individual quantum particles. One essential ingredient, in both solids and cold atom systems, is the presence of a periodic potential. In solids, such a potential occurs naturally for the electrons because of the presence of the regular array of positively charged nuclei. In cold atomic systems, it can be imposed by the presence of an optical lattice. Such a potential usually takes the form (in the direction of the lattice) (Bloch *et al.* 2008)

$$V(x) = V_0 \sin^2(kx), \tag{1.1}$$

where $k = \pi/a$, with a the lattice spacing. The presence of such a periodic potential considerably changes the properties of the system compared with one of free particles.

1.2.1 Zero kinetic energy ('atomic' limit)

Let us first analyse the effects of the periodic potential by considering a limit in which the periodic potential would be extremely large, and in particular much larger than the kinetic energy of the particles. In that case, as shown in Figure 1.1, it is a good approximation to consider that the particles remain mostly localized around one of the minima of the potential. Because in the condensed matter context this means that the particles stay essentially localized around each atom, this limit is called the

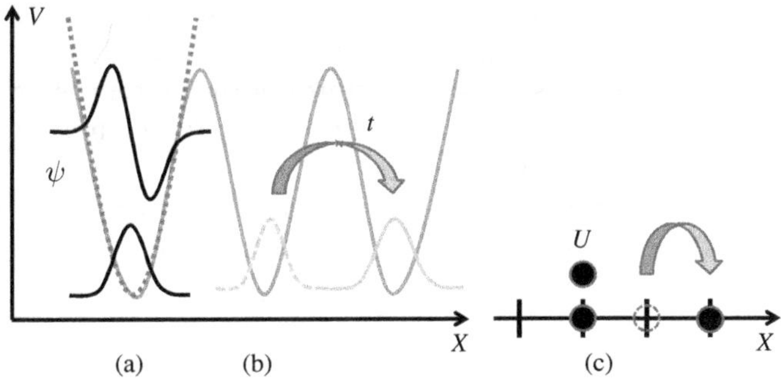

Figure 1.1 (a) If the periodic potential is extremely high compared with the kinetic energy, it is a good approximation to consider that the particles are essentially localized around the minima of the potential. In that case one has approximately the solutions of a harmonic oscillator. (b) Since the wavefunctions in different wells have a small overlap, there is a finite tunnelling amplitude t to go from one well to the next. (c) One can thus describe such a system as particles forced to be on a lattice, with a certain hopping amplitude t, which will delocalize them. In addition, if two particles are on the same site, they will feel the repulsion and pay an energy U.

'atomic limit', a somewhat confusing term in the cold atom context. Let us examine the case of the optical lattice potential, Eq. (1.1). If the particles stay around the minima we can expand the periodic potential. The Hamiltonian to be solved then becomes (for one minimum)

$$H = \frac{P^2}{2m} + V_0(kx)^2, \tag{1.2}$$

and is thus the Hamiltonian of a harmonic oscillator. As shown in Figure 1.1 around each minima, there is thus a full set of eigenstates,

$$\psi_{j,n}(r) = \psi_n(r - R_j), \tag{1.3}$$

which is centred around the jth site $R_j = aj$, and is the nth excited state, with energy

$$E_n = \omega_0 \left(n + \frac{1}{2} \right), \tag{1.4}$$

where

$$\omega_0^2 = \frac{2\, V_0 k^2}{m}. \tag{1.5}$$

Note that this frequency is associated with each well of the optical lattice in the limit of a deep lattice, and should not be confused with the frequency associated with the shallow parabolic trap usually present in those systems. If the barriers are extremely high, then the states centred around different sites j are essentially orthogonal and can thus serve as a complete basis of all the states of the system. A convenient way to represent the system is to use the second quantization representation (Mahan 2000). We introduce creation (and destruction) operators $b^\dagger_{j,n}$, which will create (destroy) a particle in the state given by Eq. (1.3). Note that this does not mean that the particle is created at the position R_j but with the wavefunction of Eq. (1.3). The Hamiltonian of the system is then

$$H = \sum_{j,n} \hbar\omega_0 \left[\frac{1}{2} + n\right] b^\dagger_{j,n} b_{j,n}. \tag{1.6}$$

Although extremely primitive, this limit already allows us to deduce a certain number of parameters. For example, one can have an estimate of the interactions among the particles. Optical lattices play, in that respect, a central role. To understand that point, let us assume that the microscopic interaction between the atoms can be described by the standard contact interaction (Pitaevskii and Stringari 2003):

$$U(r) = \frac{4\pi\hbar^2 a_s}{m}\delta(r) = g\delta(r). \tag{1.7}$$

Starting from such an interaction, one can define in the continuum a dimensionless ratio, which is the typical kinetic energy relative to the interaction energy. This ratio reads

$$\gamma = \frac{E_{\text{int}}}{E_{\text{kin}}} = \frac{gnm}{\hbar^2 n^{2/3}} \simeq 4\pi n^{1/3} a_s \tag{1.8}$$

in three dimensions, where the density of particles $n^{-1/3} = a$, the mean interparticle distance. A typical value for the parameter γ would be $\gamma = 0.02$. In other words, the interaction is normally quite weak. To see a strong interaction effect, one thus needs to reinforce it. This can be reached by either increasing the interaction itself, for example by using a Feshbach resonance (Bloch *et al.* 2008) or confinement (Olshanii 1998), or by acting on the kinetic energy via an optical lattice as we discuss next.

In the optical lattice, as we saw, wavefunctions on different sites have essentially zero overlap, which means that the interaction between particles located on different sites is essentially zero. Indeed, if we take for example the ground-state wavefunction of the harmonic oscillator:

$$\psi_0(x) = \left(\frac{m\omega_0}{\hbar\pi}\right)^{1/4} \mathrm{e}^{-\frac{m\omega_0}{2\hbar}x^2}, \tag{1.9}$$

for one dimension. In three dimensions, the wavefunction is the product of this wavefunction for each of the coordinates. Given that $\omega_0 \sim \sqrt{V_0}$ (see Eq. (1.5)) the spatial extension of the wavefunction decreases as $1/V_0^{1/4}$ and can thus be much smaller than the 'intersite' distance a (see Figure 1.1) for large potentials V_0. It means that, if we use the second quantization representation of Eq. (1.6), interaction can only involve operators on a given site.

If two particles are present on one given site, one can estimate the energy cost coming from the interactions. Let us assume that both particles are present in the lowest energy state of the harmonic oscillator (see Figure 1.1). Then the energy cost is

$$U = \frac{1}{2}\int \mathrm{d}r_1 \mathrm{d}r_2\, U(r_1 - r_2)|\psi_0(r_1)|^2|\psi_0(r_2)|^2. \tag{1.10}$$

Using the expression for the interaction, Eq. (1.7), and the wavefunction, Eq. (1.9), extended to the three-dimensional case, one obtains

$$U = \frac{g}{2\sqrt{2}}\left(\sqrt{m\omega_0 \hbar \pi}\right)^{3/2}. \tag{1.11}$$

Using the second quantization representation, and general expressions for the two-body operators, this leads to an interaction term in the Hamiltonian of the form

$$H_{\text{int}} = \frac{U}{2}\sum_j \hat{n}_{j,0}(\hat{n}_{j,0} - 1), \tag{1.12}$$

where

$$\hat{n}_{j,0} = b^\dagger_{j,0} b_{j,0} \tag{1.13}$$

is the operator counting the number of particles in the state 0 on site j. One thus sees that the higher the barriers the larger is the energy cost of having two particles or more on the same site. This is because the wavefunctions are more and more tightly confined and thus feel a local repulsion more strongly. Of course this expression, involving only one orbital, is only valid if the population of the higher levels is zero. This implies in particular that one should be in a limit where the temperature is small compared with the interlevel separation $T \ll \omega_0$ but also that the interaction parameter U is smaller than the interlevel separation $U \ll \omega_0$. Otherwise it is more favourable to promote one of the particles to a higher orbital state, which might reduce the overlap of the wavefunctions. This is energetically more favourable than paying the full repulsion price. If these conditions are not met, one needs to involve several orbitals to build the model.

1.2.2 Tight binding approximation

The approximation of the previous section essentially removes the kinetic energy of the particles, which remain localized around one site. This is clearly an oversimplification. Given the fact that there is some level of overlap of the wavefunctions on different sites, there is a finite probability of tunnelling between two sites. We can thus build a theory to include this tunnelling, starting from the basis of wavefunctions localized around each site, defined in the previous section. This method is known as a the tight-binding approximation (Ashcroft and Mermin 1976; Ziman 1972). It is specially transparent and contains all the main features of exact solutions in periodic potential that we will outline in the next section. We will thus examine it in detail.

For simplicity, let us restrict ourselves to the lowest orbital $|0\rangle$ on each site. Generalizing to several orbitals per site poses no problem. Let us take a system with N sites. We can write the full wavefunction of the problem as a linear combination of all the wavefunctions on each site, since we consider that they are essentially orthogonal,

$$\psi(r) = \frac{1}{\sqrt{N}} \sum_{j=0}^{N-1} \alpha_j \psi_0(r - R_j), \tag{1.14}$$

where the α_j are coefficients to be determined. Since we want the problem to be invariant by a translation of a, the wavefunction can only be multiplied by a phase if we translate x by a

$$\psi(x+a) = \mathrm{e}^{\mathrm{i}ka}\psi(x), \tag{1.15}$$

which defines the parameter k. This parameter, which of course depends on the wavefunction ψ, is known as the pseudo-momentum of the system. Note that this constraint is in fact an exact statement, known as the Bloch theorem. To satisfy the constraint of Eq. (1.15), it is easy to see that we can take

$$\psi_k(r) = \frac{1}{\sqrt{N}} \sum_{j=0}^{N-1} \mathrm{e}^{\mathrm{i}kR_j} \psi_0(r - R_j), \tag{1.16}$$

which also ensures the proper normalization of the wavefunction. To have *independent* wavefunctions we should not take values of k leading to the same coefficients. Since $R_j = aj$, values of k differing by $2\pi/a$ would lead to the same coefficients. We must therefore restrict the values of k to an interval of size $2\pi/a$, called the first Brillouin zone. Typically, one takes $k \in [-\pi/a, \pi/a]$. All physical quantities are thus periodic over this interval. In addition, not all values of k are allowed. Because the system is of size N, k must be quantized. The precise quantization depends on the boundary conditions. For example, for periodic boundary conditions, $\psi(x + La) = \psi(x)$ imposes that k is a multiple of an integer:

$$k = \frac{2\pi p}{N}, \quad p \in \mathbb{Z}. \tag{1.17}$$

There are thus in the first Brillouin zone exactly N values of k and so N independent functions $\psi_k(x)$.

1.2.3 More general relations

Many of the relations or properties that we have obtained within the tight binding approximation are in fact general and exact. Let us briefly review them here.

The first one is the Bloch theorem, which states that in a periodic potential there exists a quantum number k labelling the eigenfunctions, such that

$$\psi_k(r) = \mathrm{e}^{\mathrm{i}kr} u_k(r), \tag{1.18}$$

where $u_k(r)$ is a periodic function,

$$u_k(r+a) = u_k(r). \tag{1.19}$$

The constraints on the pseudo-momentum k that we have established in the previous section hold.

In the same way, the tight binding wavefunction has the right structure. One can represent the eigenfunction under a form known as a Wannier function (Ashcroft and Mermin 1976; Ziman 1972):

$$\psi_k(r) = \frac{1}{\sqrt{N}} \sum_j \phi(r - R_j). \tag{1.20}$$

The Wannier function is given by

$$\phi(r - R_j) = \phi_{R_j}(r) = \frac{1}{\sqrt{N}} \sum_k \mathrm{e}^{-\mathrm{i}kR_j} \psi_k(r). \tag{1.21}$$

Two Wannier functions centred on two different sites are *exactly* orthogonal

$$\langle \phi_{R_i} | \phi_{R_j} \rangle = \delta_{i,j}, \tag{1.22}$$

and the wavefunction $\phi_{R_j}(r)$ is essentially localized around the site R_j. We see that in the limit of high barriers, the local functions around one of the minimum of the potential provide an approximate form for the Wannier function.

Let us, for example, look at a Wannier function that would correspond to Eq. (1.18) with $u_k(r)$ independent of k. In that case the Wannier function would be (in one dimension)

$$\phi_{R_j}(x) = u(r) \frac{\sqrt{N}}{\pi} \frac{\sin(\pi(x - R_j)/a)}{x - R_j}, \tag{1.23}$$

showing the localization around the site R_j.

1.2.4 Hubbard and related models

Optical lattices thus provide a natural realization for certain models of interacting quantum systems with local interactions. In condensed matter these models are approximations of the realistic situations. Indeed, in a solid the basic interaction is normally the Coulomb interaction between the electrons. However, in a metal this interaction is screened, with a quite short screening length, of the order of the lattice spacing in a good metal (Ashcroft and Mermin 1976; Ziman 1972). It is thus tempting to replace the interaction with a local one. This is, however, in principle a caricature of reality since the screening length can vary, hence the need to take into account interactions with a range longer than a single site, etc. When comparing a certain solution of these models with reality, it is difficult to know whether the discrepancies are due to the approximations made in the solution or the approximations made in the model. Optical lattices at least provide a reasonably clean realization of such models that can be compared directly with theoretical predictions. Let us examine some of these models.

Bosonic Hubbard model We already obtained this model in Section 1.2.2. It is

$$H = -t \sum_{\langle i,j \rangle} \left(b_i^\dagger b_j + \text{h.c.} \right) + \frac{U}{2} \sum_j \hat{n}_j (\hat{n}_j - 1), \tag{1.24}$$

where $\langle\rangle$ denotes nearest neighbours, t is the hopping amplitude from one site to the next, and U the energy cost of putting two particles on the same site. This model describes quantum particles (typically bosons) hopping on a lattice and paying the interaction price U. This is essentially the simplest model that contains all the important elements of the competition between kinetic energy and interactions in a solid: (i) the kinetic energy; (ii) the notion of filling of a band (which would not be present in a continuum); (iii) the interaction. This model, known as the Hubbard model, was introduced in 1963 (Hubbard 1963) for fermions. The model of Eq. (1.24) that applies to bosons is sometimes referred to as the Bose–Hubbard model to distinguish it from its venerable ancestor. One can, of course, add several perturbations to this model. The most common ones are the confining potential or any local potential, such as disorder. This would lead to

$$H_\mu = \sum_j \mu_j \hat{n}_j. \tag{1.25}$$

For the confining potential, the chemical potential term is of the form $\mu_j \propto j^2$ and takes any suitable form depending on the perturbation. Optical lattices allow an easy control of the hopping amplitude t, while Feshbach resonance changes U (Bloch *et al.* 2008). These two methods allow for a large variation of the ratio U/t, which controls the strength of the interaction effects.

As mentioned already, Eq. (1.24) is a faithful description of the system in the optical lattice, provided the temperature T and interaction U are smaller than the distance between the lowest orbital and the first excited one, an energy of order $\hbar\omega_0$.

Otherwise, one must generalize the model to a multiorbital one. Note that if the optical lattice is not deep enough, or if the scattering length is too large, additional terms will appear in the Hamiltonian and the simple one-band Hubbard model is no longer valid: for a discussion, see e.g. (Werner *et al.* 2005).

t–V model For spinless fermions, Eq. (1.24) would not contain any interaction since the Pauli principle forbids double occupancy of a given site. For spinless fermions one can thus consider an interaction of the form

$$H_V = V \sum_{\langle i,j \rangle} \hat{n}_j \hat{n}_j. \tag{1.26}$$

In condensed matter, this is merely taking into account the long-range nature of the interactions. In cold atoms, it is rather difficult to realize but could be relevant for systems with longer-range interactions, such as dipolar molecules. The model with kinetic energy in the lattice and the interaction of Eq. (1.26) is known as the $t-V$ model and is also related to models for spins as we will see later.

Hubbard model Since electrons in solids have a spin 1/2, i.e., an internal degree of freedom, it is important to consider the generalization of this class of models to the case of two species of particle. This is the canonical Hubbard model for fermions (Hubbard 1963). Hopping conserves the internal degree of freedom (which we will call 'spin' for simplicity), while a local interaction can only exist between two opposite spins, since the Pauli principle prevents two fermions of the same spin from being on the same site. The model is thus

$$H = -t \sum_{\langle ij \rangle, \sigma=\uparrow\downarrow} \left(c^\dagger_{i\sigma} c_{j\sigma} + \text{h.c.} \right) + U \sum_i \hat{n}_{i,\uparrow} \hat{n}_{i,\downarrow}, \tag{1.27}$$

where $\uparrow, \downarrow$ denote the two eigenstates of opposite spin (for example the two eigenstates of the spin along the z-direction). In the cold atom context the 'spin' degree of freedom can denote any two possible internal states. This model contains the essential ingredients of the physics of strongly correlated quantum systems. Although it is extremely simple to write, it is extremely challenging to solve.

Generalizations Of course, this Hamiltonian can be complicated in several ways, for example, by putting state-dependent hopping amplitudes $t_\uparrow$ $t_\downarrow$ (Cazalilla *et al.* 2005), by adding longer-range interactions to the system, or by considering a larger number of internal degrees of freedom. All these models can be (or have already been) potentially realized in cold atomic systems.

In addition to the fermionic Hubbard model, cold atomic systems have also allowed the realization of bosonic systems with internal degrees of freedom. This has led to several interesting models, in particular the two-component Bose–Hubbard model. In contrast to the case of fermions, for which the Pauli principle prevents the occupation of a site by two particles of the same species, Bosons can have such terms.

The interaction term for the two-component Bose–Hubbard model thus involves three different interactions

$$H = \frac{U_{\uparrow\uparrow}}{2} \sum_i \hat{n}_{i\uparrow}(\hat{n}_{i\uparrow} - 1) + \frac{U_{\downarrow\downarrow}}{2} \sum_i \hat{n}_{i\downarrow}(\hat{n}_{i\downarrow} - 1) + U_{\uparrow\downarrow} \sum_i \hat{n}_{i\uparrow}\hat{n}_{i\downarrow}. \tag{1.28}$$

As we will see in the next section, the combination of these three interactions can lead to a wide range of physical behaviours.

With respect to these models, cold atoms, given the local nature of the interactions and the degree of control on the lattice, interactions and nature of the particles are a fantastic laboratory for realizing and testing these models. There are, however, several limitations or points to keep in mind. We have already mentioned some of them. Let us summarize them here:

1. If one wants to be able to use a single-band model, the separation between levels in one of the optical lattice wells must be larger than the interaction. This is not a major problem when the lattice is deep, and when the interaction is reasonably small, but it can become a serious limitation if the interaction is increased by a Feshbach resonance.
2. If one wants to use the optical lattice to reduce the kinetic energy in order to change the ratio of the kinetic energy to the interactions then one has to worry about the temperature. Indeed, if the kinetic energy becomes small compared with the temperature, one essentially has a classical system.
3. Finally, the confining potential, which corresponds to a locally varying chemical potential, is both an advantage and a serious limitation. Indeed, as we will discuss, the physics of such systems depends strongly on the filling. So controlling the chemical potential or the number of particles per site is, of course, crucial. Having a confining potential has the advantage that in the system there are many different values of the chemical potential and thus one does not need (it would be extremely difficult in practice) to control exactly the number of particles compared with the number of sites. On the other hand, the system is inhomogeneous, which means that most measurements will give an average response over many different phases, deeply obscuring the physics. Clearly, this question is related to the ability—or not—to probe the system locally.

1.2.5 Superexchange

The models of the previous sections describe the behaviour of itinerant quantum particles on a lattice. Particularly interesting behaviour occurs when these particles can have internal degrees of freedom, as in the Hubbard model. In that case it is possible, as we will discuss in the following sections, that because of the interactions the charge of the particles becomes localized for special filling of the lattice, for example, one particle per site (Mott transition). In such a case, as we will discuss later, the repulsion between the charges (U in the Hubbard model of Eq. (1.27) can lead to an insulating phase in the case of one particle per site (Mott transition). For the

Bose–Hubbard model with one component, such a ground state would be featureless. But for systems with two (or more) components, both fermionic and bosonic, the ground state is a priori quite complex since on each site one has to choose the state of the internal degree of freedom (which we will call spin throughout this section).

As shown in Figure 1.2, if the repulsion is very large, charge excitations that would put two particles per site would cost an energy of order U and are thus essentially forbidden. On the other hand, since U is not infinite there could be virtual processes that allow the system to benefit from the kinetic energy, while leaving the system in a sector with exactly one particle per site. These processes are the so-called superexchange processes. We will not give here the full derivation of the superexchange term; this can be found in sufficient details elsewhere (e.g., Giamarchi 2011). We simply give here a qualitative argument.

Since the charge is essentially frozen, one can stay in the Hilbert space in which each site has exactly one particle per site and only the spin degree of freedom remains. This means that on each site we require states that fully describe the Hilbert space. We can thus reduce the complete Hamiltonian of Eq. (1.27) to an *effective* Hamiltonian acting only on the spin degrees of freedom. For fermions, it is easy to see that if one has two parallel spins on neighbouring sites, no kinetic energy process can take place. On the other hand, if the spins are antiparallel, second-order perturbation theory

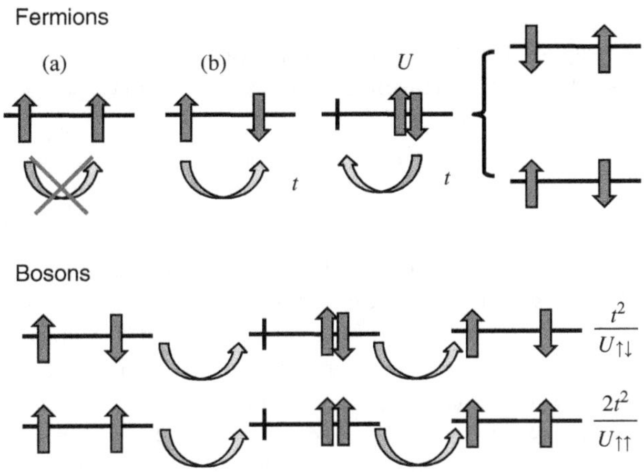

Figure 1.2 For a large repulsion U and one particle per site, charge excitations cost an energy of order U, but virtual processes allow some kinetic energy to be gained. (a) For fermions, the Pauli principle completely blocks hopping if the spins are parallel. (b) For opposite spins, virtual hopping is possible. This leads to a superexchange that is dominantly antiferromagnetic (see text). For bosons, both processes are possible and depend on the intra- and interspecies interactions. Bosonic factors favour parallel spins. Thus, if all interactions are equal for bosons, the superexchange is dominantly ferromagnetic (see text). Changing the interactions between the two types of species allows transition from the ferromagnetic exchange to the antiferromagnetic one.

(see Figure 1.2) can lead back to the initial state or lead to a state in which the two spins have been exchanged. The matrix element involved is of order $J = t^2/U$ since each hopping has an amplitude t and the intermediate state is of energy U. The first process can be described by the effective Hamiltonian (written only for two spins)

$$H_1 = JS_1^z S_2^z - \frac{J}{4}, \tag{1.29}$$

where we have introduced the spin operators $S^\alpha = \frac{1}{2}\sigma^\alpha$ and the σ^α are the three Pauli matrices. As usual, we introduce the two eigenstates of S^z and the Hermitian conjugate operators $S^+ = S^x + \mathrm{i}S^y$ and $S^- = S^x - \mathrm{i}S^y$. These operators verify

$$\begin{aligned} S^z \left|\uparrow, \downarrow\right\rangle &= \pm\frac{1}{2} \left|\uparrow, \downarrow\right\rangle, \\ S^+ \left|\downarrow\right\rangle = \left|\uparrow\right\rangle, &\quad S^+ \left|\uparrow\right\rangle = \left|\downarrow\right\rangle \end{aligned} \tag{1.30}$$

Equation (1.29) shows that the energy of two antiparallel spins is lowered by an energy $-J/2$ while one of two parallel spins remains zero. The second process leads to an exchange of the two spins and can be written as

$$H_2 = \frac{J}{2}\left[S_1^+ S_2^- + S_1^- S_2^+\right]. \tag{1.31}$$

Putting the two processes together, and taking proper care of the numerical factors, one obtains for the full effective Hamiltonian (up to a constant energy term)

$$H = \frac{J}{2}\sum_{\langle ij\rangle}\left[S_i^+ S_j^- + S_i^- S_j^+\right] + J\sum_{\langle ij\rangle} S_i^z S_j^z = J\sum_{\langle ij\rangle} \vec{S}_i \cdot \vec{S}_j, \tag{1.32}$$

where $J \simeq 4t^2/U$ for large values of U. This Hamiltonian is known as the Heisenberg Hamiltonian. We thus see that the combination of kinetic energy, Pauli principle, and interaction leads to a remarkable exchange term between the spins, which looks very much like the dipolar one that would exist for the direct magnetic exchange between magnetic moments. However, there are also remarkable differences. This exchange, nicknamed superexchange, is responsible for many of the magnetic properties of the strongly interacting quantum systems in solids (Auerbach 1998). Some noteworthy points are as follows:

1. Compared to an exchange between magnetic moments, this superexchange is isotropic in the spin variables and will not couple the lattice direction with the spin directions. In that sense, it is even simpler than a normal dipolar exchange. The spin rotation invariance of Eq. (1.32) comes, of course, from the spin rotation invariance of the original Hubbard Hamiltonian of Eq. (1.27).
2. Quite importantly, the order of magnitude of typical interactions is quite different. Direct magnetic exchanges are quite ridiculous in solids. If one takes spins on typical lattice spacing distance in a solid, one obtains a direct magnetic exchange

of less than 1 K. On the contrary, since kinetic energy is typically 1 eV and interactions of the order of ∼10 eV this leads for solids to a J of the order $J \sim 1000\,\mathrm{K}$. Superexchange is thus by far the most dominant term in solids and lies at the root of the magnetic properties that we can observe in nature. In cold atoms, the 'spin' is, of course, merely an internal degree of freedom, so the superexchange is the only term that can exist.

3. For fermions, because of the Pauli principle $J > 0$, which means that the fermionic Hubbard model leads to antiferromagnetic phases. The situation is quite different for bosons, as indicated in Figure 1.2. In that case, both species can hop, so the sign of the effective exchange J will depend on the relative values of the intra- and interspecies interactions. If the intraspecies $U_{\uparrow\uparrow}$ and $U_{\downarrow\downarrow}$ is the largest, then it is very much like the Pauli principle and one recovers an antiferromagnetic superexchange. On the contrary, if the interspecies interaction $U_{\uparrow\downarrow}$ is the largest, then one has a *ferromagnetic* (i.e., a negative J) superexchange. In the case where all the interactions are equal, the bosonic factors still favour a ferromagnetic exchange (Duan *et al.* 2003). Multicomponent bosonic systems will thus offer particularly rich physics (Kleine *et al.* 2007; Zvonarev *et al.* 2007).

1.3 The Bose–Hubbard model and the superfluid to Mott insulator transition

In this section, we make our first encounter with the Mott phenomenon: strong repulsive interactions between particles can prevent the formation of an itinerant state and favour a situation in which particles are localized (Figure 1.3). This phenomenon is of key importance to the physics of strongly correlated materials. Many remarkable physical properties are found for those materials that are close to a Mott insulating state. For example, high-temperature superconductivity is found in copper oxides when a

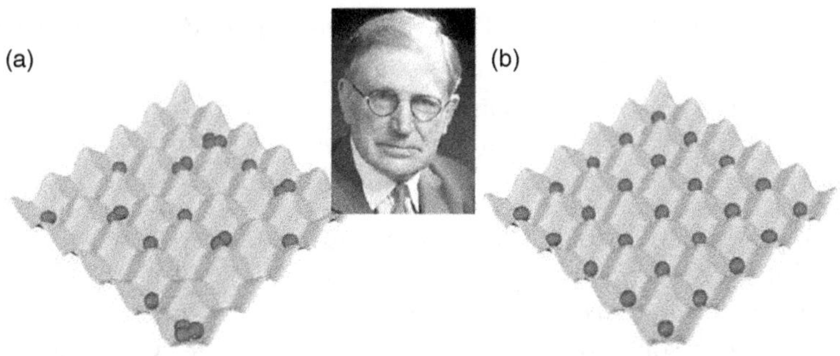

Figure 1.3 (a) Typical real-space configuration of particles in an itinerant (metallic or superfluid) state. (b) Typical real-space configuration in the Mott insulating state, in which double occupancies are strongly suppressed. (centre) Sir Nevil Mott. Adapted in part from (Bloch 2005).

metallic state is induced by introducing a relatively small amount of charge carriers into a Mott insulator.

In such circumstances, particles 'hesitate' between itinerant and localized behaviour, making quantum coherence more difficult to establish and leading to a number of possible instabilities. From a theoretical viewpoint, one of the key difficulties is that of describing consistently an entity that is behaving simultaneously in a wave-like (delocalized) and particle-like (localized) manner. Viewed from this perspective, strongly correlated quantum systems raise fundamental questions in quantum physics. Because the Mott phenomenon is so important, the theoretical proposal (Jaksch *et al.* 1998) and experimental observation (Greiner *et al.* 2002) of the Mott transition in a gas of ultracold bosonic atoms in an optical lattice have been truly pioneering works, establishing a bridge between modern issues in condensed matter physics and ultracold atomic systems.

In this section, we deal with this phenomenon in the simplest possible context: that of the Hubbard model for bosonic atoms in an optical lattice. The case of fermions will be considered later, in Sections 1.6 and 1.7. The Hamiltonian of this model reads (see also Eq. (1.24)):

$$H = -\sum_{ij} t_{ij}\, b_i^\dagger b_j + \frac{U}{2}\sum_i \hat{n}_i(\hat{n}_i - 1) + \sum_i v_{\text{trap}}(i)\,\hat{n}_i - \mu\sum_i \hat{n}_i. \tag{1.33}$$

1.3.1 General considerations: lifting a macroscopic degeneracy

Let us first consider a homogeneous system ($v_{\text{trap}} = 0$) in the limit where there is no hopping $t_{ij} = 0$ (very deep lattice), as discussed in Section 1.2.1. The Hamiltonian is then diagonal in occupation-number basis and has eigenstates $|n\rangle$ with energies $E_n^0 = \frac{U}{2}n(n-1) - \mu n$. These energy levels cross at specific values of the chemical potential $\mu_n^0 = nU$ at which $E_n^0 = E_{n+1}^0$. Hence, the nature of the ground state depends crucially on the value of the chemical potential:

- If $\mu \in \,](n-1)U, nU[$, the ground state is *non-degenerate*, with exactly n bosons on each lattice site.
- If $\mu = nU$, having n or $n-1$ bosons on each lattice site is equally probable. Hence, the ground state has a *macroscopic degeneracy* 2^{N_s} (with N_s the number of sites in the lattice).

The number of particles per site in the ground state as a function of chemical potential has the form of a 'staircase' made of plateaus of width U in which $\langle\hat{n}\rangle$ remains constant, separated by steps at $\mu_n^0 = \mu n$, up which it jumps by one unit (Figure 1.4). In the context of mesoscopic solid-state devices, this is called the 'Coulomb staircase': to increase the charge by one unit, a Coulomb charging energy must be paid, owing to the electrostatic repulsive interactions between electrons.

Within a given plateau $\mu \in \,](n-1)U, nU[$, excitation to the first excited state (at constant total particle number) involves moving one boson from one site to another

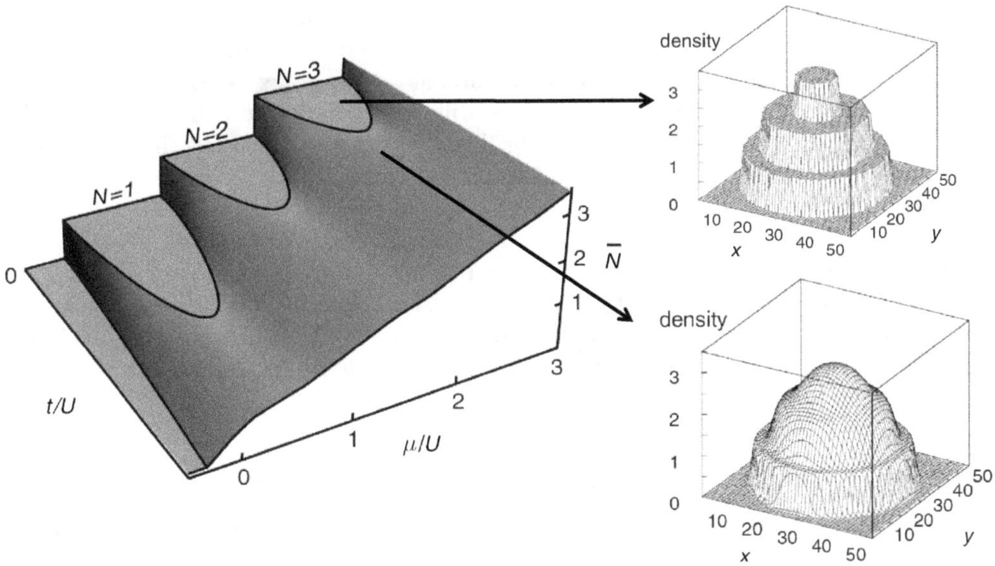

Figure 1.4 Left: phase diagram of the Bose Hubbard model as a function of chemical potential μ/U and coupling t/U. An incompressible Mott insulator is found within each lobe of integer density. Right: density profiles in a harmonic trap. The 'wedding cake' structure (see text) is due to the incompressibility of the Mott insulator (numerical calculations and figures courtesy of H. Niemeyer, H. Monien and M. Greiner).

one, leaving a site with occupancy $n-1$ and creating another one with occupancy $n+1$. The energy of this excitation is:

$$\Delta_g^0 = E_{n+1}^0 + E_{n-1}^0 - 2E_n^0 = U. \tag{1.34}$$

Hence, the ground state is separated from the first excited state by a *finite energy gap*. (In passing, we note that this gap can be written as $\Delta_g^0 = (E_{n+1}^0 - E_n^0) + (E_{n-1}^0 - E_n^0)$, which in chemist's terminology corresponds to ionization energy minus affinity). Adding or removing an electron also requires a finite amount of energy: hence the system is *incompressible*. Indeed, each plateau has a vanishing compressibility:

$$\kappa = \left(\frac{\partial^2 E}{\partial n^2}\right)^{-1} = \frac{\partial n}{\partial \mu}. \tag{1.35}$$

Having understood the zero-hopping limit, we can ask what happens when a small hopping amplitude is turned on. Obviously, a non-degenerate incompressible ground state separated by a gap from all excitations is a quite protected state. Hence, we expect that the system will remain incompressible and localized when turning on a small hopping, for values of μ well within a given charge plateau. In contrast, the hopping amplitude is likely to be a singular perturbation when starting from the macroscopically degenerate ground state at each of the degeneracy points $\mu_n^0 = nU$. One natural way for the perturbation to lift the degeneracy is to select a unique ground state that is a superposition of the different degenerate configurations, with different numbers of particles on each site. If the mixing between the different charge states corresponds to a state with small phase fluctuations (the phase is the conjugate variable to the local charge), the resulting state will be a superfluid. Hence, we expect that a superfluid state with Bose condensation will occur for infinitesimal hopping at the degeneracy points $\mu = nU$. These expectations are entirely confirmed by the mean-field theory presented in the next section. We note in passing that interesting phenomena often happen in condensed-matter physics when a perturbation lifts a large degeneracy of the ground state (the fractional quantum Hall effect is another example).

1.3.2 Mean-field theory of the bosonic Hubbard model

As is usually the case in statistical mechanics, a mean-field theory can be constructed by replacing the original Hamiltonian on the lattice by an effective single-site problem subject to a self-consistency condition. Here, this is naturally achieved by factorizing the hopping term (Fisher *et al.* 1989; Sheshadri *et al.* 1993): $b_i^\dagger b_j \to \text{const.} + \langle b_i^\dagger \rangle b_j + b_i^\dagger \langle b_j \rangle + \cdots$, in which '$\cdots$' denotes fluctuations that are neglected. Another essentially equivalent formulation is based on the Gutzwiller wavefunction (Rokhsar and Kotliar 1991; Krauth *et al.* 1992). The effective one-site Hamiltonian for site i reads:

$$h_{\text{eff}}^{(i)} = -\lambda_i b^\dagger - \lambda_i b + \frac{U}{2}\hat{n}(\hat{n}-1) - \mu\hat{n}. \tag{1.36}$$

In this expression, λ_i is a 'Weiss field', which is determined self-consistently by the boson amplitude on the other sites of the lattice through the condition

$$\lambda_i = \sum_j t_{ij} \langle b_j \rangle. \tag{1.37}$$

For nearest-neighbour hopping on a uniform lattice of connectivity z, with all sites being equivalent, this reads:

$$\lambda = z\, t\, \langle b \rangle. \tag{1.38}$$

These equations are easily solved numerically, by diagonalizing the effective single-site Hamiltonian of Eq. (1.36), calculating $\langle b \rangle$ and iterating the procedure such that Eq. (1.38) is satisfied. The boson amplitude $\langle b \rangle$ is an order parameter associated with Bose condensation in the $\vec{k} = 0$ state: it is non-zero in the superfluid phase.

For densities corresponding to an integer number n of bosons per site on average, one finds that $\langle b \rangle$ is non-zero only when t/U is larger than a critical ratio $(t/U)_c$ (which depends on the filling n). For $t/U < (t/U)_c$, $\langle b \rangle$ (and λ) vanishes, signalling a non-superfluid phase in which the bosons are localized on the lattice sites: the Mott insulator. For non-integer values of the density, the system is a superfluid for all $t/U > 0$. This fully confirms the expectations deduced on a qualitative basis at the end of the previous section.

Perturbative analysis It is instructive to analyse these mean-field equations close to the critical value of the coupling: because λ is then small, it can be treated in Eq. (1.36) as a perturbation of the zero-hopping Hamiltonian. Considering a given plateau $\mu \in](n-1)U, nU[$, the perturbed ground state reads:

$$|\psi_0\rangle = |n\rangle - \lambda \left[\frac{\sqrt{n}}{U(n-1)-\mu}|n-1\rangle + \frac{\sqrt{n+1}}{\mu - Un}|n+1\rangle \right], \tag{1.39}$$

so that:

$$\langle\psi_0|b|\psi_0\rangle = -\lambda \left[\frac{n}{U(n-1)-\mu} + \frac{n+1}{\mu - Un} \right]. \tag{1.40}$$

Inserting this in the self-consistency condition yields

$$\lambda = -z\,t\,\lambda \left[\frac{n}{U(n-1)-\mu} + \frac{n+1}{\mu - Un} \right] + \cdots, \tag{1.41}$$

where '...' denotes higher-order terms in λ. This equation can be viewed as the linear term in the expansion of the equation of state for λ. As usual, the critical value of the coupling corresponds to the vanishing of the coefficient of this linear term (corresponding to the quadratic or mass term of the expansion of the Landau free energy). Hence, the critical boundary for a fixed average (integer) density n is given by

$$\frac{zt}{U} = \frac{(n-\mu/U)(\mu/U - n + 1)}{1+\mu/U}. \tag{1.42}$$

Phase diagram This expression gives the location of the critical boundary as a function of the chemical potential. As expected, it vanishes at the degeneracy points $\mu_n^0 = nU$, where the system becomes a superfluid for infinitesimal hopping amplitude. In the $(t/U, \mu/U)$ plane, the phase diagram (Figure 1.4) consists of lobes inside which the density is an integer and the system is a Mott insulator. Outside these lobes, the system is a superfluid. The tip of a given lobe corresponds to the the maximum value of the hopping at which an insulating state can be found. For n atoms per site, this is given by

$$\left.\frac{zt}{U}\right|_{\mathrm{c},n} = \max_{x\in[n-1,n]} \frac{(n-x)[x-n+1]}{1+x} = \frac{1}{2n+1+2\sqrt{n(n+1)}}, \tag{1.43}$$

so that the critical interaction strength is $(U/zt)_{\mathrm{c}} \simeq 5.8$ for $n=1$, and increases as n increases ($(U/zt)_{\mathrm{c}} \sim 4n$ for large n).

Mott gap The gap in the Mott insulating state is, of course, reduced by the hopping from its zero hopping value $\Delta_g^0 = U$. We can obtain its mean-field value from the extension of the density plateau:

$$\Delta_g(n) = \mu_+(n) - \mu_-(n), \tag{1.44}$$

where $\mu_\pm$ are the solutions of the quadratic equation corresponding to Eq. (1.42), i.e:

$$(\mu/U)^2 - [2n - 1 - (zt/U)](\mu/U) + n(n-1) + (zt/U) = 0, \tag{1.45}$$

yielding

$$\Delta_g(n) = U\left[\left(\frac{zt}{U}\right)^2 - 2(2n+1)\frac{zt}{U} + 1\right]^{1/2}. \tag{1.46}$$

The Mott gap is $\sim U$ at large U/t and vanishes at the critical coupling ($\propto (U - U_c)^{1/2}$ within mean-field theory).

Incompressibility and 'wedding-cake' shape of the density profile in the trap The existence of a gap means that the chemical potential can be changed within the gap without changing the density. As a result, when the system is placed in a trap, it displays density plateaus corresponding to the Mott state, leading to a 'wedding cake' structure of the density profile (Figure 1.4). This is easily understood in the local density approximation, in which the local chemical potential is given by $\mu(r) = \mu - v_{\text{trap}}(r) = \mu - m\omega_0^2 r^2/2$, yielding a maximum extension of the plateau: $\sim(2\Delta_g/m\omega_0^2)^{1/2}$. Several authors have studied these density plateaus beyond the LDA by numerical simulation (see e.g (Batrouni *et al.* 2002)), and they have also been imaged experimentally, see e.g. (Fölling *et al.* 2006).

1.3.3 Mean-field theory: the wavefunction viewpoint

An alternative, but equivalent, approach to the mean-field theory is to formulate it as a variational ansatz for the ground-state wavefunction (Krauth *et al.* 1992; Rokhsar and Kotliar 1991).

In the zero-hopping limit, the ground-state wavefunction within a given density plateau reads

$$\Psi_0^{t=0} = \prod_i |n\rangle_i = \prod_i \frac{1}{\sqrt{n!}}\left(b_i^\dagger\right)^n |0\rangle. \tag{1.47}$$

In the opposite limit of a non-interacting system ($U = 0$), the ground-state wavefunction is obtained by placing all bosons in the $\vec{k} = 0$ state:

$$\Psi_0^{U=0} = \frac{1}{\sqrt{N!}}\left(b_{\vec{k}=0}^\dagger\right)^N |0\rangle = \frac{1}{\sqrt{N!}}\left[\frac{1}{\sqrt{N_s}}\sum_i b_i^\dagger\right]^N |0\rangle. \tag{1.48}$$

In the limit of large N, N_s, the ground-state wavefunction for the non-interacting case can alternatively be formulated (by letting N fluctuate) as a product of coherent states on each site,

$$\Psi_0^{U=0} = \prod_i |\alpha\rangle_i, \quad |\alpha\rangle = \mathrm{e}^{-|\alpha|^2/2} \sum_{n=0}^{\infty} \frac{\alpha^n}{\sqrt{n!}} |n\rangle, \tag{1.49}$$

with $|\alpha|^2 = \langle n \rangle = N/N_\mathrm{s}$. In this limit, the local density obeys Poisson statistics $p(n) = \mathrm{e}^{-|\alpha|^2} |\alpha|^{2n}/n! = \mathrm{e}^{-\langle n \rangle} \langle n \rangle^n / n!$.

We note that in both limits, the ground-state wavefunction is a product of individual wavefunctions over the different lattice sites. The individual wavefunctions, however, have a very different nature in each limit: they are number state for $t = 0$, but they are a phase-coherent superposition of number states in the $U = 0$ limit.

A natural variational ansatz is then to assume that the wavefunction remains an uncorrelated product over sites for arbitrary U/t, namely,

$$\Psi_0^{\mathrm{var}} = \prod_i \left[\sum_n c_n |n\rangle_i \right]. \tag{1.50}$$

The variational principle then leads to equations for the coefficients c_n, which are identical to the mean-field equations above. The trial wavefunction interpolates between the Poisson statistics $c_n = \alpha^n / \sqrt{n!}$ for $U = 0$ and the zero-fluctuation limit $c_n = \delta_{n,n_0}$ as the insulator is reached. The fact that n has no fluctuations throughout the Mott phase is, of course, an artefact of the mean-field.

The derivation of these results rests heavily on the fact that one can build a mean-field theory, and in particular that a well-defined superfluid phase, with perfect order of the phase, exists. It is thus interesting to see how this physics and competition between the superfluid and Mott insulating phase would be modified in situations where phase fluctuations are very strong and the mean-field theory is invalid. This is clearly the case as the dimension of the system becomes smaller, since in low enough dimensions it is impossible to break a continuous symmetry (the so-called Mermin–Wagner theorem (Mermin 1968)), and thus no true superfluid phase—which would correspond to a breaking of the phase symmetry of the wavefunction—can exist. Since cold-atom systems allow an excellent control on the dimensionality of the problem by changing the strength of the optical lattice, they particularly enable those questions in the one-dimensional situation to be tackled for which one can expect novel effects to occur. We will thus examine in Section 1.4 the case of one-dimensional quantum systems.

1.3.4 Probing Mott insulators: shaking of the optical lattice

To probe the physics of Mott insulators it is important to have good probes. Time-of-flight measurements, which give access to the single-particle correlations, are, of course, one method and we will examine several others in this chapter. In this section, we want to discuss a relatively simple probe, which, however, gives extremely useful information for such systems and which consists in shaking of the optical lattices.

The idea is to modulate, in a time-dependent way, the amplitude of the optical lattice (Stöferle 2004) over a given period and then to measure the energy deposited in the system by such a process, as a function of the modulation frequency. This corresponds to adding a term in the Hamiltonian of the form

$$H_L = \int \mathrm{d}x \, [V_L + \delta V_L \cos(\omega_0 t)] \cos(Qx)\rho(x). \tag{1.51}$$

The results for such an operation are shown in Figure 1.5. One sees marked differences depending on the strength of the interactions. In particular, in the Mott insulator one can recognize a peak structure.

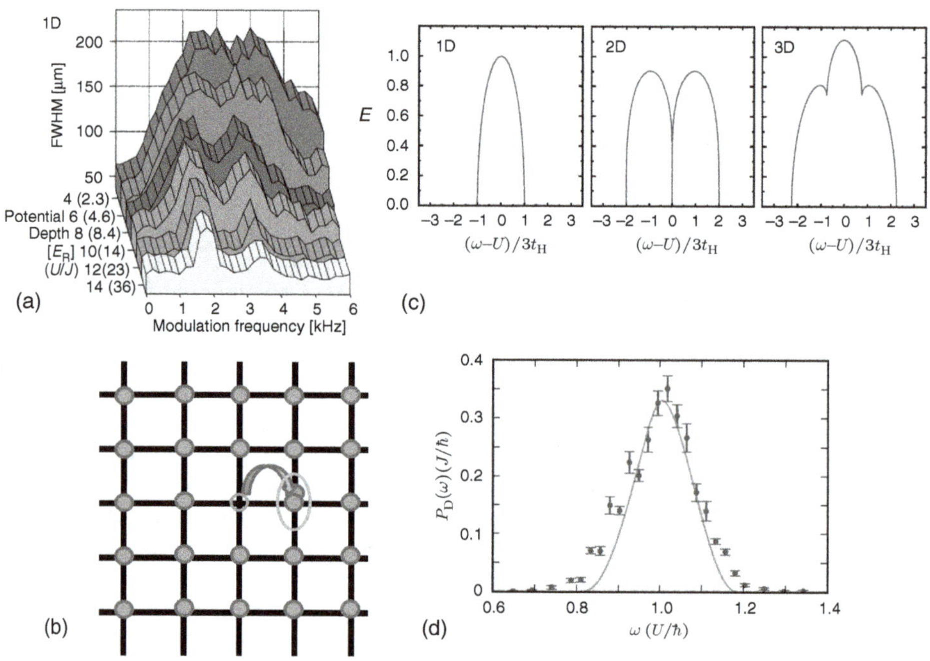

Figure 1.5 (a) Shaking of the optical lattice for a system of bosons. One sees marked differences depending on the depth of the optical lattice. In the Mott insulating phase a peak structure is observed. (After (Stöferle *et al.* 2004)). (b) Deep in the Mott phase, the structure can be explained by considering the creation of a doublon (doubly occupied site) and a holon (empty site) due to the modulation of the kinetic energy by the shaking. (c) Structure of the peak depending on the dimension. This structure is located at an energy around the Mott gap, and the width reflects the kinetic energy of the doublon and holon. (After (Tokuno and Giamarchi 2011)). (d) For fermions, similar results can be obtained by considering the creating of doubly occupied states, which makes this a very sensitive probe. Fitting to a slave boson theory gives excellent agreement with the data and can give some access to the temperature of the system. (After (Tokuno *et al.* 2011)).

Interpreting such data is, of course, not easy, since one must deal with a full time-dependent Hamiltonian, which is difficult to deal with analytically and numerically. Analytically, it is possible to use a linear response to study the effects of the shaking (Iucci *et al.* 2006). The results depend crucially on whether the lattice is weak or strong. We will concentrate here on the case of the strong lattice and refer the reader to the literature for the other limit. In that case, the main effect of modulating the lattice is to change, in the resulting effective Bose–Hubbard model of Eq. (1.24), the hopping t and the interaction U. Indeed as we saw in Section 1.2, these terms are directly determined by the shape of the wavefunctions and thus by the depth of the lattice. One can even realize that the main effect will be that on the tunnelling term (Reischl *et al.* 2005), which depends exponentially on the lattice depth. In the case of the strong lattice, the main consequence is thus a modulation of the kinetic energy in the Hubbard model:

$$H_K = H_K^0 + \delta H_K(t) = [t_0 + \delta t \cos(\omega_0 t)] \sum_{\langle i,j \rangle} \left(b_i^\dagger b_j + \text{h.c.} \right). \tag{1.52}$$

It is thus possible to study the effects of the shaking by considering the linear response in this term (Huber *et al.* 2007; Iucci *et al.* 2006; Kollath *et al.* 2006a; Tokuno and Giamarchi 2011). We will not go into the analysis in detail but give again the main ideas. In linear response, the energy absorbed is directly related to the imaginary part of the Fourier transform of the equilibrium correlation function

$$\chi(t) = -\mathrm{i}\langle[\delta H_K(t), \delta H_K(0)]\rangle. \tag{1.53}$$

We thus see that the shaking of the lattice measures the kinetic energy–kinetic energy correlations. In other words we have to consider the processes that are shown in Figure 1.5. We transfer at time zero a particle from one site to the neighbouring one; this excitation propagates and at a later time we undo it by applying the kinetic energy operator again. Deep in the Mott phase we start with one particle per site. The application of the kinetic energy term thus creates a doubly occupied site and an empty site. The energy of this excitation is of the order of the Mott gap $\Delta_\mathrm{M} \sim U$. We can thus expect that the system absorbs energy when the frequency of the modulation matches the Mott gap; $\hbar\omega_0 = \Delta_\mathrm{M}$. The shaking of the lattice thus allows us to measure the Mott gap of the system directly. In addition, the doublon and holon can propagate and thus have their own kinetic energy of the order of t_0. This will broaden the peak in a way that reflects this propagation. Such a propagation can be computed by properly taking into account the fact that the holon and doublon cannot be at the same site without recombining, giving the remarkable peak structure of Figure 1.5. Not taking such a constraint into account leads to incorrect results. Such a structure reflects the van Hove singularities in the density of state. We refer the reader to the literature for more details and references on the subject.

A variant of the shaking of the optical lattice, namely a modulation of the phase of the lattice rather than its amplitude, can be treated using similar methods. Quite remarkably, modulating the phase leads to the current–current correlation function instead of the kinetic energy–kinetic energy one. It thus gives direct access to the frequency-dependent conductivity of the system (Tokuno and Giamarchi 2011), something that allows us to make a direct connection with comparable experiments done in the condensed matter context. It will be interesting to implement such a probe practically.

The shaking is thus an extremely useful probe for Mott insulating physics. One drawback for the bosons is that measuring the energy absorbed is difficult. As a result, the modulation must be of a relatively large intensity, which takes the system out of the linear response regime. To describe the absorption in this limit it is thus necessary to perform a numerical analysis of the system, something not trivial given the out-of-equilibrium nature of the problem. In one dimension, density matrix renormalization group (DMRG) studies that fully utilize the time-dependent Hamiltonian have been performed (Kollath *et al.* 2006a) and have enabled elucidation of the nature of the higher peaks in the experimental data, as shown in Figure 1.5. To circumvent the difficulty caused by the measure of the energy (and in particular, as we will see in Section 1.6.2, for fermions this is extremely difficult), it has been suggested (Kollath *et al.* 2006b) that one measure instead the production of the doubly occupied states as a function of time. This allows for much more precise measurements. We will come back to this point in Section 1.6.2.

1.4 One-dimensional bosons and bosonization

Let us now turn to one-dimensional systems, for which very special effects arise. Indeed, as we discussed already for the case of bosons, and will see in Section 1.5 for fermions, the effects of interactions are crucial. For bosons, interactions lead both to the superfluid state and to the Mott insulating one. As one can naively expect in one dimension, the effects of interactions will be maximum since the particles cannot avoid each other, while in three dimensions one can naively expect that particles will 'see' each other much less. In addition, as we already mentioned, it is impossible to break a continuous symmetry, so even at $T = 0$ a true ordered superfluid ground state cannot exist. However, a bosonic system will still retain strong superfluid tendencies. One can thus expect that quantum systems in one dimension exhibit a radically different physics than is seen for their higher-dimensional counterparts. Cold atomic systems have been remarkable in showing such physics, given the remarkable control over dimension and interactions.

We will examine some of the aspects of this novel physics in this section. Of course, there is much too much to be examined in these few pages. This section will thus simply be a general presentation, and will not pretend to be exhaustive. The interested reader can find much more detail in a whole book on the subject of one-dimensional systems (Giamarchi 2004), where a complete description of the various one-dimensional systems and physical effects and methods is given. In addition, for the specific case of bosons in cold atoms, several volumes of lecture notes also contain complementary material (Giamarchi 2006; Giamarchi 2011). Finally, this chapter will not attempt to give a comprehensive list of references, since an extensive review on the subject of one-dimensional bosons exists (Cazalilla *et al.* 2011).

1.4.1 Peculiarities of one dimension

Before we embark on the one-dimensional world, let us briefly recall some of the points of the typical solution for a bosonic system in higher dimensions. As discussed in Section 1.3 for a high (meaning $d > 1$) dimensional system, one can expect that there is a well-defined superfluid order. As a result, the wavefunction can be written as

$$\psi(x) = \sqrt{\rho(x)}\mathrm{e}^{\mathrm{i}\theta(x)}, \tag{1.54}$$

where $\rho(x)$ is the density of particles at point x and $\theta(x)$ the phase of the wavefunction at the same point. The presence of superfluid order implies that we can use $\rho(x) \to \rho_0$ and $\theta(x)$ acquires a finite expectation value $\theta(x) \to \theta_0$, so that the wavefunction has a coherent phase through the whole sample. Fluctuations above this ground state can be described by the Bogoliubov theory (Pitaevskii and Stringari 2003). We will not repeat the theory here but just give the results. The Bogoliubov spectrum is linear at small k with a velocity u of the excitations, which represent the Goldstone modes of the superfluid. The velocity u depends on the interactions among the particles. This linear mode is the hallmark of the superfluidity in the system. At larger k the dispersion gives back the k^2 dispersion of free particles:

$$E(k) = \sqrt{u^2k^2 + (k^2/(2m))^2}. \tag{1.55}$$

An important point is that the mode is a well-defined dispersive mode, which characterizes excitations that have a well-defined relation between their momentum and energy.

Given the superfluid order, the single-particle correlation function tends to a constant,

$$g_1(r) = \lim_{r\to\infty} \langle \psi(r)\psi^\dagger(0)\rangle \to \text{const.} \tag{1.56}$$

and as a result the occupation factor $n(k)$, which is the Fourier transform of the correlation function

$$n(k) = \int \mathrm{d}r g_1(r), \tag{1.57}$$

has a δ-function divergence at $k = 0$.

These results are summarized in the Figure 1.6. We will contrast them with the results in one dimension in the following sections.

1.4.2 Realization of one-dimensional systems

The possibility of obtaining 'one-dimensional' systems is deeply rooted in the quantum nature of the problem. Indeed, the objects themselves are much smaller than the possibility of confining them, so one could naively think that it is always possible for them to avoid each other. The answer comes from the quantization of the wavefunction.

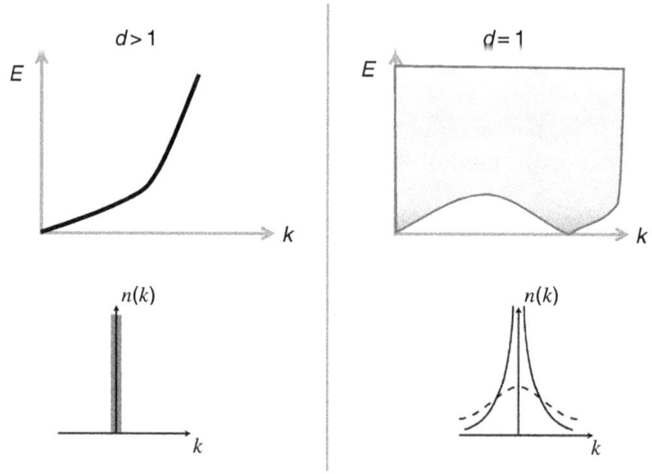

Figure 1.6 (left) Behaviour in high dimension ($d > 1$). One expects an ordered superfluid state for which the phase of the wavefunction is well defined (see text). The excitation spectrum is made of Bogoliubov excitations with a linear dispersion at small k. The single-particle correlation $g_1(k)$ (see text) has a divergent δ peak at $k = 0$. (right) In $d = 1$, no state with a fully ordered phase can exist and correlation functions usually decay as power laws at $T = 0$ and exponentially at finite T. The spectrum has a continuum of excitations and low-energy modes at $k = 2\pi\rho_0$, where ρ_0 is the average density. The single-particle correlation has (at $T = 0$) a power law divergence that characterize the quasi-long range order of the superfluid. At finite temperatures this turns into an exponential decay and thus a Lorentzian-like behaviour for $n(k)$.

In the presence of an optical lattice, one has the wavefunction of Eq. (1.9) with one frequency ω_0 in the longitudinal direction and $\omega_\perp$ in the two other directions (see Figure 1.7). If the confinement along the longitudinal direction is very weak, one can consider that the wavefunction is essentially a plane wave in the longitudinal direction, leading to a wavefunction of the form

$$\psi(x, r_\perp) = \mathrm{e}^{\mathrm{i}kx}\phi(r_\perp), \tag{1.58}$$

where ϕ depends on the precise form of the confining potential. For an infinite well, as shown in Figure 1.7, ϕ is $\phi(y) = \sin((2n_y + 1)\pi y/l)$, whereas it would be a Gaussian function (Eq. (1.9)) for a harmonic confinement. The energy is of the form

$$E = \frac{\hbar^2 k^2}{2m} + \hbar\omega_\perp\left(n_\perp + \frac{1}{2}\right). \tag{1.59}$$

Owing to the narrowness of the transverse channel l, the transverse quantization energy is sizeable, while the energy along the longitudinal direction is nearly continuous. This leads to minibands, as shown in Figure 1.7. If the distance between the minibands is larger than the temperature or interaction energy, one is in a situation where only one

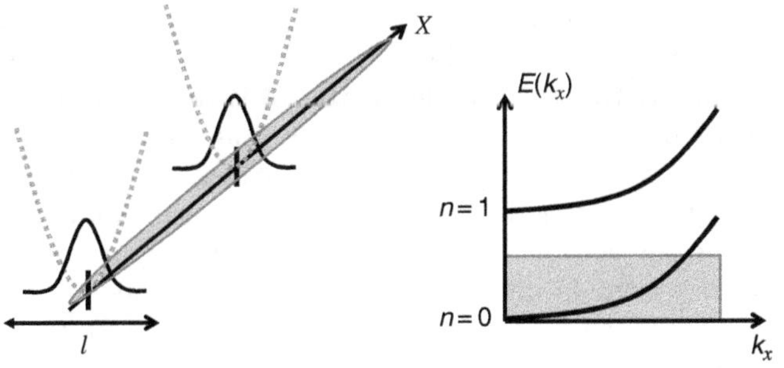

Figure 1.7 (left) Confinement of the electron gas in a one-dimensional tube of transverse size l; x is the direction of the tube. Only one transverse direction of confinement has been shown for clarity. Owing to the transverse confining potential, the transverse degrees of freedom are strongly quantized. (right) Dispersion relation $E(k)$. Only half of the dispersion relation is shown for clarity; k is the momentum parallel to the tube direction. The degrees of freedom transverse to the tube direction lead to the formation of minibands, labelled by a quantum number n. If only one miniband is populated, as represented by the grey box, the system is equivalent to a one-dimensional system where only longitudinal degrees of freedom can vary.

miniband can be excited. The transverse degrees of freedom are thus frozen and only k_x matters. The system is a one-dimensional quantum system.

This is quite similar to the conditions established at the end of Section 1.2 for the use of a single-band model. In addition to the cold-atom situation, similar conditions have been met in condensed matter systems in a variety of problems, such as spin chains and ladders, organic superconductors, nanotubes, edge states in the quantum Hall effect, quantum wires in semiconducting structures, Josephson junction arrays, and helium in nanopores. For more details on these systems, we refer the reader to (Giamarchi 2004). Quantum systems thus allow us to realize situations where, although of course the physical system is three-dimensional, all the important properties can be described purely in one dimension. Solving one-dimensional problems is thus not just a theorist's game but has deep consequences for a large number of physical systems. Let us note that in addition to realizing purely one-dimensional systems, by including a larger and larger number of minibands, an intermediate world can be modelled between the one-dimensional world and the two-dimensional one.

1.4.3 One-dimensional techniques

Treating interacting particle in one dimension is a quite difficult task, since we lose most of the techniques (mean field theory, for example) that we used to have to handle higher-dimensional systems. Fortunately, there are some techniques that have proven very efficient and which, when combined, allow us to make significant progress in our understanding of such systems. We will, of course, not detail these techniques and refer the reader to (Giamarchi 2004; Cazalilla *et al.* 2011) for details and references. Here is a brief summary.

Exact solutions Some models in one dimension are exactly solvable by a technique known as the Bethe ansatz (BA). This technique is limited to special models. For example, the fermionic Hubbard model or the $t-V$ model are BA solvable, but the bosonic Hubbard model is not. These exact solutions allow one to extract the spectrum of excitations relatively easily, and thus with some effort the thermodynamic properties of the system. It is a Herculean task to go beyond this and in particular to compute the correlation functions. Fortunately, significant progress could be accomplished in this domain and some correlation functions have been obtained by BA in recent years. This technique can be potentially extended to out-of-equilibrium situations as well.

Numerical techniques Numerical techniques to deal with quantum interacting particles suffer from notorious convergence problems (specially for fermions) or have a hard time dealing with real-time dynamics. Fortunately, in one dimension, a special technique, the density matrix renormalization group technique, introduced by S. White in the 1990s, produces extremely precise results without suffering from essential convergence problems. Recently, this technique has been extended to deal with dynamical correlation functions as well. It is thus a method of choice for tackling one-dimensional systems. As with any numerical techniques, it is well adapted to give short- and intermediate-range physics, but has the advantage of being able to deal with additional complications, such as the trap or other modifications of the model without too much problem.

Low-energy techniques As for high-dimensional materials (see in particular Section 1.5 for the Fermi liquid theory) there is in one dimension a way to extract a universal description of the physical properties of the problem at low energy. This technique, resting on something called bosonization, is thus complementary to the two above-mentioned techniques. It allows us, right from the start, to obtain the asymptotic properties of the system, as a function of space, time at zero or finite temperature. It also provides a nice framework for understanding the new physical properties of one-dimensional systems. It depends on parameters that can be efficiently determined by the two above techniques or extracted directly from experiments.

These three lines of approach are directly complementary. In this section, we will mostly discuss the bosonization technique, since it is the one that gives the most direct physical representation of the physics of the problem.

The idea behind the bosonization technique is to re-express the excitations of the system in a basis of collective excitations. Indeed, in one dimension it is easy to realize that single particle excitations cannot really exist. One particle when moving will push its neighbours and so on, which means that any individual motion is converted into a collective one. Collective excitations should thus be a good basis for representing a one-dimensional system.

To exploit this idea, let us start with the density operator

$$\rho(x) = \sum_i \delta(x - x_i), \tag{1.60}$$

where x_i is the position operator of the ith particle. We write the position of the ith particle as an 'equilibrium' position R_i^0 that the particle would occupy if the particles were forming a perfect crystalline lattice, and the displacement u_i relative to this equilibrium position. Thus,

$$x_i = R_i^0 + u_i. \tag{1.61}$$

If ρ_0 is the average density of particles, $d = \rho_0^{-1}$ is the distance between the particles. Then, the equilibrium position of the ith particle is

$$R_i^0 = di. \tag{1.62}$$

Note that at this stage it is not important whether we are dealing with fermions or bosons. The density operator written as Eq. (1.60) is not very convenient. To rewrite it in a more pleasant form we introduce a labelling field $\phi_l(x)$ (Haldane 1981a). This field, which is a continuous function of the position, takes the value $\phi_l(x_i) = 2\pi i$ at the position of the ith particle. It can thus be viewed as a way to number the particles. Since in one dimension, unlike higher dimensions, one can always number the particles in a unique way (e.g., starting at $x = -\infty$ and proceeding from left to right), this field is always well defined. Some examples are shown in Figure 1.8. Using this labelling field, one can rewrite the density as

$$\begin{aligned}\rho(x) &= \sum_i \delta(x - x_i) \\ &= \sum_n |\nabla\phi_l(x)|\delta(\phi_l(x) - 2\pi n).\end{aligned} \tag{1.63}$$

It is easy to see from Figure 1.8 that $\phi_l(x)$ can always us be taken as an increasing function of x, which allows us to drop the absolute value in Eq. (1.63). Using the Poisson summation formula, this can be rewritten as

$$\rho(x) = \frac{\nabla\phi_l(x)}{2\pi} \sum_p e^{ip\phi_l(x)}, \tag{1.64}$$

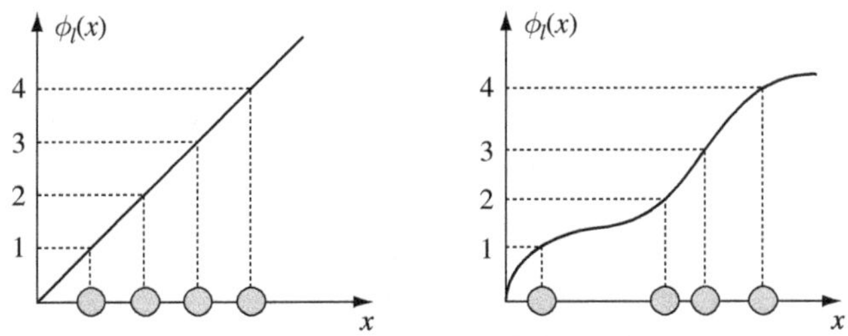

Figure 1.8 Some examples of the labelling field $\phi_l(x)$. If the particles form a perfect lattice of lattice spacing d, then $\phi_l^0(x) = 2\pi x/d$, and is just a straight line. Different functions $\phi_l(x)$ allow us to put the particles at any position in space. Note that $\phi(x)$ is always an increasing function, regardless of the position of the particles. (From (Giamarchi 2004)).

where p is an integer. It is convenient to define a field ϕ relative to the perfect crystalline solution and to introduce

$$\phi_l(x) = 2\pi\rho_0 x - 2\phi(x). \tag{1.65}$$

The density becomes

$$\rho(x) = \left[\rho_0 - \frac{1}{\pi}\nabla\phi(x)\right] \sum_p e^{i2p(\pi\rho_0 x - \phi(x))}. \tag{1.66}$$

Since the density operators at two different sites commute, it is reasonable to expect that the field $\phi(x)$ commutes with itself. Note that if one averages the density over distances that are large compared with the interparticle distance d, all oscillating terms in Eq. (1.66) vanish. Thus, only $p = 0$ remains and this smeared density is

$$\rho_{q\sim 0}(x) \simeq \rho_0 - \frac{1}{\pi}\nabla\phi(x). \tag{1.67}$$

The formula in Eq. (1.66) has the following semiclassical interpretation: the field $\phi(x)$ is essentially the displacements of the particles from a perfect crystalline order with a distance $a = \rho_0^{-1}$. The $p = 0$ term is essentially the standard elastic representation of the density of particles. In addition, the density is composed of density waves with wavevectors $2\pi\rho_0 p$ (the lowest of these is simply the one corresponding to a maximum on each particle). The field $\phi(x)$ gives the phase of these density waves.

We can now write the single-particle creation operator $\psi^\dagger(x)$. Such an operator can always be written (note the similarity with Eq. (1.54)) as

$$\psi^\dagger(x) = [\rho(x)]^{1/2} e^{-i\theta(x)}, \tag{1.68}$$

where $\theta(x)$ is some operator. In the case where one would have Bose condensation, θ would just be the superfluid phase of the system. The commutation relations between the ψ impose some commutation relations between the density operators and the $\theta(x)$. For bosons, the condition is

$$\left[\psi_B(x), \psi_B^\dagger(x')\right] = \delta(x - x'). \tag{1.69}$$

If we assume, quite reasonably, that the field θ commutes with itself ($[\theta(x), \theta(x')] = 0$), a sufficient condition for satisfying Eq. (1.69) is thus

$$\left[\rho(x), e^{-i\theta(x')}\right] = \delta(x - x') e^{-i\theta(x')}. \tag{1.70}$$

It is easy to check that if the density is only the smeared density of Eq. (1.67) then Eq. (1.70) is obviously satisfied if

$$\left[\frac{1}{\pi}\nabla\phi(x), \theta(x')\right] = -i\delta(x - x'). \tag{1.71}$$

One can show that this is indeed the correct condition to use (Giamarchi 2004). Equation (1.71) proves that θ and $\frac{1}{\pi}\nabla\phi$ are canonically conjugate. Note, for the moment, that this results from totally general considerations and does not rest on a given microscopic model. Such commutation relations are also physically very reasonable, since they encode the well-known duality relation between the superfluid phase and the total number of particles. Integrating Eq. (1.71) by parts shows that

$$\pi\Pi(x) = \hbar\nabla\theta(x), \tag{1.72}$$

where $\Pi(x)$ is the canonically conjugate momentum to $\phi(x)$. To obtain the single-particle operator, one can substitute Eq. (1.66) into Eq. (1.68). Since the square root of a delta function is also a delta function up to a normalization factor, the square root of ρ is identical to ρ up to a normalization factor that depends on the ultraviolet structure of the theory. Thus,

$$\psi_{\mathrm{B}}^{\dagger}(x) = \left[\rho_0 - \frac{1}{\pi}\nabla\phi(x)\right]^{1/2} \sum_p \mathrm{e}^{\mathrm{i}2p(\pi\rho_0 x - \phi(x))}\mathrm{e}^{-\mathrm{i}\theta(x)}, \tag{1.73}$$

where the index B emphasizes that this is the representation of a *bosonic* creation operator.

The fact that all operators are now expressed in terms of variables describing *collective* excitations is at the heart of the use of such representations, since, as already pointed out, excitations in one dimension are necessarily collective as soon as interactions are present. In addition, the fields ϕ and θ have a very simple physical interpretation. If one forgets their canonical commutation relations, order in θ indicates that the system has a coherent phase as indicated by Eq. (1.73), which is the signature of superfluidity. On the other hand, order in ϕ means that the density is a perfectly periodic pattern, as can be seen from Eq. (1.66). This means that the system has 'crystallized'. The representation of Eq. (1.66) and Eq. (1.73), and the commutation relation of Eq. (1.71) is thus a dictionary allowing one to re-express every term in any interacting one-dimensional bosonic problem in terms of the new collective variables θ and ϕ. Although this does not solve the problem, but simply re-expresses it, because these are the excitations of the system, we can expect the theory to be much simpler in these variables. We will see that this is indeed the case, and that we can extract some universal physical behaviour.

1.4.4 Universal physics: Luttinger liquids (LL)

To determine the Hamiltonian in the bosonization representation we use Eq. (1.73) in the kinetic energy of bosons. It becomes

$$\begin{aligned} H_K &\simeq \int \mathrm{d}x \frac{\hbar^2\rho_0}{2m}\left(\nabla \mathrm{e}^{\mathrm{i}\theta}\right)\left(\nabla \mathrm{e}^{-\mathrm{i}\theta}\right) + \frac{\hbar^2(\nabla\rho(x))^2}{2m\rho(x)} \\ &= \int \mathrm{d}x \frac{\hbar^2\rho_0}{2m}(\nabla\theta)^2 + \frac{\hbar^2}{2m\pi^2\rho_0}\left(\nabla^2\phi(x)\right)^2. \end{aligned} \tag{1.74}$$

The first part is the part coming from the single-particle operator, which contains fewer powers of $\nabla\phi$ and is thus the most relevant. We have also kept here the second (less relevant term), which allows us to make the connection with Bogoliubov's theory. Using Eq. (1.66), the interaction term becomes

$$H_{\text{int}} = \int \mathrm{d}x V_0 \frac{1}{2\pi^2}(\nabla\phi)^2 \tag{1.75}$$

plus higher-order operators. Keeping only the lowest order shows that the Hamiltonian of the interacting bosonic system can be rewritten as

$$H = \frac{\hbar}{2\pi}\int \mathrm{d}x \left[\frac{uK}{\hbar^2}(\pi\Pi(x))^2 + \frac{u}{K}(\nabla\phi(x))^2\right], \tag{1.76}$$

where we have put back the $\hbar$ for completeness. This leads to the action

$$S/\hbar = \frac{1}{2\pi K}\int \mathrm{d}x\ \mathrm{d}\tau \left[\frac{1}{u}(\partial_\tau\phi)^2 + u(\partial_x\phi(x))^2\right]. \tag{1.77}$$

This Hamiltonian is a standard sound wave one. The fluctuation of the phase ϕ represents the 'phonon' modes of the density wave, as given by Eq. (1.66). One immediately sees that this action leads to a dispersion relation, $\omega^2 = u^2k^2$, i.e., to a linear spectrum, where u is the velocity of the excitations. Note that keeping the second term in Eq. (1.74) gives the dispersion

$$\omega^2 = u^2k^2 + Ak^4, \tag{1.78}$$

which is exactly similar to the Bogoliubov dispersion relation. Note, however, that the theory is quite different from the Bogoliubov one, given the highly non-linear representation of the operators in terms of the fields θ and ϕ.

K is a dimensionless parameter, whose role will become apparent. The parameters u and K are used to parameterize the two coefficients in front of the two operators. In the above expressions, they are given by

$$\begin{aligned} uK &= \frac{\pi\hbar\rho_0}{m}, \\ \frac{u}{K} &= \frac{V_0}{\hbar\pi}. \end{aligned} \tag{1.79}$$

This shows that for weak interactions $u \propto (\rho_0 V_0)^{1/2}$, while $K \propto (\rho_0/V_0)^{1/2}$. In establishing the above expressions we have thrown away the higher-order operators, which are less relevant. The important point is that these higher-order terms will not change the form of the Hamiltonian (like making cross terms between ϕ and θ appears, etc.) but *only* renormalize the coefficients u and K (for more details see (Giamarchi 2004)).

The low-energy properties of interacting quantum fluids are thus described by a Hamiltonian of the form of Eq. (1.76), *provided* the proper u and K are used.

These two coefficients *totally* characterize the low-energy properties of massless one-dimensional systems. The bosonic representation and Hamiltonian of Eq. (1.76) play the same role for one-dimensional systems that the Fermi liquid theory that will be discussed in Section 1.5 plays for higher-dimensional systems. It is an effective low-energy theory that is the fixed point of any massless phase, regardless of the precise form of the microscopic Hamiltonian. This theory, which is known as Luttinger liquid theory (Haldane 1981a; 1981b), depends only on the two parameters u and K. Provided that the correct values of these parameters are used, *all* asymptotic properties of the correlation functions of the system can then be obtained *exactly* using Eqs (1.66) and (1.73) or (1.113).

The Luttinger liquid coefficient can be computed very efficiently. For small interactions, perturbation theory such as Eq. (1.79) can be used. More generally, one just needs two relations involving these coefficients to obtain them. These could be, for example, two thermodynamic quantities, which makes the coefficient easy to extract from either Bethe-ansatz solutions, if the model is integrable, or numerical solutions. The Luttinger liquid theory thus provides, coupled with the numerics, an incredibly accurate way to compute correlations and physical properties of a system (see (Klanjšek *et al.* 2008) for a remarkable example). For more details on the various procedures and models see (Giamarchi 2004; Cazalilla *et al.* 2011). But, of course, the most important use of Luttinger liquid theory is to justify the use of the boson Hamiltonian and fermion–boson relations as starting points for any microscopic model. The Luttinger parameters then become most effective. They can be taken as input, based on general rules (e.g., for bosons $K=\infty$ for non-interacting bosons and K decreases as the repulsion increases; for other general rules see (Giamarchi 2004)), without any reference to a particular microscopic model. This removes part of the caricatural aspects of any modelization of a true experimental system. The Luttinger liquid theory is thus an invaluable tool for tackling the effects of perturbations on an interacting one-dimensional electron gas (such as the effect of lattice, impurities, coupling between chains, etc.). We refer the reader to (Giamarchi 2004) for more on those points.

Let us now examine in detail the physical properties of such a Luttinger liquid. For this, we need the correlation functions. We just give the results here. More detailed calculations and functional integral methods are given in (Giamarchi 2004).

The density–density and the single particle correlations are given by

$$\begin{aligned}\langle T_\tau \psi(r)\psi^\dagger(0)\rangle &= A_1\left(\frac{\alpha}{r}\right)^{\frac{1}{2K}} + \cdots \\ \langle T_\tau \rho(r)\rho(0)\rangle &= \rho_0^2 + \frac{K}{2\pi^2}\frac{y_\alpha^2 - x^2}{(y_\alpha^2+x^2)^2} + A_3\cos(2\pi\rho_0 x)\left(\frac{1}{r}\right)^{2K} + \cdots,\end{aligned} \tag{1.80}$$

where $r=\sqrt{x^2+y^2}$ and $y=u\tau$, with τ the standard imaginary time. Here, the shortest distance in the theory is $\alpha \sim \rho_0^{-1}$. The amplitudes A_i are non-universal objects. They depend on the precise microscopic model, and even on the parameters of the model. These amplitudes can be computed either by BA or by the DMRG calculations. Contrary to the amplitudes A_n, which depend on the precise microscopic model, the power-law decays of the various terms are *universal*. They *all* depend on the unique Luttinger coefficient K. The long-wavelength fluctuations decay with a universal power law. These fluctuations correspond to the hydrodynamic modes of the interacting quantum fluid. The fact that the fluctuations decay very slowly is the signature that there are massless modes present. This corresponds to the sound waves of density described by Eq. (1.76). However, the particle density also has higher Fourier harmonics. The corresponding fluctuations also decay very slowly but this time with an interaction-dependent exponent that is controlled by the LL parameter K. This is also the signature of the presence of a *continuum* of gapless modes, which exists for Fourier components around $Q = 2n\pi\rho_0$. For bosons, K goes to infinity when the interaction goes to zero, which means that the correlations in the density decay increasingly quickly with smaller interactions. This is consistent with the idea that the system, in becoming more and more superfluid, increasingly smears its density fluctuations. This is shown in Figure 1.6.

The single-particle correlation function decays with distance. This reflects the fact that no true superfluid order exists. For the non-interacting system, $K = \infty$ and we discover that the system has off-diagonal long-range order, since the single-particle Green's function does not decay with distance. The system has condensed in the $k = 0$ state. As the repulsion increases (K decreases), the correlation function decays more quickly and the system tends less and less towards superconductivity. The occupation factor $n(k)$ thus has no delta function divergence but a power law one, as shown in Figure 1.6. Note that the presence (or not) of the condensate is not linked directly to the question of superfluidity. The fact that the system is a Luttinger liquid with a finite velocity u implies that in one dimension an interacting-boson system always has a linear spectrum $\omega = uk$, contrary to a free-boson system, where $\omega \propto k^2$. Such a system is thus a *true* superfluid at $T = 0$, since superfluidity is the consequence of the linear spectrum (Mikeska and Schmidt 1970). Of course, when the interaction tends to zero, $u \to 0$, as it should, reinstating the quadratic dispersion of free bosons.

Correlation functions can be computed as easily at finite temperatures, using either standard methods or a conformal mapping. We refer the reader to (Giamarchi 2004) for these calculations. Essentially, the correlation functions now decrease exponentially as $\mathrm{e}^{-C\beta x}$, where β is the inverse temperature and C some constant related to the velocity u and the LL parameter K. This will transform the occupation factor into a Lorentzian one, as shown in Figure 1.6.

Note that one very often encounters an approximation called 'quasi-condensates'. This approximation involves assuming that the density is essentially $\rho(x) = \rho_0$ but that the phase can fluctuate. As is obvious from the representations of Section 1.4.3, this is an approximation compared with the true LL representation. It is a very accurate one in the limit where K is large (small interactions), since in that case the density–density correlation decays extremely quickly with distance. However, for larger interactions the density fluctuations affect the decay of the correlation functions, as described by Eq. (1.80), and the full theory must be retained.

One specially interesting limit to investigate is the so-called Tonks–Girardeau limit (Girardeau 1960; Lieb and Liniger 1963), for which the repulsion between bosons tends

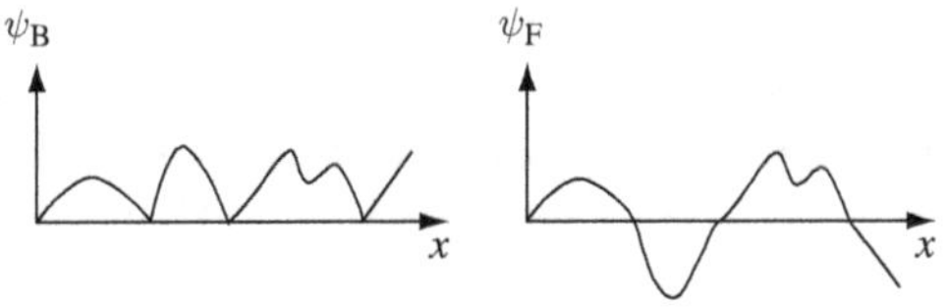

Figure 1.9 If the repulsion between bosons is infinite, one can replace this problem with a free-fermion problem, since the Pauli principle will impose a node at the position of each particle. However, the two problems differ by the sign of the wavefunction across each particle. The properties depending on the square of the wavefunctions are thus identical between the two problems, while single particle properties are quite different.

to infinity. In that case, the repulsion between the bosons acts as a constraint forbidding two fermions from being at the same point. The wavefunction of each particle thus has a node at the position of each other particle. We can thus imagine replacing the repulsion with the Pauli principle of a fake spinless-fermion problem and thus map the infinitely repulsive boson problem onto a free-fermion one. The price to pay for this, as shown in Figure 1.9, is that the wavefunction of the first problem is totally symmetric, while that of the second is totally antisymmetric. Thus, the sign of the wavefunction differs, between each particle. This means that properties that only depend on the square of the wavefunction—such as the thermodynamics and the density correlations—are the same, while those directly depending on the wavefunction (such as the single particle correlations) will, of course, be more complicated to compute. There are direct methods for exploiting this limit. Let us see here how the LL theory allows us to have the correlations. We see that choosing $K = 1$ ensures that the density–density correlations decay as $1/r^2$ and have oscillations at $2\pi\rho_0$. This is exactly what one expects for a free-fermion system (Ashcroft and Mermin 1976; Ziman 1972). The mapping on the free-fermion problem allows us here to fix unambiguously the LL parameter K to $K = 1$. Of course, a direct determination as a function of the interaction also shows that this is a good limit for this parameter when the interaction becomes infinite. The single-particle correlation function is not easy to obtain even in the Tonks–Girardeau limit, given the change of sign of the wavefunction, and one must use rather sophisticated techniques (see, e.g., (Cazalilla *et al.* 2011) for more details). However, the LL theory directly gives that the single-particle correlation decays as $1/\sqrt{r}$. We see in this particularly clear example the universal features that one can extract for the physics of one-dimensional interacting systems.

The Luttinger liquid theory has been checked in various contexts in both condensed matter and cold-atom systems. In condensed matter, the first evidence of a LL power law was obtained for organic superconductors (Schwartz *et al.* 1998), followed by experiments on nanotubes (Yao *et al.* 1999). Many additional tests have been made for other systems, see (Giamarchi 2004; Cazalilla *et al.* 2011) for more systems and references. Recently spin-ladder systems have provided remarkable opportunities for quantitatively test the exponents (Klanjšek *et al.* 2008). In cold atomic systems beautiful experiments could probe one-dimensional interacting bosonic systems. Coupled one-dimensional tubes were obtained (Stöferle *et al.* 2004), in which the role of the superfluid–Mott transition was investigated and the single-particle correlation function measured. The existence of the Tonks–Girardeau limit could be checked by investigating the thermodynamics of the system on a single tube (Kinoshita *et al.* 2004). In such a system, the interaction was raised by using the transverse confinement (Olshanii 1998). The Tonks–Girardeau limit was also observed in systems with optical lattices (Paredes *et al.* 2004). In such systems, the ratio between the kinetic energy and the interactions was controlled by the optical lattice. The single-particle correlation functions were measured and the results were roughly compatible with the $n(k) \sim 1/\sqrt{k}$ that one would expect. However, the inhomogeneities of density, both within a single tube and between the tubes, makes the comparison more complicated. For the particular case of $U = \infty$, the mapping to free fermions allows averaging, allowing a reasonable fit to the experiment. It would, however, be very interesting to have local measurements of single tube systems, to check the intermediate interaction regimes for which no comparison with the LL theory has yet been made. Finally, a remarkable system to test such predictions is provided by atom chips. Indeed, in such systems, the homogeneity is very good, and one can make measurements on a single tube. Interference experiments (Gritsev *et al.* 2006; Hofferberth *et al.* 2008) on condensates have shown excellent agreements with the LL theory, both for the correlation functions and even the full counting statistics. Unfortunately, the interactions are small so that K remains very large and it is difficult to distinguish between LL and simple quasi-condensates. Clearly, further experiments will prove interesting in this remarkable experimental system.

1.4.5 Mott insulators in one dimension

Let us finally examine how the Mott transition can take place in one dimension, and compare this with the results of Section 1.3. Although we have already shown that the superfluid phase is quite different in one dimension and in higher dimensions, we can certainly expect the basic arguments in favour of the Mott transition of Section 1.3 to be still valid. We can thus expect the existence of a Mott transition in one dimension as well. One would then go from a quasi-long-range order of the phase (a power law decay of the superfluid correlations) to a system with one (or an integer number of) boson per site, which would be an insulator. The LL formalism provides a remarkable way of studying such a transition. As in the previous section, we only sketch the solution and refer the reader to (Giamarchi 2004; Cazalilla *et al.* 2011) for the gist of the calculations and for references.

In the absence of a lattice, the interacting one-dimensional system is described by the quadratic action of Eq. (1.77). To determine the effect of a lattice, we just have to add the interaction with a lattice to this action. If we represent the lattice by the potential $V(x) = V_0 \cos(Qx)$, then such a term is

$$H_V = -V_0 \int \mathrm{d}x \cos(Qx)\rho(x). \tag{1.81}$$

We can then use the representation of the density, Eq. (1.66), to see that terms of the form

$$V_0 \int \mathrm{d}x \mathrm{e}^{\mathrm{i}(Q-2p\pi\rho_0)x} \mathrm{e}^{-\mathrm{i}2p\phi(x)} \tag{1.82}$$

appear from Eq. (1.81). We thus see that there are two possibilities. The first possibility is that the wavevector Q of the periodic potential is not commensurate with the density of particles, $Q \neq 2\pi\rho_0$. In that case, one does not have exactly one particle per site; the terms in the integral of Eq. (1.82) oscillate fast and essentially kill the extra term in the action. It follows that the lattice potential is irrelevant and one recovers a LL (superfluid) phase with renormalized parameters u and K. This is exactly similar to the case described in Section 1.3, where the Mott phase could only occur for one particle per site. The second possibility is that the Mott phase can potentially appear if Q is commensurate with the particle density. If $Q = 2\pi\rho_0$, this means, as shown in Figure 1.10, that there is exactly one particle per site. In this case, the oscillations go away and Eq. (1.81) becomes

$$H_V = -V_0 \int \mathrm{d}x \cos(2\phi(x)). \tag{1.83}$$

There are potentially terms with higher p, which correspond to higher commensurabilities (one particle every two sites, etc.). We will not deal with them here and refer the reader to (Giamarchi 2004; Cazalilla *et al.* 2011) for these cases.

The effect of the term in Eq. (1.83) is quite remarkable. There is on the one hand the quadratic action of Eq. (1.77), which allows the field ϕ to fluctuate. These

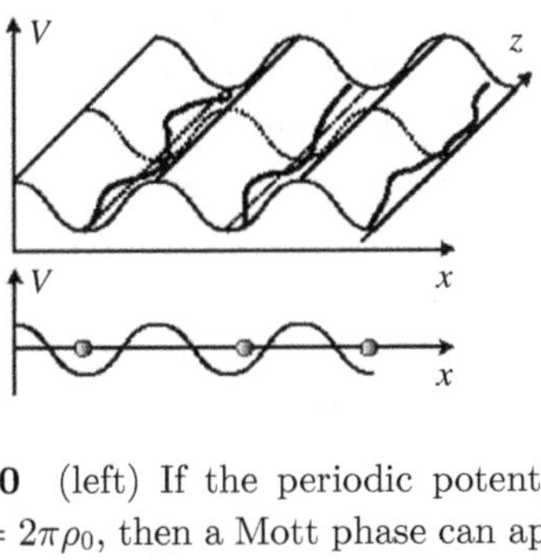

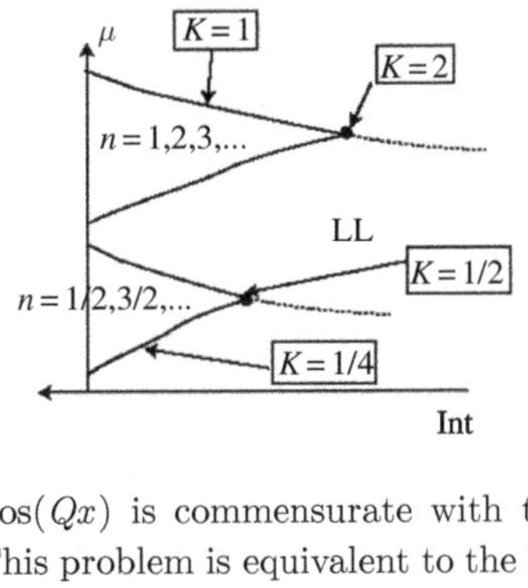

Figure 1.10 (left) If the periodic potential $V_0 \cos(Qx)$ is commensurate with the particle density $Q = 2\pi\rho_0$, then a Mott phase can appear. This problem is equivalent to the localization of elastic lines in a periodic potential or to the Berezinskii–Kosterlitz–Thouless transition in a two-dimensional xy model. It occurs for sufficiently repulsive interactions $K \leq 2$. (right) The phase diagram showing the universal value taken by the LL parameter at the transition. Note the presence of two transitions: the Mott-U transition at constant density upon variation of the interactions; the Mott-δ transition, where the interactions are fixed and the system is doped. These correspond to different universality classes. (After (Giamarchi 2004).)

fluctuations are responsible for the decay of the density correlations. On the other hand, Eq. (1.83) wants to pin the field ϕ to one of the minima of the cosine. If the field ϕ is pinned it means, firstly, that the density does not fluctuate any more. We thus have a phase with one particle per site, which is the Mott phase; secondly, the field θ that is conjugate will fluctuate wildly, with the result that the superfluid correlations are killed exponentially quickly. We thus see that the combination of Eqs (1.77) and (1.83), known as the sine–Gordon model, is the model giving the description of the Mott transition in one dimension. This model has connections with several other models (Giamarchi 2004; Cazalilla *et al.* 2011). As shown in Figure 1.10, it is connected to the fluctuations of classical lines, in a tin-roof potential. It is also connected, in a less obvious way, to the classical Beresinskii–Kosterlitz–Thouless transition in the xy model, with the operator $\cos(2\phi)$ being the vortex creation operator in such a model. We will not examine the connection between these models and refer the reader to (Giamarchi 2004; Cazalilla *et al.* 2011). The transition can occur if the strength of the potential V_0 increases beyond a certain value or if the interaction becomes large enough. In particular, one can show that if the fluctuations are small enough, i.e., if $K \leq 2$, even an *infinitesimal* V_0 is able to pin the field ϕ and one goes into the Mott phase. This is a quite remarkable feature, since it shows that a large enough repulsion between the particles can lead to a Mott phase even if the lattice is very weak. This can be viewed as the pinning of the charge–density wave of the bosons by the periodic potential of the lattice, and is a true quantum effect. Of course, if the lattice is deep, we also recover our usual intuition of the Mott transition.

We thus see that the Mott transition in one dimension is quite similar to its higher-dimensional counterpart. One important difference is that it can also occur for weak lattices, provided that the repulsion is large enough. At the transition, as indicated in Figure 1.10, K takes the *universal* value $K = 2$. Here, the transition is in the universality class of the two-dimensional xy model. This feature persists even to higher dimensions (Fisher *et al.* 1989). The fact that in one dimension we have the bosonized representation of the Hamiltonian allows us to compute all the correlation functions, in both the superfluid and the Mott phase. We refer the reader to (Giamarchi 2004; Cazalilla *et al.* 2011) for further information on that point. An important point to note is that there are in fact two types of Mott transition (Giamarchi 1997) (see also Figure 1.4).

1. One can keep commensurate values for the density, and vary the interactions. This Mott-U transition is in the universality class of the d-dimensional xy model. In one dimension, it is described by the sine–Gordon theory and leads to the universal values shown in Figure 1.10.
2. One can have interactions corresponding to being inside the Mott phase, but dope the system, i.e., vary the density. This Mott-δ transition is in a different universality class. In one dimension, it corresponds to a universality class known as the commensurate–incommensurate phase transition (Giamarchi 2004; Cazalilla *et al.* 2011) and leads to the different critical exponents indicated in Figure 1.10.

In one dimension, these two universality classes have been confirmed by DMRG calculations, where the phase diagram and the LL exponents have been obtained (Kühner *et al.* 2000). In the cold-atom context, the existence of the Mott transition

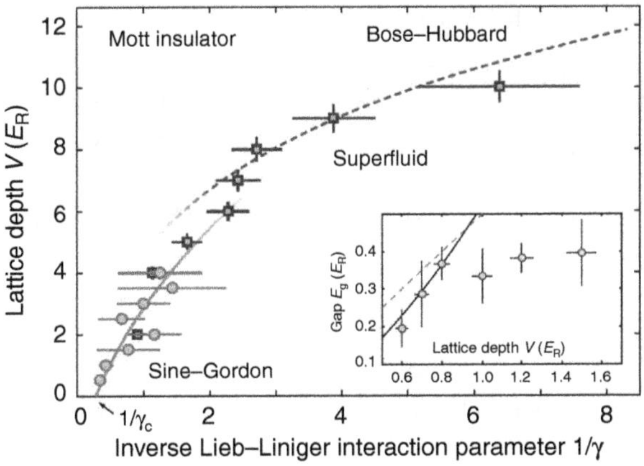

Figure 1.11 Phase diagram of a one-dimensional system gap as a function of the interaction parameter γ (relative to the kinetic energy) and the strength of the optical lattice V, as probed by a shaking of the optical lattice. The finite gap indicates the existence of a Mott phase. One sees that regardless of the strength of the lattice, a Mott transition can occur provided that the repulsion is large enough, in agreement with the LL predictions. After (Haller *et al.* 2010).

in one dimension for an arbitrarily small lattice with sufficiently repulsive interactions has been checked in a remarkable experiment (Haller *et al.* 2010). The gap of the Mott phase, probed by the shaking method described in Section 1.3.4, is shown in Figure 1.11.

The term in Eq. (1.83) has another remarkable consequence. The object that is ordered is not simply the density but the field ϕ itself. Given the relation of Eq. (1.67) the field ϕ is, in a way, the integral of the density, and changes in density (δ-function peaks at the particle positions) correspond to kinks in the field ϕ. The fact that ϕ itself is ordered means that any function of the form $e^{i\alpha\phi}$ tends to a constant. This is making a much stronger statement than merely imposing the density fixed on each site. In fact, the order of ϕ can be traced to the existence of non-local string order parameters (Berg *et al.* 2009). Such string orders, being non-local, are, of course, notoriously difficult to measure. However, the recent possibilities of local addressability in cold atomic systems has allowed us to probe such string orders directly, and a recent experiment has shown the existence of such an order parameter for the Mott transition (Endres *et al.* 2011).

1.5 From free fermions to Fermi liquids

This section is based on graduate courses given in Geneva (together with C. Berthod, A. Iucci, and P. Chudzinski) and in Paris (together with O. Parcollet). For more details, see the course notes:
http://dpmc.unige.ch/gr_giamarchi/Solides/solides.html
and
www.cpht.polytechnique.fr/cpht/correl/teaching/teaching.htm.

1.5.1 Non-interacting fermions

Let us start by recalling some well-known but important facts about non-interacting fermion systems. We shall state these facts without detailing the calculations, since they can be found in every textbook on solid-state physics (Ashcroft and Mermin 1976; Ziman 1972).

We consider independent electrons described by the Hamiltonian

$$H_{\text{kin}} = \sum_{\vec{k}\nu\sigma} \varepsilon_{\vec{k}\nu} c^{\dagger}_{\vec{k}\nu\sigma} c_{\vec{k}\nu\sigma}. \tag{1.84}$$

When considering fermions in a lattice, the sum over $\vec{k}$ runs over the first Brillouin zone, and ν is a band index (which we shall sometimes omit when focusing on a single band). It is important to keep in mind that the creation and destruction operators in this expression refer to *single-particle wavefunctions*. In a lattice, those wavefunctions are of the form (Bloch's theorem): $\phi_{\vec{k}\nu}(\vec{r}) = u_{\vec{k}\nu}(\vec{r})\mathrm{e}^{\mathrm{i}\vec{k}\cdot\vec{r}}$, where $u_{\vec{k}\nu}$ is a Bloch function having the periodicity of the lattice, while $\phi_{\vec{k}} = \mathrm{e}^{\mathrm{i}\vec{k}\cdot\vec{r}}/\sqrt{\Omega}$ in the continuum. The fermion-creation field operator at point $\vec{r}$ is expanded onto these wavefunctions as $\psi^{\dagger}_{\sigma}(\vec{r}) = \sum_{\vec{k}\nu} \phi^{*}_{\vec{k}\nu}(\vec{r}) c^{\dagger}_{\vec{k}\nu\sigma}$.

The eigenstates of Eq. (1.84) are Slater determinants of single-particle wavefunctions, of the form $\det\{\phi_{\vec{k}_i\nu_i}(\vec{r}_j)\}$, which can conveniently be represented in an occupation number basis (Fock representation) as $|\{n_{\vec{k}\nu\sigma}\}\rangle$ with $n_{\vec{k}\nu\sigma}$ equal to zero when the single-particle state is empty, and 1 when it is occupied. The ground state for N fermions corresponds to filling all single-particle states with those fermions, starting from the lowest possible single-particle energy and placing two fermions with opposite spin per state. Hence, the ground state is the 'Fermi sea':

$$|\mathrm{FS}\rangle = \prod_{\vec{k}\nu,\varepsilon_{\vec{k}\nu}<\varepsilon_{\mathrm{F}}} c^{\dagger}_{\vec{k}\nu\uparrow} c^{\dagger}_{\vec{k}\nu\downarrow} |\emptyset\rangle. \tag{1.85}$$

In this expression, ε_{F} is the *Fermi energy*: the largest single-particle energy corresponding to an occupied state. It is also the zero-temperature limit of the chemical potential (assuming a metallic, or liquid, state): $\mu(T \to 0) = \varepsilon_{\mathrm{F}}$. One often incorporates the chemical potential in the energy and defines $\xi^{0}_{\vec{k}\nu} = \varepsilon_{\vec{k}\nu} - \mu$.

In momentum space, the condition $\varepsilon_{\vec{k}\nu} = \varepsilon_{\mathrm{F}}$ ($\xi^{0}_{\vec{k}\nu} = 0$) defines the *Fermi surface*. It has important physical significance, since it defines the loci in momentum space of *zero-energy excitations*. Hence, the presence of a Fermi surface (FS) is a distinctive aspect of a metallic (or liquid) state, in which excitations with arbitrarily low energy are present, in contrast to an insulating state, in which the ground state is separated from excited states by an energy gap.

In the context of cold fermionic atoms in optical lattices, direct imaging of the Fermi surface is possible, as first demonstrated in a remarkable experiment by M. Köhl and coworkers (Köhl *et al.* 2005) for a two-dimensional lattice. The idea is to switch off the lattice potential adiabatically, so that the quasi-momentum of a fermion in a given quasi-momentum state in the lattice is transferred to the corresponding momentum in the continuum (i.e., quasi-momentum is conserved to a good approximation) (Figure 1.12). Time-of-flight expansion and absorption imaging are performed, to obtain the starting quasi-momentum distribution of the atoms inside the lattice (Figure 1.13). By changing the lattice depth, a gradual increase in the size of the FS was observed, resulting at some point into a transition from a metal to a band-insulator, in which the FS coincides exactly with the first Brillouin zone of the lattice. Note that this effect is a consequence of the presence of a confining potential. Indeed, in a homogeneous system, the FS is entirely determined by the number of particles present in the system, and, for a given N, will remain unchanged if the depth of the lattice (hopping amplitude t) is varied. In contrast, in the presence of a harmonic potential of frequency ω_0, the effective density is $\rho = N(a/l)^3$ with $l = \sqrt{t/m\omega_0^3}$, and ρ can be changed by changing t.

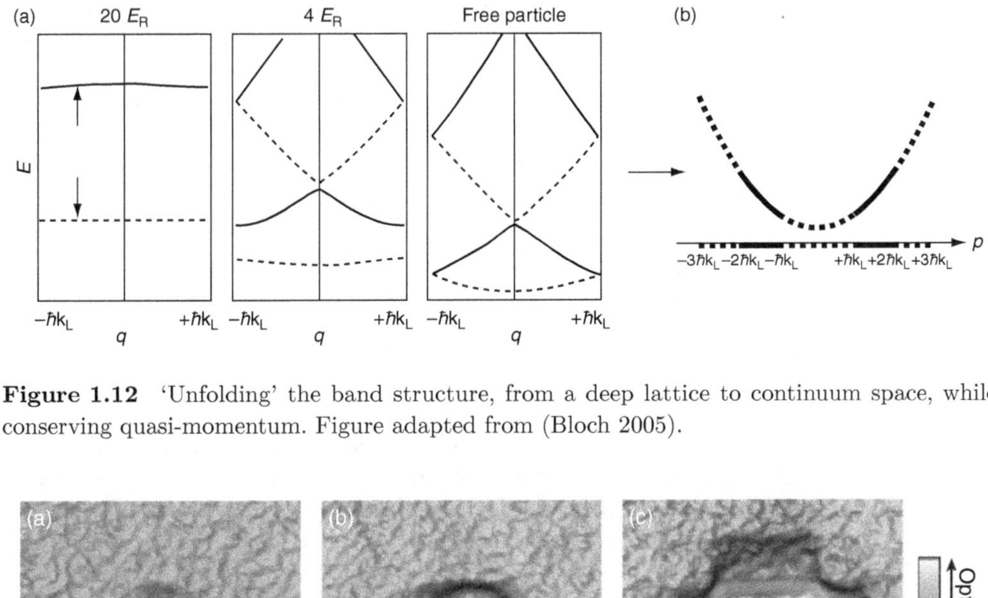

Figure 1.12 'Unfolding' the band structure, from a deep lattice to continuum space, while conserving quasi-momentum. Figure adapted from (Bloch 2005).

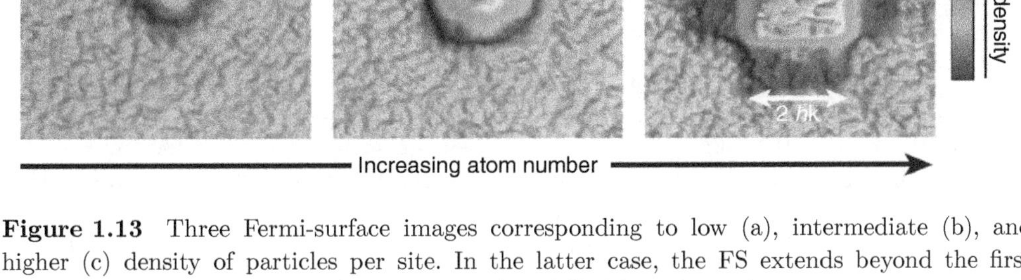

Figure 1.13 Three Fermi-surface images corresponding to low (a), intermediate (b), and higher (c) density of particles per site. In the latter case, the FS extends beyond the first (square) Brillouin zone. From (Esslinger 2010) and (Köhl *et al.* 2005).

It is important, at this stage, to have in mind the order of magnitude of key physical parameters (such as density and mass of particles) relevant to different physical systems of interest, namely: liquid ^{3}He (for which Landau Fermi-liquid theory was first developed historically), electrons in solids, and cold atomic gases. These are summarized in Table 1.1. This table illustrates the fact that the study of degenerate quantum Fermi gases can be undertaken in physical systems with widely different values of the key energy scale: the Fermi energy. While the Fermi temperature ($= \varepsilon_F/k_B$) is a few kelvin in ^{3}He, it is several tens of thousands of kelvin for electrons in solids (ε_F is a few eV) and of the order of a micro- or a nanokelvin for cold atomic gases in optical lattices! The natural unit in the latter case is the recoil energy of the atoms in the lattice laser beams $\hbar^2 k_L^2/2m$, typically of the order of a μK. By these standards, it is considerably more difficult to reach the low-temperature (quantum degenerate) regime $T \ll T_F$ for 'ultracold' atomic gases than for electrons in solids, for which T/T_F is always of order 10^{-2} or less in usual conditions. This observation can also be turned into an advantage: while the high-temperature crossover between a quantum degenerate and a classical gas cannot be observed easily in solids, it is easily observed with atomic gases. The theoretical description of this crossover (and of the 'incoherent' regime $T \gtrsim T_F$ requires us to handle not only very low-energy excitations but also higher-energy excited states, which as explained next, are outside the scope of low-energy effective theories, such as Landau Fermi-liquid theory.

At finite temperature, single-particle states of a non-interacting Fermi gas are occupied with a probability given by the Fermi factor (Figure 1.14):

$$f\left(\xi_{\vec{k}}^0\right) = \left\langle \mathrm{FS}|c_{\vec{k}}^\dagger c_{\vec{k}}|\mathrm{FS}\right\rangle = \frac{1}{e^{\beta\xi_{\vec{k}}^0}+1}. \tag{1.86}$$

In the quantum degenerate regime $T \ll T_F$, the broadening of the Fermi distribution is extremely small. The important states are thus the ones in which particles are excited in a tiny shell close to the Fermi level, as shown in Figure 1.14. The other excitations are completely blocked by the Pauli principle. This strong constraint on available excited states is, of course, what confers to fermionic systems their unique properties, and makes them so different from a classical system, or from a bosonic quantum system. As a consequence, the specific heat $C = \mathrm{d}U/\mathrm{d}T = T\mathrm{d}S/\mathrm{d}T$ is linear

Table 1.1 Typical values of physical parameters for three physical systems. Note that, in the absence of a lattice, $\varepsilon_F \propto n^{2/3}/m$.

	Mass	Lattice spacing	Density	$T_F = \varepsilon_F/k_B$
Liquid ^{3}He	5×10^{-27} kg	–	$\sim 2 \times 10^{22}$ cm^{-3}	A few K
Electrons in solids	9.1×10^{-31} kg	A few Å – $\sim 10^{-10}$ m	10^{21}–10^{23} cm^{-3}	A few $\times 10^4$ K
Cold atoms	m(^{40}K) $\sim$ 66×10^{-27} kg	$\sim$μm	$\sim 10^{11}$–10^{12} cm^{-3}	$\sim$nK–μK

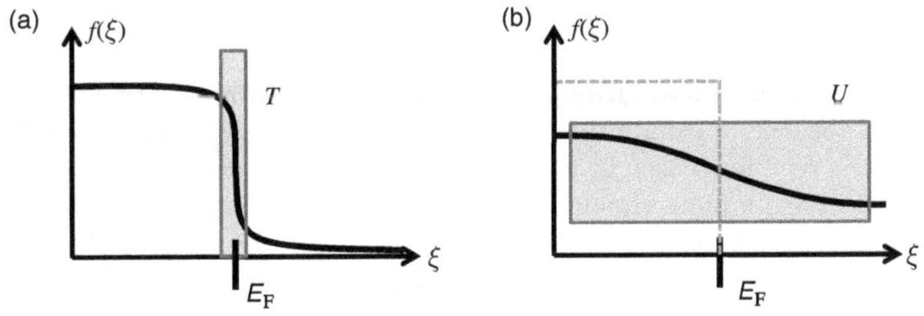

Figure 1.14 Broadening of the Fermi distribution due to temperature, T. (a) When $T \ll T_F$, only a tiny fraction of particles close to the Fermi level can be excited. These low-energy particle-hole excitations control all the physical properties of the system. (b) In the regime $T \gg T_F$, the Fermi factor is broad and high-energy excitations become relevant.

with temperature (contrarily to the case of a classical gas, for which it would be a constant),

$$C(T) \propto k_B^2 \mathcal{N}(\varepsilon_F) T, \quad (T \ll T_F), \tag{1.87}$$

where $\mathcal{N}(\varepsilon_F)$ is the density of states at the Fermi level. The compressibility of the fermion gas,

$$\kappa \propto \frac{\partial n}{\partial \mu} \propto \mathcal{N}(\varepsilon_F), \quad (T \ll T_F), \tag{1.88}$$

reaches a constant value in the limit $T \to 0$ (in contrast to a band insulator, for which ε_F lies within a gap and is hence incompressible $\kappa \propto \mathcal{N}(\varepsilon_F) = 0$). Similarly, the spin susceptibility, which measures the magnetization M of the electron gas in response to an applied magnetic field H, reaches a constant value (Pauli behaviour):

$$\chi = \left(\frac{\partial M}{\partial H}\right)_T \propto \mathcal{N}(\varepsilon_F), \quad (T \ll T_F). \tag{1.89}$$

Note that a system made of independent spins would have instead a divergent spin susceptibility $\chi \propto 1/T$ when $T \to 0$ (Curie behaviour) instead of a constant one. In a non-interacting Fermi gas, the slope of the specific heat, the compressibility and the spin susceptibility are all controlled by the same quantity, namely the density of states at the Fermi level.

1.5.2 Quasi-particles and the $(N \pm 1)$-particle problem

We now consider the effect of interactions between fermions. We focus on a system that is in a compressible liquid (or metallic) state, with no symmetry breaking of any sort (apart from the translational symmetry breaking due to the lattice). Possible transitions into a (Mott) insulating state induced by interactions, as well as magnetic ordering, will be discussed in Section 1.6.

The first important observation is that the ground-state wavefunction, which was simple in the absence of interactions (the Fermi sea), now becomes exceedingly complex. There are very few cases in which this ground-state wavefunction can be found exactly (one such example is the one-dimensional Hubbard model, thanks to the Bethe ansatz). Even numerically, the problem is very difficult. The size of the Hilbert space grows exponentially with the size of the system (for example, for a single-band Hubbard model, it has dimension 4^{N_s}, with N_s the number of lattice sites). Hence, exact diagonalization (e.g., Lanczos) methods can only handle small systems (say, $N_s \lesssim 12$). As for quantum Monte-Carlo simulations, they are faced with the infamous 'minus-sign problem', which severely limits their use, at least in direct simulations without further approximations.

The second important observation is that we may not care so much, after all, about the detailed form of the ground-state wavefunction. What most experiments actually probe are the *excitations* above the ground state (Nozieres 1961). Furthermore, if the system is at low temperature ($T \ll T_F$) and probed in a gentle-enough manner, only low-energy excitations matter. So what we really want is a description of these low-energy excitations. This is fortunate, since general wisdom (backed up by renormalization-group ideas) teaches us that low-energy phenomena (or, equivalently, phenomena involving long timescales) have a large degree of *universality*. Hence, the nature of the low-energy excitations may not depend on all the microscopic details of the specific problem at hand, and a universal effective theory of those low-energy excitations may be in sight. For interacting fermions in more than one dimension, this universal theory is Landau's Fermi-liquid theory (Landau 1957a; 1957b). For one-dimensional systems, it is Luttinger liquid theory (Sections 1.4 and 1.7).

A word of warning, however: being effective theories of low-energy excitations, their applicability is limited to ... low-energy. Hence, these descriptions come with a characteristic scale, above which they are no longer valid and a more detailed quantitative description is required. This scale is associated, as we shall see, with the lifetime of quasi-particles: for energies above a certain coherence scale, long-lived quasi-particles no longer make sense and Landau Fermi liquid theory does not apply. In strongly correlated systems, this coherence scale may be quite low, making the range of validity of effective theories too limited to explain all experimental observations. Also, experiments that perturb the system too strongly (e.g., in a pump-probe experiment) require tools beyond low-energy effective theories.

Quasi-particles: qualitative picture The basic idea behind Landau Fermi liquid theory is that low-energy excited states can be constructed by combining together elementary excitations, with combination rules and quantum numbers identical to those of free particles.

For free fermions, this is clearly the case: the simplest low-energy excitation (at constant particle number N) is obtained by considering a single Slater determinant, which is obtained from the Fermi sea by exciting an electron from a state just below the FS to a state just above it. Hence, a 'particle-hole' excitation has been created, which is a combination of a hole-like excitation (removing a particle) and a particle-like excitation (adding a particle). In a free system, adding a particle in an empty state yields an eigenstate: such an excitation does not care about the presence of all the other electrons in the ground state (other than via the Pauli principle, which prevents it from being created in an already occupied state).

In the presence of interactions, this will not be the case and the added particle interacts with the existing particles in the ground state. For example, for repulsive interactions, one can expect that this excitation repels other electrons in its vicinity. This is represented schematically in Figure 1.15. On the other hand, if one is at low temperature (compared with the Fermi energy) there are very few such excitations and one can thus neglect their mutual interactions (or rather treat them in a mean-field manner). This defines a new composite object (fermion or hole surrounded by its own polarization cloud). This complex object essentially behaves as a particle, with the same quantum numbers (charge, spin) as the original fermion, albeit with renormalized parameters, for example its mass. This image thus strongly suggests that, even in the presence of interactions, good elementary excitations, appearing like free particles, still exist. These particle resemble free fermions but with a renormalized excitation energy $\xi_{\vec{k}}$, which is different from $\xi^0_{\vec{k}}$.

Since our system is gapless, the excitation energy $\xi_{\vec{k}}$ must vanish on a certain surface in momentum space. This defines the Fermi surface *of the interacting system*. Note that, in contrast to the free system, we have established no connection between the FS

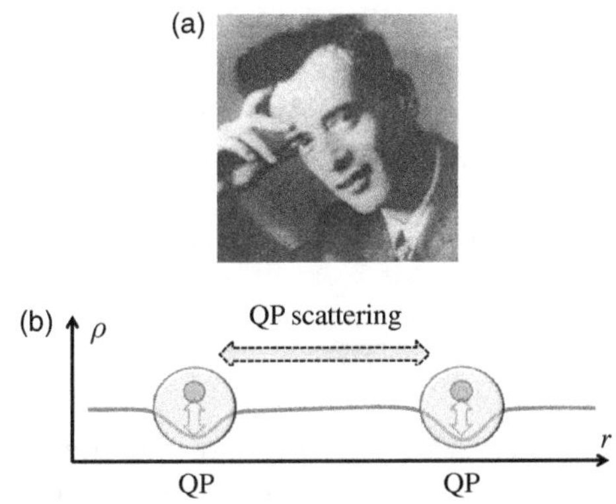

Figure 1.15 (a) Lev Landau, the man behind Fermi liquid theory (and many other things). (b) In a Fermi liquid, the few excitations above the ground state can interact strongly with all the other electrons present in the ground state. The effect of such interactions is strong and leads to a strong change of the parameters entering the low-energy effective theory, compared with free electrons. The combined object behaves as a long-lived particle, named a quasi-particle: its characteristics depend strongly on interactions. However, the scattering of the quasi-particles is blocked by the Pauli principle, leaving a very small phase space for scattering. The lifetime of the quasi-particles is thus extremely large. Furthermore, a low-energy excited state involves a low density of excited quasi-particles. This is the essence of the Fermi liquid theory.

and the ground-state wavefunction: we have been referring only to excitation energies. Under quite general assumptions, a remarkable property does hold, however (even when quasi-particles do not exist in the Landau sense): the momentum-space volume encompassed by the FS is identical to that of the free system (*Luttinger theorem*), and hence entirely fixed by the number of particles. The *shape of the FS*, however, can be changed by interactions (except in the continuum, where it is a sphere specified only by its radius $k_{\rm F}$, given by $2 \times \frac{4}{3}\pi k_{\rm F}^3/h^3 = \frac{N}{\Omega}$).

The dispersion relation $\xi_{\vec{k}}$ specifying the quasi-particle's excitation energy can be expanded around a given point $\vec{k}_{\rm F}$ of the FS as

$$\xi_{\vec{k}} = \nabla_{\vec{k}}\xi|_{\vec{k}_{\rm F}} \cdot (\vec{k} - \vec{k}_{\rm F}) + \cdots , \tag{1.90}$$

which defines a renormalized Fermi velocity of the quasi-particles at this point: $\vec{v}^*(\vec{k}_{\rm F}) = \nabla_{\vec{k}}\xi|_{\vec{k}_{\rm F}}/\hbar$. In the continuum, the Fermi velocity is identical on all points of the spherical FS (by isotropy) and it is customary to define the *effective mass* of quasi-particles by analogy to the free-particle dispersion relation $\xi^0_{\vec{k}} = \hbar^2k^2/2m - \hbar^2k_{\rm F}^2/2m \sim \hbar k_{\rm F}(k - k_{\rm F})/m + \cdots$,

$$\xi_{\vec{k}} = \frac{\hbar k_{\rm F}}{m^*}(k - k_F) + \cdots \tag{1.91}$$

The effective mass controls the low-temperature behaviour of the specific heat of a Landau Fermi liquid, which has the same linear temperature dependence as a free fermion gas:

$$\frac{(C/T)}{(C/T)_0} = \frac{\mathcal{N}_{\rm QP}}{\mathcal{N}}|_{\rm FS} = \frac{m^*}{m}. \tag{1.92}$$

While the low-temperature compressibility and susceptibility are again constant,

$$\frac{\kappa}{\kappa_0} = \frac{m^*/m}{1 + F_0^s}, \quad \frac{\chi}{\chi_0} = \frac{m^*/m}{1 + F_0^a}, \tag{1.93}$$

but involve distinct renormalizations, which parameterize the effective interaction between quasi-particles in the low-energy effective theory (Landau parameters F_0^s and F_0^a). Note that Eqs (1.92 and 1.93) have been written for the isotropic case (continuum). In an anisotropic lattice case, the Landau parameters acquire a more complex angular dependence and these expressions have less predictive power.

The $N \pm 1$-particle problem: Green's function, spectral function and self-energy Of course, those are qualitative ideas. Let us now turn to a more formal treatment. To this aim, we imagine probing the excitations of the N-particle system by removing a fermion from the system (at, say, point $\vec{r}$ and time 0) and sending it to the outside vacuum, or by injecting a fermion from the outside (Figure 1.16). Starting from the (complicated) ground state $|\Psi_0\rangle$ of the N-particle system, we thus prepare the

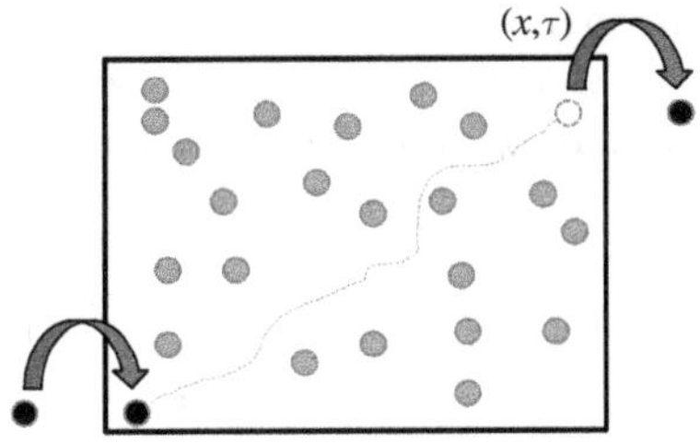

Figure 1.16 A *gedanken* experiment probing the excitations of an interacting system. A particle is injected from outside at point $\vec{r}=0$ and time $t=0$. It propagates through the system and partially decays by creating excitations. The particle is removed at position $\vec{r'}(=x)$ and time $t=\tau$. The amplitude of this process contains a wealth of information about single-particle excitations of the system, encapsulated in the spectral function $A(\vec{k},\omega)$. 'Resemblance' of the final state to the original bare particle signals the existence of long-lived quasi-particle excitations.

wavefunction (for simplicity, we consider a one-band system and one spin component, so that we omit both indices ν, σ):

$$|\Psi(\vec{r},0)\rangle \equiv \psi^\dagger(\vec{r},t)|\Psi_0\rangle = \sum_{\vec{k}} \phi^*_{\vec{k}}(\vec{r})\, c^\dagger_{\vec{k}}|\Psi_0\rangle. \tag{1.94}$$

In a non-interacting system, the wavefunctions on the right-hand side are eigenstates of the system with $N+1$ particles. But this is not the case in the presence of interactions. Hence, to understand the time-evolution of this wavefunction, we expand it onto the *exact eigenstates* of the $N+1$-particle system:

$$|\Psi(\vec{r},0)\rangle = \sum_{\vec{k}} \phi^*_{\vec{k}}(\vec{r}) \sum_A \langle\Psi_A|c^\dagger_{\vec{k}}|\Psi_0\rangle|\Psi_A\rangle. \tag{1.95}$$

We then time-evolve this state with the evolution operator $\exp -\frac{i}{\hbar}(\hat{H}-\mu\hat{N})$, with $\hat{H}$ the full (interacting) many-body Hamiltonian, yielding the state $|\Psi(\vec{r},t)\rangle$. We compare this state with the state obtained by injecting a fermion into the ground state directly at time t, and point $\vec{r'}$, $\psi^\dagger(\vec{r'})|\Psi_0(t)\rangle$, by forming the overlap of the two states $\langle\Psi_0(t)\psi(\vec{r'})|\Psi(\vec{r},t)\rangle$. This overlap can also be viewed as the amplitude for injecting a particle at $\vec{r}$ at time 0 and removing it at $\vec{r'}$ at time t (Figure 1.16). It reads

$$\langle\Psi_0|\psi(\vec{r'},t)\psi^\dagger(\vec{r},0)|\Psi_0\rangle = \sum_{\vec{k}} \phi_{\vec{k}}(\vec{r'})\phi^*_{\vec{k}}(\vec{r}) \sum_A |\langle\Psi_A|c^\dagger_{\vec{k}}|\Psi_0\rangle|^2\, e^{-\frac{i}{\hbar}[E_A-(E_0+\mu)]t}. \tag{1.96}$$

Note that $\mu = \partial E_0/\partial N$ and thus the ground-state energy of the $N+1$-particle system is (for a large gapless system) $E_0^{N+1} \simeq E_0+\mu$. Hence, the frequencies appearing in the time evolution on the r.h.s involve the excitation energies $\hbar\omega = E_A - E_0^{N+1} > 0$ of the $N+1$-particle system. We see that this *gedanken experiment* provides information on the excitation of the system; more precisely on the excited states to which $|\Psi_0\rangle$ couples by injecting a particle. It is very useful to introduce the *one-particle spectral function*, which condenses all this information, and is defined (at $T = 0$) as:

$$\begin{aligned} A(\vec{k},\omega) &\equiv \sum_{A(N+1)} |\langle\Psi_A|c^\dagger_{\vec{k}}|\Psi_0\rangle|^2\,\delta\left[\omega - \frac{1}{\hbar}(E_A - E_0 - \mu)\right] \quad , \quad (\omega > 0), \\ &\equiv \sum_{B(N-1)} |\langle\Psi_B|c_{\vec{k}}|\Psi_0\rangle|^2\,\delta\left[\omega - \frac{1}{\hbar}(E_0 - \mu - E_B)\right] \quad , \quad (\omega < 0). \end{aligned} \tag{1.97}$$

It can easily be checked that the spectral function is normalized over frequencies for each value of the momentum: $\int_{-\infty}^{+\infty} A(\vec{k},\omega)\mathrm{d}\omega = 1$ and that the quasi-momentum distribution of particles in the ground state is given by $N(\vec{k}) \equiv \langle\Psi_0|c^\dagger_{\vec{k}}c_{\vec{k}}|\Psi_0\rangle = \int_{-\infty}^{0} A(\vec{k},\omega)\mathrm{d}\omega$. The spectral function can also be related to the Fourier transform of the *retarded Green's function*, defined as

$$G(\vec{k},t) = -\mathrm{i}\,\theta(t)\langle\Psi_0|[c_{\vec{k}}(t), c^\dagger_{\vec{k}}(0)]_+|\Psi_0\rangle, \tag{1.98}$$

by

$$A(\vec{k},\omega) = -\frac{1}{\pi}\,\mathrm{Im}\,G(\vec{k},\omega). \tag{1.99}$$

In Figure 1.17, we display a cartoon of the spectral function of a Fermi liquid. For momenta not too far from the Fermi surface, the liquid can be decomposed into two spectral features: a narrow peak corresponding to quasi-particle excitations, and a broad continuum corresponding to incoherent excitations. The narrow peak is centred at the excitation frequency $\omega = E_{\vec{k}} - \mu = \xi_{\vec{k}}$ corresponding to the quasi-particle dispersion. It has a spectral weight $Z_{\vec{k}} \leq 1$, while its width $\gamma_{\vec{k}} = \hbar/\tau_{\vec{k}}$ corresponds to the inverse lifetime of quasi-particle excitations and can be approximated by a Lorentzian:

$$A_{\mathrm{QP}}(\vec{k},\omega) \simeq Z_{\vec{k}}\,\frac{\gamma_{\vec{k}}/\pi}{(\omega - \xi_{\vec{k}})^2 + \gamma_{\vec{k}}^2}. \tag{1.100}$$

Correspondingly, the Green's functions can be separated into two components, involving very different timescales:

$$G(\vec{k},\omega) \simeq Z_{\vec{k}}\,\mathrm{e}^{-t/\tau_{\vec{k}}}\,\mathrm{e}^{-\mathrm{i}\xi_{\vec{k}}\,t/\hbar} + G_{\mathrm{inc}}(\vec{k},t). \tag{1.101}$$

The notion of quasi-particle excitations makes sense because their lifetime $\tau_{\vec{k}}$ becomes very large as $\vec{k}$ approaches the Fermi surface, for the phase-space reasons detailed in the next section. As a result, the first term decays very slowly, while the second 'incoherent' one decays quickly (corresponding to a broad frequency spectrum).

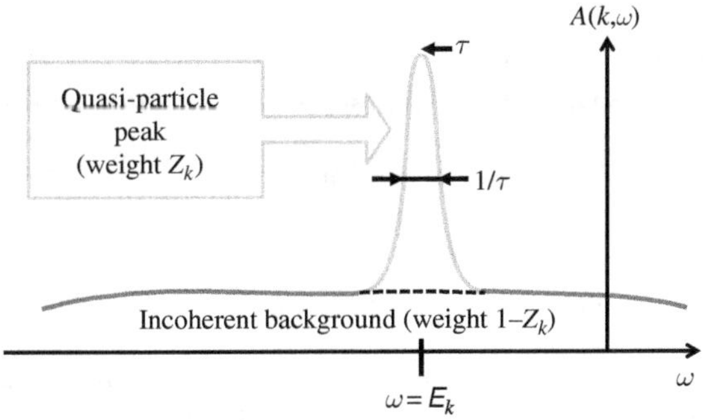

Figure 1.17 A cartoon of the spectral function for interacting particles. One can recognize several features. There is a continuous background of excitations of total weight $1 - Z_{\vec{k}}$. This part of the spectrum corresponds to incoherent excitations, which are not associated with quasi-particles. In addition to this continuous background, there is a quasi-particle peak. The total weight of the peak $Z_{\vec{k}}$ is determined by the real part of the self-energy. The centre of the peak is at a frequency $\xi_{\vec{k}}$, the renormalized quasi-particle dispersion. The quasi-particle peak has a Lorentzian line shape that reflects the finite lifetime of the quasi-particles, and is inversely proportional to the imaginary part of the self-energy.

A very useful quantity is the *self-energy*, which is a measure of the difference between the Green's function of the interacting system and that of the free system. It is defined by (with $\xi^0_{\vec{k}} = \varepsilon_{\vec{k}} - \mu$):

$$G(\vec{k}, \omega) = \frac{1}{\omega - \xi^0_{\vec{k}} - \Sigma(\vec{k}, \omega)}. \tag{1.102}$$

By expanding this expression close to the FS $\vec{k} \simeq \vec{k}_F$ and at low-frequency $\omega \simeq 0$, we find that the key quantities characterizing quasi-particles can be read off from the self-energy. The FS of the interacting system are formed by the quasi-momenta, which satisfy

$$\varepsilon_{\vec{k}_{\rm F}} + \Sigma(\vec{k}_{\rm F}, 0) = \mu, \tag{1.103}$$

in which μ in the r.h.s should be viewed as a function of the particle density n and, of course, of the interaction strength. The quasi-particle spectral weight, dispersion $\xi_{\vec{k}} = \vec{v}^*(\vec{k}_{\rm F}) \cdot (\vec{k} - \vec{k}_{\rm F})$, and inverse lifetime are given by:

$$\begin{aligned} Z_{\vec{k}} &= \left[1 - \frac{\partial \Sigma'}{\partial \omega}|_{\omega=0}\right]^{-1}, \\ \vec{v}^*(\vec{k}_{\rm F}) &= Z_{\vec{k}_{\rm F}} \left[\nabla_{\vec{k}} \xi^0_{\vec{k}} + \nabla_{\vec{k}} \Sigma'\right]_{\omega=0, \vec{k}=\vec{k}_{\rm F}}, \\ \gamma_{\vec{k}} &= Z_{\vec{k}} \Sigma''(\vec{k}, \omega = \xi_{\vec{k}}). \end{aligned} \tag{1.104}$$

In these expressions, Σ' and Σ'' stand for the real and imaginary parts of the retarded self-energy, respectively. As expected, the inverse quasi-particle lifetime is related to the latter (but also involves the weight Z, which in contrast would not appear in the scattering rate measured from transport or optical conductivity). For an isotropic system, this leads to the following expression for the effective mass:

$$\frac{m}{m^*} = Z\left[1 + \frac{m}{\hbar k_F}\frac{\partial \Sigma'}{\partial k}|_{\omega=0,k=k_F}\right]. \quad (1.105)$$

Note that the quasi-particle weight is related only to the frequency dependence of the self-energy, while the effective mass involves both the frequency and momentum dependence. Only when the self-energy is momentum-independent (as, e.g., in the limit of large dimensionality, or within the dynamical mean-field theory approximation) do we have $m^*/m = 1/Z$.

On general grounds, the following phenomena are clear signatures of strong correlations (and need not necessarily occur together):

- A small quasi-particle weight, Z,
- A large effective mass (low $\vec{v}_F^*$),
- A short quasi-particle lifetime (large $\gamma_{\vec{k}}$).

Lifetime of quasi-particles: phase-space constraints To estimate the lifetime of a quasi-particle, let us look at the scattering of a particle from a state $\vec{k}$ to another state. Let us start from the non-interacting ground state in the spirit of a perturbative calculation in the interactions. As shown in Figure 1.18, a particle entering the system with an energy ω and a momentum $\vec{k}$ can excite a particle–hole excitation, taking a particle below the Fermi surface with an energy ω_1 and putting it above the Fermi level with an energy ω_2. The process is possible if the initial state is occupied and the final state is empty. One can estimate the probability of transition using the Fermi golden rule. The probability of the transition directly gives the inverse lifetime of the particle, and thus the imaginary part of the self-energy. We will not care here about the matrix elements of the transition, assuming that all possible transitions will effectively happen with some matrix element. The probability of transition is thus the sum over all possible

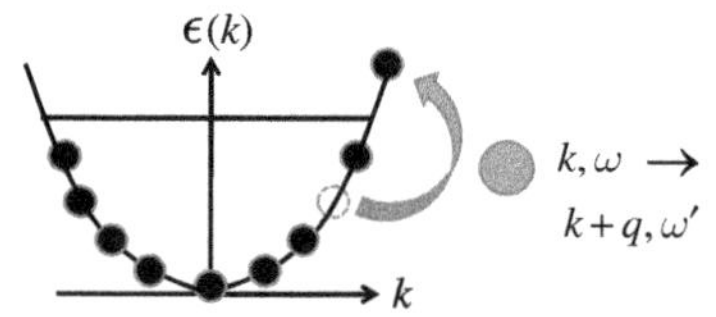

Figure 1.18 Cartoon of the process, giving the lifetime of a particle with energy ω. The ground state of the free system has all its single-particle states filled below the Fermi energy ε_F. The excitations are thus particle–hole excitations, where a particle is promoted from below the Fermi level to above the Fermi level. Owing to the presence of the sharp Fermi level, the phase space available for making such a particle–hole excitations is severely restricted.

initial states and final states that respect the constraints (energy conservation and initial state occupied, final state empty). Since the external particle has an energy ω, it can give at most ω in the transition. Thus $\omega_2 - \omega_1 \leq \omega$. This also directly implies that the initial state cannot go deeper below the Fermi level than ω, otherwise the final state would also be below the Fermi level and the transition would be forbidden. The probability of transition is thus

$$P \propto \int_{-\omega}^{0} \mathrm{d}\omega_1 \int_{0}^{\omega+\omega_1} \mathrm{d}\omega_2 = \frac{1}{2}\omega^2. \tag{1.106}$$

One thus has the remarkable result that because of the *discontinuity* due to the Fermi surface and the Pauli principle that only allows transitions from below to above the Fermi surface, the inverse lifetime behaves as ω^2. This has drastic consequences, since it means that contrarily to naive expectations, when one considers a quasi-particle at the energy ω, the lifetime grows much *faster* than the period $\tau_\omega = 2\pi/\omega$ characterizing the oscillations of the wavefunction (Figure 1.19). In fact,

$$\frac{\tau_{\mathrm{QP}}}{\tau_\omega} \propto \frac{1}{\omega} \to \infty \tag{1.107}$$

when one approaches the Fermi level. In other words, the Landau quasi-particles become *better and better defined* as one gets closer to the Fermi level. This is a remarkable result, since it confirms that we can view the system as composed of single-particle excitations that resemble the original electrons, but with renormalized parameters (effective mass m^*, quasi-particle weight Z_k, etc.).

Probing quasi-particles photoemission and out-coupling spectroscopies Experimental spectroscopic techniques are available, which to a good approximation realize in

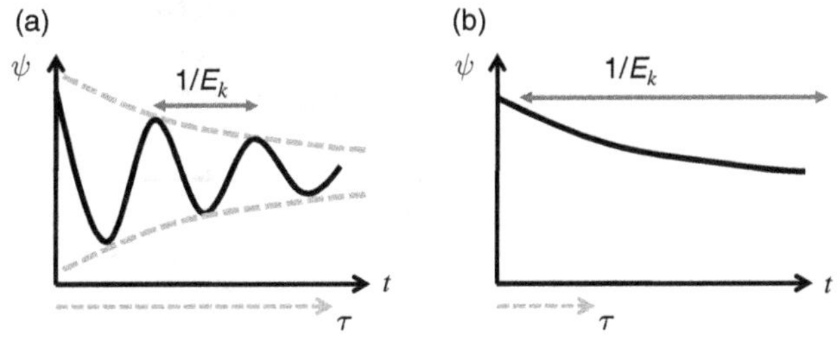

Figure 1.19 For particles with energy E_k and a finite lifetime τ, the energy controls the oscillations in time of the wavefunction. (a) To identify properly an excitation as a particle, it is mandatory that the wavefunction oscillates several times before being damped by the lifetime. (b) In contrast, if the damping is too fast, it is impossible to define the frequency of the oscillations precisely, and thus a precise excitation energy associated with a long-lived quasi-particle.

practice the *gedanken* experiment of Figure 1.16, and hence allow for a direct imaging of quasi-particle excitations.

In the solid-state context, angular-resolved photoemission spectroscopy (ARPES) is a remarkable experimental method, which has undergone considerable development over the past two decades (stimulated to a large extent by the study of high-T_c superconductors) see, e.g., (Damascelli *et al.* 2003; Damascelli 2004). The basic principle of this method is illustrated on Figure 1.20. Under certain conditions and approximations, the measured photoemission intensity is given by

$$I(\vec{k},\omega) = M(\vec{k},\omega)\, A(\vec{k},\omega)\, f(\omega). \tag{1.108}$$

In this expression, M is a matrix element, $A(\vec{k},\omega)$ is the one-particle spectral function introduced already, and $f(\omega)$ is the Fermi distribution. In addition, because of the finite energy resolution, the measured signal is a convolution of Eq. (1.108) with a Gaussian of a certain width. Currently available energy resolutions greatly depend on the incident photon energy: this is typically of order 50−100 meV when using X-rays with energies of several hundred eV at the synchrotron, of order 5−10 meV for laboratory sources, such as a helium lamp (~21 eV), and as low as a fraction of a meV for the recently developed laser-based photoemission ($h\nu \sim 6$ eV). These different sources provide complementary information, since there is a trade-off between bulk versus surface sensitivity, energy resolution, and the momentum-space constraints limiting the area of the Brillouin zone that can be probed.

The Fermi function appears in Eq. (1.108) because this spectroscopy measures the probability of extracting an electron from the system, and hence *mostly probes hole-like excitations*. Momentum-resolved spectroscopies of particle-like excitations,

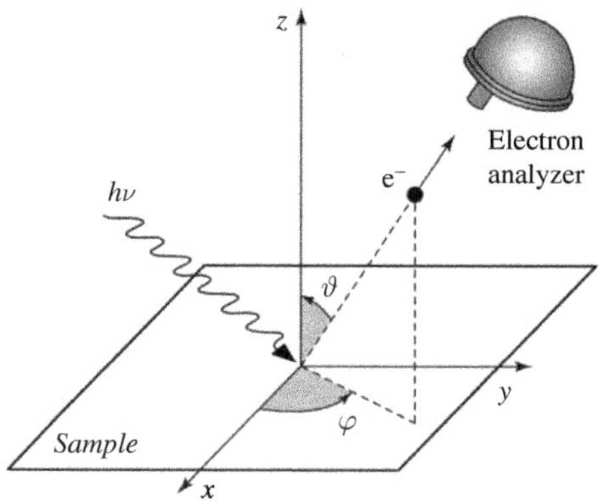

Figure 1.20 Basic principle of photoemission spectroscopy. A photon beam is directed at the (carefully cleaved) surface of the sample. An electron is extracted (photoelectric effect) and its energy and momentum is recorded from the electron analyser. (Adapted from (Damascelli 2004)).

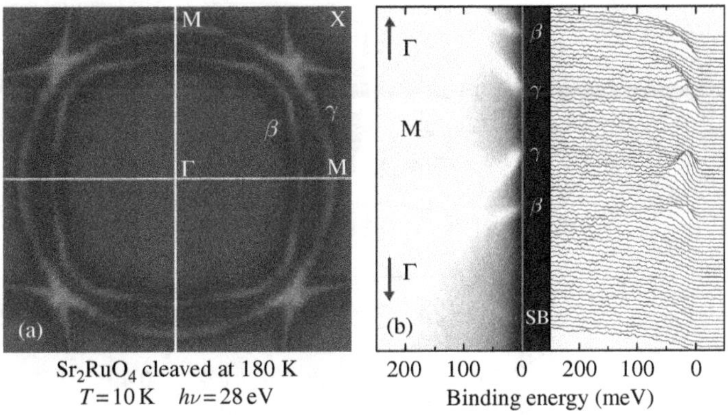

Figure 1.21 ARPES spectroscopy of Sr_2RuO_4. (a) ARPES intensity map providing a determination of the Fermi surface, which has three sheets (α, β, γ). (b) Energy-dependence of the photoemission signal (energy-distribution curves, or EDCs) for several momenta along the Γ–M–Γ-direction in the Brillouin zone. Clear quasi-particle peaks are seen when approaching the FS crossing of the β- and γ-sheets. After (Damascelli *et al.* 2000).

unfortunately, have a much poorer resolution. Scanning tunnelling microscopy (STM), in contrast, does probe both $\omega < 0$ and $\omega > 0$, but in a momentum-integrated way.

As an example, Figure 1.21 displays ARPES measurements on Sr_2RuO_4, a two-dimensional transition-metal oxide with strong electronic correlations. This material has a three-sheeted FS, which can be beautifully imaged with ARPES (as well as with other techniques, such as quantum oscillations in a magnetic field, with good agreement between these two determinations of the FS). In Figure 1.21b, the photoemission signal is displayed along a certain cut $(M - \Gamma)$ in momentum-space, which reveals quasi-particle peaks corresponding to two of these FS sheets. For momenta $\vec{k}$ far from the FS, only a broad incoherent signal is seen. With $\vec{k}$ approaching $\vec{k}_F$, a peak develops, revealing the quasi-particles. When $\vec{k}$ crosses the FS into empty states, the signal disappears because of the Fermi factor. Careful examination of these spectra show that the quasi-particle peak becomes more narrow as the FS is approached, as expected from our (Landau) phase-space arguments.

In the context of cold atomic gases, an analogue of photoemission spectroscopy can also be performed (Dao *et al.* 2007; Stewart *et al.* 2008); see also (Chen *et al.* 2009; Dao *et al.* 2009) and references therein. The idea is to trigger the conversion of one of the hyperfine states (say, $|1\rangle$) present in the system of interest into an out-coupled state (say, $|3\rangle$). This can be achieved either by exciting the system with radio frequencies (rf spectroscopy) or by inducing a stimulated Raman transition using two laser beams (Figure 1.22). A time-of-flight measurement can then be performed, which allows for a determination of the initial momentum $\vec{k}$ of the out-coupled atom. When studying, for example, an interacting mixture of two hyperfine species, $|1\rangle, |1'\rangle$, one would ideally like to pick the out-coupled state, $|3\rangle$, such that it has only very weak interactions with

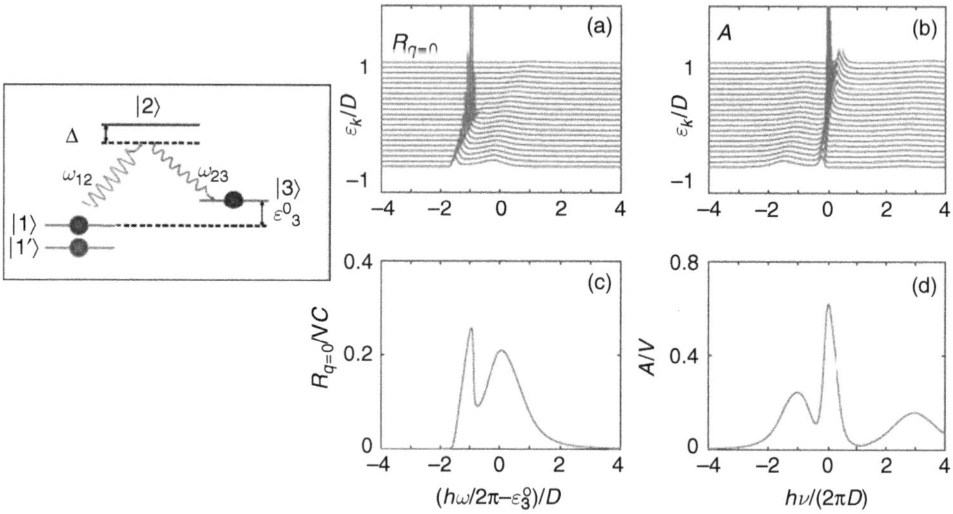

Figure 1.22 Left: Raman out-coupling process. Atoms in hyperfine state $|1\rangle$ are transferred to state $|3\rangle$ using two laser beams with frequencies ω_{12}, ω_{23}. Right: Theoretical spectra for a homogeneous Hubbard model in a strongly correlated regime $U/W = 1.75$ (with W the bandwidth) and total density per site $n = 0.85$, as obtained from dynamical mean-field theory. (a,b): momentum-resolved rf- spectra (a) and spectral function (b). (c,d): momentum-integrated rf-spectrum (c) and spectral function (d). Three main features are seen in the spectral function: upper and lower Hubbard bands corresponding to incoherent, high-energy, quasi-local excitations, and a sharp dispersing quasi-particle peak near the Fermi level. Both the quasi-particle peak and lower Hubbard band are seen in the out-coupled spectra. From (Bernier *et al.* 2010).

either $|1\rangle$ or $|1'\rangle$. Under such conditions, the production rate of out-coupled atoms is obtained from Fermi's golden-rule as:

$$R_{\vec{q}}(\vec{k},\omega) = \frac{2\pi}{\hbar}\sum_{\mathbf{r}} W_{\vec{k}}^{\vec{q}}|\Omega(\vec{r})|^2 f\left(\varepsilon_{3,\vec{k}}^{\vec{r}} - \hbar\omega - \mu_0\right) \times A\left(\vec{k}-\vec{q}, \varepsilon_{3,\vec{k}}^{\vec{r}} - \mu_0 - \hbar\omega; \mu_{\vec{r}}\right). \tag{1.109}$$

In this expression, $\vec{q} = \vec{k}_1 - \vec{k}_2$ is the momentum difference between the two laser beams in a Raman set-up (the case of rf spectroscopy simply amounts to setting $\vec{q} = \vec{0}$). $\vec{r}$ denotes a given point in the trap, with $\mu_{\vec{r}} = \mu_0 - V_1(\vec{r})$ the local chemical potential (using the LDA approximation) and μ_0 the chemical potential at the centre of the trap. $\Omega(\vec{r})$ is the Rabi frequency of the transition and $\omega = \omega_{12} - \omega_{23}$ (Figure 1.22). $W_{\vec{k}}^{\vec{q}}$ is a matrix element involving Wannier functions in the lattice and $\varepsilon_{3,\vec{k}}^{\vec{r}} = \varepsilon_{3,\vec{k}} + V_3(\vec{r})$ is the dispersion of the out-coupled atom corrected by its trapping potential.

The main message of this expression is that, as in photoemission spectroscopy, measuring the out-coupling rate provides access to the spectral function (provided

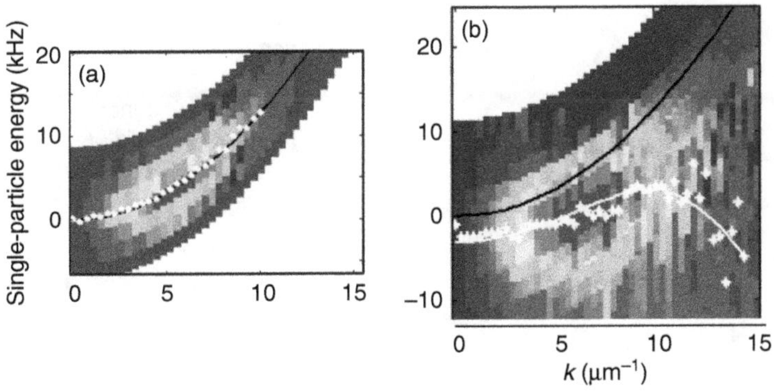

Figure 1.23 'Photoemission' spectroscopy of a two-component trapped gas of ^{40}K fermionic atoms. Displayed are intensity maps obtained by energy- and momentum-resolved rf spectroscopy. (a) Data for a very weakly interacting gas, showing the expected parabolic dispersion of excitations. (b) Data close to unitarity $1/k_F a \sim 0$. From (Stewart *et al.* 2008).

that certain conditions are met). In Figure 1.22, we display theoretical results for the fermionic Hubbard model in a strongly correlated regime, which demonstrate that the key features seen in the spectral function (quasi-particle peak and incoherent lower Hubbard band) can be detected by rf or Raman out-coupling spectroscopy.

A few years ago, the JILA group performed a beautiful experiment (Stewart *et al.* 2008) in which the single-particle excitations of a trapped fermionic gas were measured using energy- and momentum- resolved rf spectroscopy, hence demonstrating the usefulness of such spectroscopic probes. Some of their results are reproduced on Figure 1.23.

1.6 Mott transition of fermions: three dimensions

In this section, we consider again the physics of Mott localization, this time in the context of a two-component gas of fermions in an optical lattice with a repulsive interaction. In comparison to the bosonic case considered in Section 1.3, a major novelty here is the existence of an internal degree of freedom (the two hyperfine states, or the spin in the case of electrons in a solid). This leads to the possibility of long-range 'magnetic' ordering. Even so, it is important to keep in mind that the basic physics behind the Mott localization of fermions is identical to the bosonic case, at least at strong coupling $U \gg t$. Namely, the repulsive interaction makes it unfavourable for particles to hop, resulting in an incompressible state with suppressed density fluctuations.

1.6.1 Homogeneous system: the half-filled Hubbard model

In Figure 1.24, we display the phase diagram of the fermionic Hubbard model for a three-dimensional cubic lattice and a homogeneous density of one particle per site on

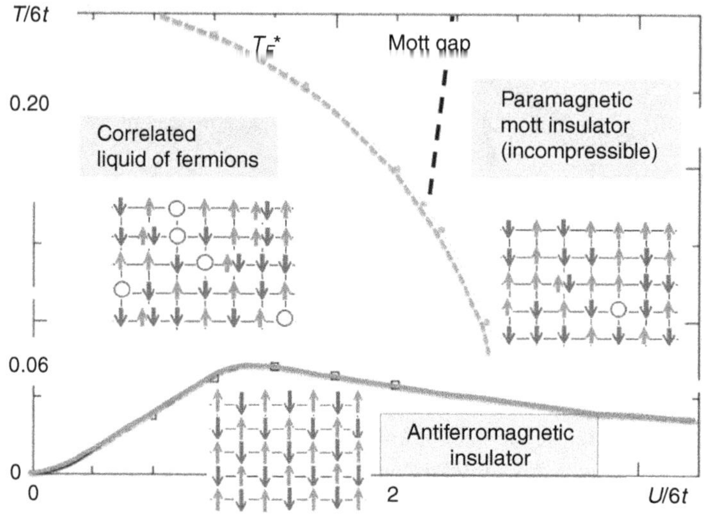

Figure 1.24 Phase diagram of the homogeneous Hubbard model, for a three-dimensional cubic lattice with one particle per site on average. The solid line denotes the phase transition into a long-range ordered antiferromagnet (Néel temperature). The long-dashed line denotes the Mott gap: to the right of this line the paramagnetic phase behaves as an incompressible Mott insulator. The short-dashed line denotes the quasi-particle coherence scale. To the right of this line, the paramagnetic phase behaves as an itinerant fermionic liquid with long-lived quasi-particles. Typical snapshots of the wavefunction in real space are displayed for each regime.

average ('half-filled band'). From the point of view of symmetry breaking and long-range order, there are only two phases. Below the Néel temperature $T_N(U)$ (solid line), antiferromagnetic long-range order occurs, in which spin and translational symmetries are broken. At $T = 0$ this phase has a gap and is insulating. The phase $T > T_N$ is a paramagnet (no long-range spin ordering). However, physically important *crossovers* take place within this phase. The short-dashed line in Figure 1.24 denotes the coherence scale of quasi-particles $T_F^*(U)$. For $T < T_F^*$ (i.e., low-enough temperature and weak-enough coupling) one has an itinerant fermionic liquid, with long-lived quasi-particles (essentially a Fermi liquid, apart from possible subtleties associated with perfect nesting). Another key energy scale is the Mott gap (long-dashed line), which is of order $\Delta_g \sim U$ at large U. For $T < \Delta_g$, one has an essentially incompressible Mott insulator (up to very rare thermal excitations), with frozen density fluctuations and a high spin entropy, i.e., a localized paramagnet.

Hence, when increasing the strength of the repulsive coupling U/t, one crosses over from a fermionic liquid to an incompressible localized paramagnet (through an intermediate incoherent state, which is a 'bad metal' or a poor insulator). In a situation where magnetic long-range order is suppressed (e.g., owing to geometrical frustration of the lattice), this crossover may be replaced by a true phase transition. Because no symmetry breaking distinguishes a metal from an insulator at finite temperature, this transition is expected to be first-order at $T \neq 0$, similar to a liquid–gas transition. Precise description of this crossover or transition is not so easy theoretically. Indeed, by contrast with the phase transition between a superfluid and a Mott insulator of bosons, there is no evident order parameter associated with a static correlation function, which discriminates between a metal and a paramagnetic Mott insulator of fermions. Possible order parameters are all related to frequency-dependent (dynamical) response functions: for example the Drude weight associated with the $\omega \to 0$ component of the ac conductivity, or the quasi-particle weight Z introduced earlier and associated with the low-frequency behaviour of the one-particle Green's function. For this reason, a mean-field theory of this crossover or transition must focus on one- or two-particle response functions. Currently, the most complete approach of this kind is dynamical mean-field theory (DMFT), which has allowed for many successes in understanding strongly correlated fermion systems. For brevity, we refer the reader to review articles for a presentation of this theoretical approach.

From Mott to Slater The transition into the antiferromagnetic state deserves some further remarks. At strong coupling $U/t \gtrsim u^*$, there is a clear separation of energy scales: the gap Δ_{g} ($\sim U$ at large U) is much larger than the antiferromagnetic superexchange $J_{\text{AF}} \sim t^2/U$, which also controls $T_N \propto J_{\text{AF}}$. Hence, for $T \ll \Delta_{\text{g}}$, density fluctuations are frozen out, particles are localized into a Mott insulating state and only the spin degrees of freedom that are described by an effective Heisenberg model are active. In this Mott regime, localization precedes spin ordering, which is a low-energy instability of the insulating paramagnet. In contrast, at weak coupling $U/t \ll u^*$, long-range magnetic order and the blocking of translational degrees of freedom cannot be distinguished: in this regime the opening of a gap is intimately connected to spin ordering and can be described using a simple spin-density-wave mean-field theory (Slater regime). The characteristic coupling u^* separating these two regimes is also the one at which the crossover from a liquid to an insulating state takes place in the paramagnetic state. The Slater and Mott–Heisenberg regimes are connected by a smooth crossover.

From repulsion to attraction There is actually a direct formal analogy between this physics and that of the BCS–BEC crossover in the Hubbard model with an *attractive interaction*. Indeed, on a bipartite lattice (i.e., a lattice made of two sublattice A and B) with nearest-neighbour hopping, one can perform the following symmetry operation:

$$c_{i\uparrow} \rightarrow \widetilde{c}_{i\uparrow}, \quad c_{i\downarrow} \rightarrow (-1)^i \, \widetilde{c}^{\dagger}_{i\downarrow}, \tag{1.110}$$

with $(-1)^i = +1$ on the A sublattice and $= -1$ on the B sublattice. At half-filling, this symmetry simply changes the sign of the coupling U, hence establishing an exact connection between the two cases. Long-range AF order along the x- or y-axis is mapped onto superconducting long-range order, while AF order along the z-axis is mapped onto a charge–density wave state, in which pairs reside preferentially on one sublattice. The two types of ordering are degenerate at half-filling in the attractive case. The BCS regime of the attractive case maps onto the Slater regime of the repulsive one, and the BEC regime onto the Mott–Heisenberg one. In fact, the symmetry also maps the attractive model away from half-filling onto the repulsive model at half filling in a uniform magnetic field (Ho *et al.* 2009).

1.6.2 Trapped system

In the presence of a trapping potential $V(\vec{r}) = V_t\,(r/a)^2$, the local density changes as one moves away from the trap centre, so that different phases can coexist in the system. When the trap potential varies slowly, the density profile is accurately predicted within the local density approximation (LDA), which relates the local state of the system to that of the homogeneous system with a chemical potential $\mu(\vec{r}) = \mu_0 - V(\vec{r})$, so that the local density reads: $n(\vec{r}) = n_{\text{hom}}[\mu = \mu_0 - V(\vec{r})]$. Furthermore, for a large system, one can replace the summation over lattice sites by an integral over the chemical potential, so that the relation between the total particle number N and the chemical potential at the centre of the trap reads:

$$\rho \equiv N\left(\frac{V_t}{6t}\right)^{3/2} = 2\pi \int_{-\infty}^{\mu_0^*} \mathrm{d}\mu^* (\mu_0^* - \mu^*)^{1/2}\, n_{\text{hom}}[\mu^*], \tag{1.111}$$

where μ^* is the chemical potential normalized to the half bandwidth of the lattice $\mu^* = \mu/(6t)$. From this expression, we see that the state diagram of the system can be discussed in terms of the scaled particle number $\rho = N\,(V_t/6t)^{3/2}$: increasing the number of particles or compressing the system by increasing V_t accordingly has the same effect.

In Figure 1.25, we display the state diagram of a two-component fermionic gas confined to a cubic optical lattice in a harmonic trap, as a function of ρ and interaction strength $u = U/6t$. This state diagram was obtained (De Leo *et al.* 2008; De Leo *et al.* 2011) using DMFT calculations for the homogeneous Hubbard model. Different temperatures in the currently accessible range are considered. At still lower temperatures (not displayed in Figure 1.25), antiferromagnetic long-range order will occur in the regimes with a commensurate Mott plateau, as discussed. The state diagram displays four characteristic regimes (labelled L, B, Mc, and Ms). Three of them are illustrated with the corresponding density profiles $n(r)$ calculated at representative points. For a low interaction strength (regime L), the density profile adjusts to the trapping profile and the system remains a Fermi liquid everywhere in the trap. For very large values of the scaled particle number ρ, a band insulator with $n = 2$ forms in the centre of the trap (regime B). The pinning to $n = 2$, and hence the band insulator, is destroyed by increasing the temperature. For larger interaction strengths (regime Mc) a Mott-insulating region appears in the centre of the trap, in which the density is pinned to $n = 1$ particle per site. Close to the boundary of the trap, the Mott insulating region is surrounded by a liquid region. Increasing the number of atoms in the trap at large interaction strengths can increase the pressure exerted on the atoms, and can cause the occurrence of a liquid region with filling larger than one in the centre, surrounded by a shell of Mott insulator with $n = 1$ (regime Ms).

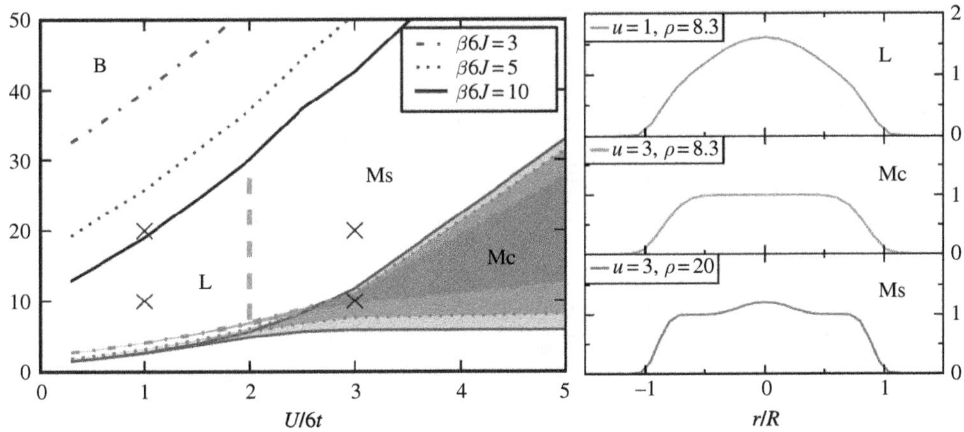

Figure 1.25 State diagram of two-component repulsive fermions in a cubic optical lattice with parabolic confinement, for different temperatures $\beta = 1/k_B T$ (DMFT). The four characteristic regimes (see text) are labelled: B (band insulator in the centre of the trap), Mc (Mott insulator in the centre of the trap, shaded areas), Ms (shell of Mott insulator away from the centre) and L (liquid state). For each temperature, the (crossover) lines indicate, from bottom to top, the ρ values at which the central density takes the values 0.995, 1.005, and 1.995. The grey dashed line marks the crossover from the liquid to the Mott state with increasing interaction. The crosses indicate points at which the density profiles are plotted (right). After (De Leo *et al.* 2008; De Leo *et al.* 2011).

Recently, as displayed on Figure 1.26, experiments have reported the observation of the Mott insulating region for fermionic atoms (Jördens *et al.* 2008; Schneider *et al.* 2008). Figure 1.26 displays a comparison between experimental data and theoretical calculations. In the left panel (Jördens *et al.* 2010), the measured double occupancy as a function of atom number (Jördens *et al.* 2008) is compared with theoretical calculations performed at constant entropy (assuming that turning on the optical lattice corresponds to an adiabatic process). High-temperature series expansions were actually sufficient for this comparison, with DMFT yielding identical results. Fitting theory to experiment allows for a determination of the actual value of the entropy, and ultimately of the temperature attained after the lattice is turned on, for a given particle number. This analysis reveals that the lowest temperature that was reached in this experiment (at small atom number) is comparable to the hopping amplitude ($T \sim t$). The regime with very small double occupancy at the two largest values of $U/6t$ actually corresponds to the formation of a Mott plateau in the centre of the trap. This is more clearly revealed in the measurement of the cloud size as a function of trap compression (right panel) (Schneider *et al.* 2008), see also (Scarola *et al.* 2009), where a plateau signals the onset of an incompressible regime for the largest value of U displayed. These experiments provide us with an 'analogue quantum simulator' validation of theoretical methods for strongly correlated fermions (such as DMFT or high-temperature series), admittedly still in a rather high-temperature regime.

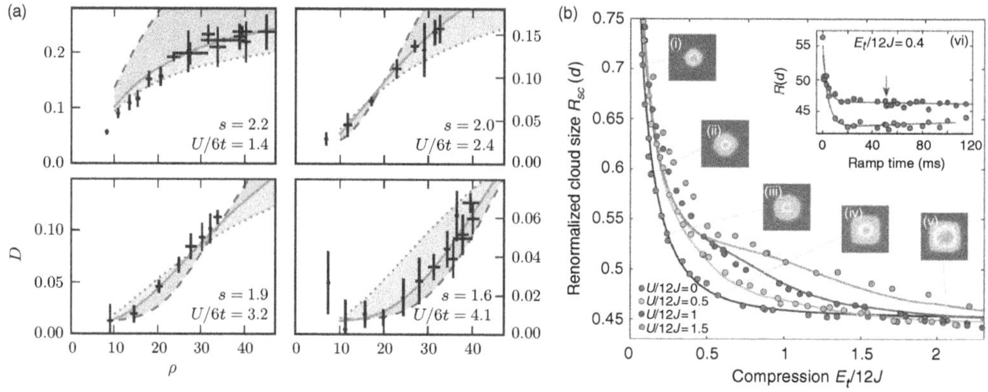

Figure 1.26 Experiments on cold fermionic atoms with repulsive interactions in a three-dimensional optical lattice, revealing the crossover into a Mott insulating regime. (a) Double occupancy: experiment versus theory. Points and error bars are the mean and standard deviation of at least three experimental runs. The solid curve in each panel is the best fit of the second-order high-temperature series to the experimental data and yields specific entropies of $s = 2.2(2), 2.0(5), 1.9(4), 1.6(4)$ for the different interactions strengths of $U/6t = 1.4(2), 2.4(4), 3.2(5), 4.1(7)$. Curves for $s = 1.3$ (dashed curve) and 2.5 (dotted curve) represent the interval of specific entropy measured before and after the ramping of the lattice. Reproduced from (Jördens *et al.* 2010). (b) Cloud sizes versus compression. Measured cloud size R_{sc} in a $V_{\text{lat}} = 8\,E_r$ deep lattice as a function of the external trapping potential for various interactions (from bottom to top) $U/12t = 0$, $U/12t = 0.5$, $U/12t = 1$, $U/12t = 1.5$—in this figure the hopping is designated by J. Dots denote single experimental shots, lines the theoretical expectation from DMFT for $T/T_{\text{F}} = 0.15$ prior to loading. The insets (i–v) show the quasi-momentum distribution of the non-interacting clouds (averaged over several shots). (vi) Resulting cloud size for different lattice ramp times at $E_t/12t = 0.4$ for a non-interacting and an interacting Fermi gas. The arrow marks the ramp time of 50 ms used in the experiment. Reproduced from (Schneider *et al.* 2008).

Shaking of the lattice For the case of bosons (see Section 1.3.4), one can also probe the physics of the fermionic Mott insulator using the shaking of the optical lattice. This probe is complementary to the other spectroscopy probes discussed in this section. For fermions, the major difficulty, compared with the scheme exposed in Section 1.3.4, is that of measuring the energy absorbed. Indeed, for bosons, this could be done by releasing the trap and determining the width of the central peak. For fermions, $n(k)$ is a step and determining how the step is broadened by the absorbed energy is a difficult proposition, given the other sources of broadening. Fortunately, one can proceed differently and it was shown (Kollath *et al.* 2006b) that a measure of the rate of creation of doubly occupied sites (doublon production rate, DPR) would give essentially the same information as a measure of the absorbed energy. Furthermore, the total weight of the peak at the Mott gap U was shown to be directly related to the degree of short-range antiferromagnetic correlations in the system, making the shaking probe a useful probe for antiferromagnetic correlations as well. This last property can easily be understood by using the same arguments as those leading to the superexchange (see Figure 1.2). If two neighbours have parallel spins, then the kinetic-energy term is blocked and thus the perturbation cannot lead to any absorption or DPR. However, if two neighbours have opposite spins, and thus short-range antiferromagnetic order, the transition can take place and absorption of energy of DPR occurs.

The proposal of this new method of probing the system by measuring the DPR was very successful, since the the counting of the doubly occupied sites can be achieved with great accuracy. This allowed implementation of this probe with the modulation amplitude kept small enough that the response stayed in the linear regime (Greif *et al.* 2011), greatly simplifying the theoretical analysis of this probe and allowing a much simpler and efficient comparison between theory and experiment. Although the position of the peak is clearly at the Mott gap Δ_{M} (Kollath *et al.* 2006b), computing the shape of the peak is much more complicated than for the bosons. Indeed, as shown in Figure 1.5 for the bosons, the doublon and holon moved in a featureless environment of singly occupied sites. On the contrary, for the fermions, these two excitations propagate in an antiferromagnetic background, scrambling the spin environment in the process. To compute their propagation it was thus necessary to use approximations of such an antiferromagnetic background. Fortunately, such approximations existed in the condensed matter context, and were made more efficient by the relatively high temperature present in the cold atomic systems. The simplest version is the so-called retraceable approximation, where the holon and doublon simply retrace their steps to return to their point of origin (Sensarma *et al.* 2009). More recently, a more sophisticated approximation using slave boson techniques allowed both the effects of temperature and the trapping to be treated and a very good comparison with the experimental data to be made, as shown in Figure 1.5. Since the shaking amplitude depends on the temperature, this allows to be used the shaking as a thermometer as well.

Reaching the antiferromagnetic state While these experiments have demonstrated the crossover into a paramagnetic Mott insulator, reaching the phase with antiferromagnetic long-range order (Figure 1.24) requires further cooling. To estimate how much further effort is needed, and assuming an adiabatic process, an analysis of the entropy of that phase in the trap is needed. As pointed out in (Werner *et al.* 2005), an important consideration in this respect is the entropy per site of the homogeneous half-filled Hubbard model on the Néel critical line. This quantity is very small at small U/t, passes through a shallow maximum for $U/t \simeq u^*$ (owing to additional density fluctuations) and reaches a finite value $\simeq \ln 2/2$ in the strong-coupling Heisenberg limit (Fuchs *et al.* 2011; Werner *et al.* 2005; Wessel 2010). Note that, in contrast, the Néel temperature becomes very small at large U/t, illustrating the importance of thinking rather in terms of entropy. In the trap, the entropy of liquid wings (in the Mc regime of Figure 1.25) needs to be taken into account as well. Theoretical studies (De Leo *et al.* 2008; Fuchs *et al.* 2011) indicate that, in the favourable case of intermediate coupling, the trapped system must be cooled down to an entropy per atom of order $s = S/N \simeq 0.66$, in order to reach the antiferromagnetic state in the centre of the

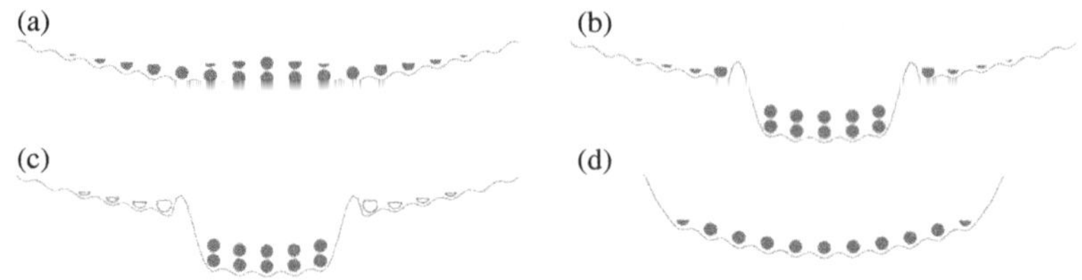

Figure 1.27 Cooling scheme by trap shaping, following (Bernier *et al.* 2009). (a) The atoms trapped in a parabolic profile are loaded into an optical lattice. (b) A band insulator (hence with a very low entropy) is created in a dimple at the centre of the trap. This core region is isolated from the rest of the system, the storage region, by rising potential barriers. (c) If needed, the storage region is removed from the system. (d) The band insulator is relaxed adiabatically (hence preserving the low entropy) to the desired quantum phase, e.g., a Mott insulator, by flattening the dimple and turning off or pushing outwards the barriers.

trap, about three times smaller than the entropy that was attained in the experiments described previously.

Obviously, cooling further fermionic atoms trapped in an optical lattice is currently a key challenge. Several proposals have been put forward to this effect, e.g. in (Bernier *et al.* 2009; Ho and Zhou 2009) A discussion and a number of relevant references on the issue of cooling can be found in those articles, as well as in (De Leo *et al.* 2011). In Figure 1.27, we display the basic idea behind the proposal for cooling by shaping the trap potential made in (Bernier *et al.* 2009).

1.7 One-dimensional fermions

In a similar approach to that used for the bosons, let us examine the case of one-dimensional fermions.

As one can easily guess, there will be no Fermi liquid in one dimension. Indeed the Fermi liquid theory relies on the fact that individual excitations very similar to those of free fermions exist. Clearly, this cannot be the case in 1D where, there can only be collective excitations. Once again, only the general idea will be given and the reader referred to (Giamarchi 2004) for more details and references.

1.7.1 Luttinger liquid and Mott insulators

The bosonization formulae of Section 1.4.3 can easily be modified to deal with bosons. The density is strictly identical and can obviously be expressed in the same way in terms of the field ϕ. For the single-particle operator, one has to satisfy an anticommutation relation instead of Eq. (1.69). We thus have to include in Eq. (1.68) something that introduces the proper minus sign when the two fermion operators are commuted. This is known as a Jordan–Wigner transformation. Here, the operator to be added can easily be guessed. Since the field ϕ_{l} has been constructed to be a multiple of 2π

at each particle, $e^{i\frac{1}{2}\phi_l(x)}$ oscillates between ±1 at the location of consecutive particles. The Fermi field can thus be easily constructed from the boson field of Eq. (1.68) by

$$\psi_F^\dagger(x) = \psi_B^\dagger(x) e^{i\frac{1}{2}\phi_l(x)}. \tag{1.112}$$

This can be rewritten in a form similar to Eq. (1.68) as

$$\psi_F^\dagger(x) = \left[\rho_0 - \frac{1}{\pi}\nabla\phi(x)\right]^{1/2} \sum_p e^{i(2p+1)(\pi\rho_0 x - \phi(x))} e^{-i\theta(x)} \tag{1.113}$$

For fermions, note that the least oscillating term in Eq. (1.113) corresponds to $p = \pm1$. This leads to two terms oscillating with a period $\pm\pi\rho_0$, which is nothing but $\pm k_F$. These two terms thus represent the fermions leaving around their respective Fermi points $\pm k_F$, also known as right movers and left movers.

The action keeps exactly the same form as Eq. (1.77). The important difference is that since the single particle operator already contains ϕ and θ at the lowest order (see Eq. (1.113)) the kinetic energy alone leads to $K = 1$ and interactions perturb around this value, while non-interacting bosons correspond to $K = \infty$. Attraction corresponds to $K > 1$, while repulsion leads to $K < 1$. The correlation functions can thus easily be obtained. For the density–density correlations, we have exactly the same form as for the bosons, Eq. (1.80), the only difference being the different potential values for the LL parameter K. In particular, for the non-interacting fermions, $K = 1$, and one recovers the universal $1/r^2$ decay of the Friedel oscillations in a free electron gas. For repulsive interactions $K < 1$, and density correlations decay more slowly, while for attractive interactions, $K > 1$, they will decay more quickly, being smeared by the superconducting pairing.

The situation is different for the single particle correlations. Contrarily to the case of bosons, for fermions the correlation contains the terms $p = \pm1$, corresponding to fermions close to $\pm k_F$. If we compute the correlation for the right movers we get

$$\begin{aligned} G_R(x,\tau) &= -e^{ik_F x}\langle T_\tau e^{i(\theta(x,\tau)-\phi(x,\tau))} e^{-i(\theta(0,0)-\phi(0,0))}\rangle \\ &= e^{ik_F x} e^{-\left[\frac{K+K^{-1}}{2}\log(r/\alpha) - i\mathrm{Arg}(y+ix)\right]}. \end{aligned} \tag{1.114}$$

The single-particle correlation thus decays as a non-universal power law whose exponent depends on the Luttinger liquid parameter. For free particles ($K = 1$) one recovers

$$G_R(r) = -e^{ik_F x} e^{-\log[(y_\alpha - ix)/\alpha]} = -ie^{ik_F x}\frac{1}{x + i(v_F\tau + \alpha\,\mathrm{Sign}(\tau))}, \tag{1.115}$$

which is the normal function for ballistic particles with velocity u. For interacting systems, $K \neq 1$, the decay of the correlation is always faster, which shows that single-particle excitations do not exist in the one-dimensional world, and thus, of course, that no Fermi liquid can exist.

One important consequence is the occupation factor $n(k)$, which is given by the Fourier transform of the equal-time Green's function

$$n(k) = \int \mathrm{d}x\, \mathrm{e}^{-\mathrm{i}kx} G_R(x, 0^-) = -\int \mathrm{d}x\, \mathrm{e}^{\mathrm{i}(k_\mathrm{F}-k)x} \left(\frac{\alpha}{\sqrt{x^2+\alpha^2}}\right)^{\frac{K+K^{-1}}{2}} \mathrm{e}^{\mathrm{i}\,\mathrm{Arg}(-\alpha+\mathrm{i}x)}. \tag{1.116}$$

The integral can easily be determined by simple dimensional analysis. It is the Fourier transform of a power law and thus

$$n(k) \propto |k - k_\mathrm{F}|^{\frac{K+K^{-1}}{2}-1}. \tag{1.117}$$

The occupation factor is shown in Figure 1.28. Instead of the discontinuity at k_F that signals in a Fermi liquid that fermionic quasi-particles are sharp excitations, one thus finds in one dimension an essential power-law singularity. Formally, this corresponds to $Z = 0$, another signature that all excitations are converted to collective excitations and that new physics emerges compared with the Fermi liquid case.

In practice, this difference on $n(k)$ is relatively difficult to see unless the interaction is quite large, since the discontinuity of $n(k)$ is smeared by the temperature. There are thus better ways to check for the LL properties for fermions (Giamarchi 2004).

In a similar method as for bosons, one can add a lattice to the problem and check for the presence of a Mott insulator. The problem and properties are essentially the same as for bosons and we will not repeat the analysis here, but refer the reader to (Giamarchi 1997; Giamarchi 2004). The essential difference comes again from the different values of the LL parameter K for the two systems. So, for example, for the Hubbard model, the Mott insulator can be obtained for any values of $K < 1$, i.e., for any repulsive interaction. This is very similar to what happens in higher dimensions (see Figure 1.24). The perfect antiferromagnetic order is replaced by a power law decay of the antiferromagnetic correlation functions. As for the case of bosons, string-order parameters can exist.

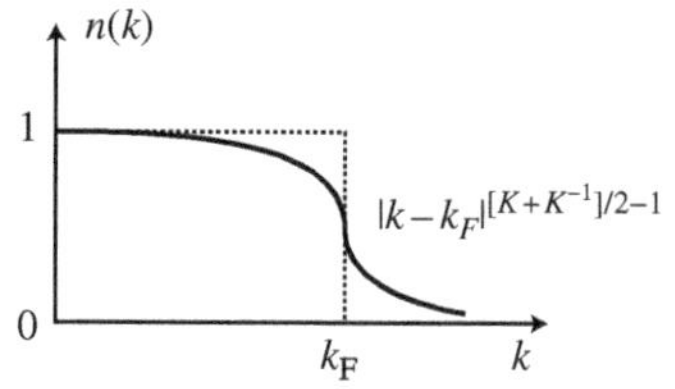

Figure 1.28 The occupation factor $n(k)$. Instead of the usual discontinuity at k_F for a Fermi liquid, it has a power-law essential singularity. This is the signature that fermionic quasi-particles do not exist in one dimension. Note that the position of the singularity is still k_F. This is a consequence of Luttinger's theorem, which states that the volume of the Fermi surface cannot be changed by interactions.

1.7.2 Two-component fermions: spin–charge separation

A very interesting property of one-dimensional systems can be seen in two-component systems (such as, e.g., the Hubbard model). In that case, one can represent the excitations by introducing collective variables for each component of the spins. One thus has four collective variables, $(\phi_\uparrow, \theta_\uparrow)$ and $(\phi_\downarrow, \theta_\downarrow)$. However, one can see that something remarkable happens. If one introduces the variables

$$\begin{aligned} \phi_\rho(x) &= \frac{1}{\sqrt{2}}[\phi_\uparrow(x) + \phi_\downarrow(x)], \\ \phi_\sigma(x) &= \frac{1}{\sqrt{2}}[\phi_\uparrow(x) - \phi_\downarrow(x)], \end{aligned} \tag{1.118}$$

the first variable represents fluctuations of the total density $\rho_\uparrow(x) + \rho_\downarrow(x)$ while the second represents fluctuations of the spin density $\rho_\uparrow(x) + \rho_\downarrow(x)$. In terms of these variables the interaction in the Hubbard model completely decouples. Indeed,

$$\begin{aligned} H = U \sum_i \hat{n}_{i\uparrow} \hat{n}_{i\downarrow} &\to \frac{U}{\pi^2} \int \mathrm{d}x (\nabla \phi_\uparrow(x))(\nabla \phi_\downarrow(x)) \\ &= \frac{U}{2\pi^2} \int \mathrm{d}x \left[(\nabla \phi_\rho(x))^2 - (\nabla \phi_\sigma(x))^2 \right] \end{aligned} \tag{1.119}$$

A similar decoupling occurs for the kinetic energy (Giamarchi 2004). This means that the full Hilbert space of the problem decouples into two sectors, one sector only involving *charge* excitations, and another involving *spin* excitations. It immediately shows that a single-particle excitation, such as the Fermi liquid quasi-particle, which carries charge *and* spin cannot exist. It shows that in one dimension what we could naively think as of an elementary excitation, namely an electron that carries both a charge and a spin, is in fact not the most elementary one. The electron fractionalizes into two more elementary excitations: (a) a holon, which carries a charge but no spin; (b) a spinon, which carries a spin but no charge. These excitations are directly linked to the fields ϕ_ρ and ϕ_σ. Such a fractionalization is thus one of the most important hallmarks of the one-dimensional world. It occurs in a variety of systems and contexts (Giamarchi 2004). One can draw a cartoon to visualize the case of fermions with spin. Such a cartoon is indicated in Figure 1.29. We also see that such a mechanism does not occur naturally in higher dimensions. An important consequence of the spin–charge separation would be the occurrence in photoemissions of a double-singularity structure at the energies of the holon and the spinon and not the single one that one expects in a Fermi liquid (see Figure 1.17). Probing for such an effect is thus extremely interesting and challenging. In the condensed matter context, only one experiment performing tunnelling between two quantum wires could observe such a spin–charge separation (Auslaender *et al.* 2005; Tserkovnyak *et al.* 2002). Cold atoms could thus be a very nice system to observe this effect. For fermions, the temperature is still an issue, thus proposals to use two components bosonic systems have instead been put forward (Kleine *et al.* 2007; Kleine *et al.* 2008) and remain to be tested.

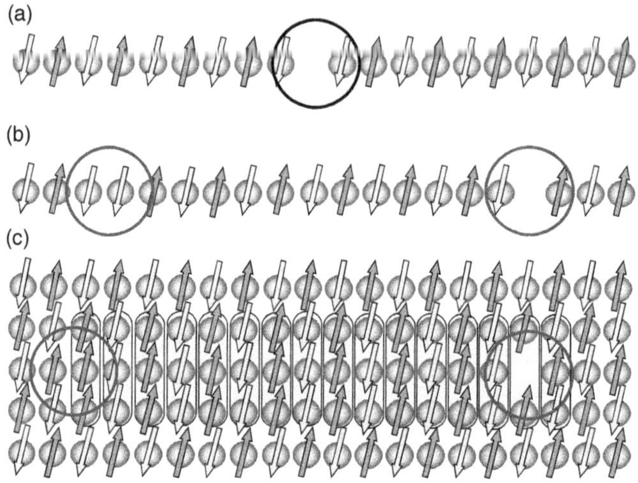

Figure 1.29 A cartoon of the spin–charge separation (fractionalization of excitations) that naturally occurs in one dimension; (a) one removes a particle, which carries a spin and a charge; (b) after the excitations have propagated, we see that there is a place in the system where two parallel spins exist but no charge is missing. This is the spinon, which carries spin but no charge. There is also a hole that has no accompanying distortion of the surrounding antiferromagnetic environment. This is the holon with a charge but no spin. The particle has thus fractionalized into to more elementary (collective) excitations. (c) Unlike what happens in 1D, in higher dimensions the holon and the spinon are held together by a series of frustrated bonds. They are thus bound and form the Fermi liquid quasi-particle.

1.8 Conclusion

This concludes our brief tour of interacting quantum fluids. We have presented the basic concepts that underlie our understanding of quantum interacting systems, both fermionic and bosonic. Two major cornerstones are the Fermi liquid and the Luttinger liquid theories, which are effective theories of the low-energy excitations of the system. They apply in two and higher dimensions, and in one dimension, respectively. They constitute references with which any novel properties or novel system must be compared. Important effects of the interactions, such as the superfluid, Mott-insulating and antiferromagnetic phases have been discussed, and are at the forefront of current research.

A full solution or complete understanding of interacting quantum models beyond these low-energy effective models is still a tremendously difficult task today, although recent years have witnessed significant progress in the field. Indeed, the arsenal of tools at our disposal to tackle such questions, on both the analytic and the numerical side, has increased considerably, and those tools have undergone considerable development. Notwithstanding, the physics of such a simple model as the Hubbard model is still a formidable challenge, especially in two dimensions. Cold atomic systems in optical lattices have provided a remarkable realization of such models, and it is certain that the 'quantum simulators' realized in this novel experimental set-up will help in driving the field forward.

Of course, many more challenges remain and this chapter cannot even list all the exciting new subjects that are connected to this physics. It is clear that questions such as cooling, thermometry, and new experimental probes or spectroscopies are of central interest in order to make progress. Cold atoms, through the control one can exert on the dimensionality of the lattice, have also opened the way to the study of dimensional crossovers between low- and higher-dimensional situations. For example, the passage from a one-dimensional situation to a two- or three-dimensional one remains a challenge that is, of course, of direct interest to many systems in condensed matter physics. In a similar way, cold atomic systems have opened up the possibility of tackling much richer situations involving several internal degrees of freedom, e.g., bosons with two 'spin' components, Bose–Fermi mixtures, multicomponent pairing states, etc. All these systems potentially display very rich and novel physics. Cold atoms have also provided remarkable isolated quantum systems, enabling physicists to tackle in a different system than in condensed matter the question of the out-of-equilibrium behaviour of interacting quantum systems. They also open the possibility of dealing in a controlled manner with the influence of an external dissipative bath. Last but not least, and because of the extreme control on the properties of the system they have allowed the study, in a controlled way, of the influence of disorder and the combined effects of disorder and interactions.

All these subjects go far beyond, but build upon, the material outlined here and constitute the heart of the research on strongly correlated quantum systems. Cold atoms have opened all these avenues and new frontiers for us. We are only at the beginning of the trip, and we can surely expect beautiful surprises and discoveries in the years to come.

Acknowledgements

We are especially grateful to: J.-S. Bernier, I. Bloch, I. Carusotto, M. Cazalilla, V. Cheianov, J. Dalibard, T. L. Dao, L. De Leo, E. Demler, T. Esslinger, M. Ferrero, F. Gerbier, A. F. Ho, A. Iucci, C. Kollath, M. Köhl, O. Parcollet, C. Salomon, U. Schollwoeck, A. Tokuno, and M. Zvonarev for discussions and collaborations. We acknowledge the support of the Agence Nationale de la Recherche, France (under programs GASCOR, FABIOLA and FAMOUS), the Swiss National Science Foundation under MaNEP and Division II, and the Army Research Office (DARPA-OLE program).

References

Ashcroft, NW and Mermin, ND (1976). *Solid State Physics*. Saunders College, Philadelphia.

Auerbach, A (1998). *Interacting Electrons and Quantum Magnetism*. Springer, Berlin.

Auslaender, OM, Steinberg, H, Yacoby, A, *et al.* (2005). *Science*, **308**, 88.

Batrouni, GG, Rousseau, V, Scalettar, RT, *et al.* (2002). *Phys. Rev. Lett.*, **89**(11), 117203.
Berg, E, Dalla Torre, EG, Giamarchi, T, and Altman, E (2009). *Phys. Rev. B*, **77**, 245119.
Bernier, J-S, Kollath, C, Georges, A, *et al.* (2009). *Phys. Rev. A*, **79**(6), 061601.
Bernier, J, Dao, T-L, Kollath, C, Georges, A, and Cornaglia, PS (2010). *Phys. Rev. A*, **81**(6), 063618.
Bloch, I (2005). *Nat. Phys.*, **1**, 23.
Bloch, I, Dalibard, J, and Zwerger, W (2008). *Rev. Mod. Phys.*, **80**, 885.
Cazalilla, MA, Ho, AF, and Giamarchi, T (2005). *Phys. Rev. Lett.*, **95**, 226402.
Cazalilla, MA, Citro, R, Giamarchi, T, Orignac, E, and Rigol, M (2011). *Rev. Mod. Phys.*, **83**, 1405.
Chen, Q, He, Y, Chien, C-C, and Levin, K (2009). *Rep. Prog. Phys.*, **72**(12), 122501.
Damascelli, A (2004). *Phys. Scripta*, **T109**, 61.
Damascelli, A, Lu, DH, Shen, KM, *et al.* (2000). *Phys. Rev. Lett.*, **85**(24), 5194.
Damascelli, A, Hussain, Z, and Shen, Z-X (2003). *Rev. Mod. Phys.*, **75**, 473.
Dao, T-L, Georges, A, Dalibard, J, Salomon, C, and Carusotto, I (2007). *Phys. Rev. Lett.*, **98**(24), 240402.
Dao, T-L, Carusotto, I, and Georges, A (2009). *Phys. Rev. A*, **80**(2), 023627.
De Leo, L, Kollath, C, Georges, A, Ferrero, M, and Parcollet, O (2008). *Phys. Rev. Lett.*, **101**(21), 210403.
De Leo, L, Bernier, J-S, Kollath, C, Georges, A, and Scarola, VW (2011). *Phys. Rev. A*, **83**, 023606.
Duan, L-M, Demler, E, and Lukin, MD (2003). *Phys. Rev. Lett.*, **91**, 090402.
Endres, M, Cheneau, M, Fukuhara, T, *et al.* (2011). *Science*, **334**(6053), 200.
Esslinger, T (2010). *Annu. Rev. Cond. Mat. Phys.*, **1**, 129.
Fisher, MPA, Weichman, PB, Grinstein, G, and Fisher, DS (1989). *Phys. Rev. B*, **40**(1), 546.
Fölling, S, Widera, A, Müller, T, Gerbier, F, and Bloch, I (2006, August). *Phys. Rev. Lett.*, **97**(6), 060403.
Fuchs, S, Gull, E, Pollet, L, *et al.* (2011). *Phys. Rev. Lett.*, **106**(3), 030401.
Giamarchi, T (1997). *Physica B*, **230–232**, 975.
Giamarchi, T (2004). *Quantum Physics in One Dimension.* Vol. 121, International Series of Monographs on Physics. Oxford University Press, Oxford, UK.
Giamarchi, T (2006). In *Lectures on the Physics of Highly Correlated Electron Systems X*, p. 94. AIP Conference Proceedings. arXiv:cond-mat/0605472.
Giamarchi, T (2011). In *Ultracold Gases and Quantum Information* (ed. C Miniatura *et al.*), Volume XCI, Les Houches 2009, p. 395. Oxford. arXiv:1007.1030.
Girardeau, M (1960). *J. Math. Phys.*, **1**, 516.
Greif, D, Tarruell, L, Uehlinger, T, Jördens, R, and Esslinger, T (2011). *Phys. Rev. Lett.*, **106**, 145302.
Greiner, M, Mandel, O, Esslinger, T, Hänsch, TW, and Bloch, I (2002). *Nature*, **415**, 39.
Gritsev, V, Altman, E, Demler, E, and Polkovnikov, A (2006). *Nat. Phys.*, **2**, 705.
Haldane, FDM (1981a). *Phys. Rev. Lett.*, **47**, 1840.

Haldane, FDM (1981b). *J. Phys. C*, **14**, 2585.
Haller, E, Hart, R, Mark, MJ, *et al.* (2010). *Nature (London)*, **466**, 497.
Ho, T-L and Zhou, Q (2009). *P. Nat. Acad. Sci.*, **106**, 6916.
Ho, AF, Cazalilla, MA, and Giamarchi, T (2009). *Phys. Rev. A*, **79**, 033620.
Hofferberth, S, Lesanovsky, I, Schumm, T, *et al.* (2008). *Nat. Phys.*, **4**, 489.
Hubbard, J (1963). *Proc. Roy. Soc. A*, **276**, 238.
Huber, SD, Altman, E, Büchler, HP, and Blatter, G (2007). *Phys. Rev. B*, **75**, 085106.
Iucci, A, Cazalilla, MA, Ho, AF, and Giamarchi, T (2006). *Phys. Rev. A*, **73**, 41608.
Jaksch, D, Bruder, C, Cirac, JI, Gardiner, CW, and Zoller, P (1998). *Phys. Rev. Lett.*, **81**, 3108.
Jördens, R, Strohmaier, N, Günter, K, Moritz, H, and Esslinger, T (2008). *Nature*, **455**, 204.
Jördens, R, Tarruell, L, Greif, D, *et al.* (2010). *Phys. Rev. Lett.*, **104**(18), 180401.
Kinoshita, T, Wenger, T, and Weiss, DS (2004). *Science*, **305**, 1125.
Klanjšek, M, Mayaffre, H, Berthier, C, *et al.* (2008). *Phys. Rev. Lett.*, **101**, 137207.
Kleine, A, Kollath, C, McCulloch, IP, Giamarchi, T, and Schollwoeck, U (2007). *Phys. Rev. A*, **77**, 013607.
Kleine, A, Kollath, C, McCulloch, IP, Giamarchi, T, and Schollwoeck, U (2008). *New J. Phys.*, **10**, 045025.
Köhl, M, Moritz, H, Stöferle, T, Günter, K, and Esslinger, T (2005). *Phys. Rev. Lett.*, **94**, 080403.
Kollath, C, Iucci, A, Giamarchi, T, Hofstetter, W, and Schollwöck, U (2006a). *Phys. Rev. Lett.*, **97**, 050402.
Kollath, C, Iucci, A, McCulloch, IP, and Giamarchi, T (2006b). *Phys. Rev. A*, **74**, 041604(R).
Krauth, W, Caffarel, M, and Bouchaud, J-P (1992). *Phys. Rev. B*, **45**, 3137.
Kühner, TD, White, SR, and Monien, H (2000). *Phys. Rev. B*, **61**(18), 12474.
Landau, LD (1957a). *Sov. Phys. JETP*, **3**, 920.
Landau, LD (1957b). *Sov. Phys. JETP*, **5**, 101.
Lieb, EH and Liniger, W (1963). *Phys. Rev.*, **130**, 1605.
Mahan, GD (2000). *Many-Particle Physics* (3rd edn). Physics of Solids and Liquids. Kluwer Academic/Plenum Publishers, New York.
Mermin, ND (1968). *Phys. Rev.*, **176**, 250.
Mikeska, HJ and Schmidt, H (1970). *J. Low Temp. Phys.*, **2**, 371.
Nozieres, P (1961). *Theory of Interacting Fermi Systems*. WA Benjamin, New York.
Olshanii, M (1998). *Phys. Rev. Lett.*, **81**, 938.
Paredes, B, Widera, A, Murg, V, *et al.* (2004). *Nature (London)*, **429**, 277.
Pitaevskii, L and Stringari, S (2003). *Bose–Einstein Condensation*. Clarendon Press, Oxford.
Reischl, A, Schmidt, KP, and Uhrig, GS (2005). *Phys. Rev. A*, **72**, 063609.
Rokhsar, DS and Kotliar, BG (1991). *Phys. Rev. B*, **44**, 10328.
Scarola, VW, Pollet, L, Oitmaa, J, and Troyer, M (2009). *Phys. Rev. Lett.*, **102**(13), 135302.
Schneider, U, Hackermüller, L, Will, S, *et al.* (2008). *Science*, **322**, 1520.

Schwartz, A, Dressel, M, Grüner, G, Vescoli, V, Degiorgi, L, and Giamarchi, T (1998). *Phys. Rev. B*, **58**, 1261.

Sensarma, R, Pekker, D, Lukin, MD, and Demler, E (2009). *Phys. Rev. Lett.*, **103**, 035303.

Sheshadri, K, Krishnamurthy, HR, Pandit, R, and Ramakrishnan, TV (1993). *Europhys. Lett.*, **22**, 257.

Stewart, JT, Gaebler, JP, and Jin, DS (2008). *Nature*, **454**, 774.

Stöferle, T, Moritz, H, Schori, C, Köhl, M, and Esslinger, T (2004). *Phys. Rev. Lett.*, **92**, 130403.

Tokuno, A and Giamarchi, T (2011). *Phys. Rev. Lett.*, **106**, 205301.

Tokuno, A, Demler, E, and Giamarchi, T (2012). *Phys. Rev. A*, **85**(5), 053601.

Tserkovnyak, Y, Halperin, BI, Auslaender, OM, and Yacoby, A (2002). *Phys. Rev. Lett.*, **89**, 136805.

Werner, F, Parcollet, O, Georges, A, and Hassan, SR (2005). *Phys. Rev. Lett.*, **95**, 056401.

Wessel, S (2010). *Phys. Rev. B*, **81**(5), 052405.

Yao, Z, Postma, HWC, Balents, L, and Dekker, C (1999). *Nature (London)*, **402**, 273.

Ziman, JM (1972). *Principles of the Theory of Solids*. Cambridge University Press, Cambridge.

Zvonarev, M, Cheianov, VV, and Giamarchi, T (2007). *Phys. Rev. Lett.*, **99**, 240404.

2
Ultracold atoms in optical lattices

Immanuel BLOCH

Max-Planck-Institute of Quantum Optics, Hans Kopfermann Str. 1,
85748 Garching, Germany and Ludwig-Maximilians University,
Schellingstr. 4/II, 80798 Munich, Germany

Chapter Contents

2.1 Overview

This work provides an introduction into the field of strong correlation physics with ultracold atoms in optical lattices. After a basic introduction to the single-particle band structure and lattice configurations, the effect of strong interactions in the Hubbard model is discussed. Detection methods are introduced, which allow one to reveal in-trap density and (quasi)-momentum distributions, as well as correlations between particles on the lattice. The fundamental phases of the bosonic and fermionic Hubbard model are discussed. Superexchange spin–spin interactions that form the basis of quantum magnetism are introduced and the current status on observing such magnetic phenomena is highlighted. Finally, novel possibilities for detecting single-site and single-atom resolved quantum gases are outlined.

2.2 Introduction

Ultracold quantum gases in optical lattices have become a major research area in the field of atomic and molecular quantum gases. Not only do they allow one to investigate prototypical condensed matter Hamiltonians, but they also offer the possibility of reaching novel physical regimes, which are in many cases unattainable in real condensed matter systems (Jaksch and Zoller 2005; Lewenstein *et al.* 2007; Bloch *et al.* 2008). Optical lattices are generally formed by interfering several laser beams; the resulting intensity of the interference pattern is experienced by the atoms as a light-induced potential, via their polarizability. The dimensionality of such a lattice and its geometry can be fully controlled through the amplitude and the directions of the interfering laser beams. Interactions can be tuned via Feshbach resonances (Chin *et al.* 2010) or by quenching kinetic energy when using deeper lattices. Ultracold atoms in optical lattices have thus become a powerful setting for the investigation of strongly correlated quantum gases. They also offer a new window for the observation of non-equilibrium dynamics in quantum many-body systems (Polkovnikov *et al.* 2011), where timescales and dynamical evolutions are conveniently in the regime of milliseconds rather than femtoseconds.

This chapter represents an introduction to the field of optical lattices and covers some of the basics in (a) the generation of optical lattices and (b) the detection methods employed to image *in-situ* and momentum distributions. Some of the many-body phases that have been investigated in this context will be discussed. In the last part of the chapter, novel single-site and single-atom resolved detection techniques are introduced, which have opened a new window for investigating quantum gases in optical lattices with unprecedented resolution, down to the level of revealing individual thermal or quantum fluctuations in the system. This chapter is far from a complete survey of what has been done in the field, but rather represents a selection of introductory topics, with references to more detailed discussions throughout the text.

2.3 Basics of optical lattices

Atoms may be trapped in standing-wave light fields by exploiting the so-called optical dipole force. The basic mechanism underlying such dipole traps may be explained via a simple semiclassical picture: the oscillating electric field of a laser induces an oscillating electric dipole in an atom. This atomic dipole can again interact with the external field, thereby creating a potential for an atom that is proportional to the intensity of the applied laser light. For a frequency below an atomic transition frequency $\omega_L < \omega_0$ (red detuning), the atomic dipole oscillates in phase with the applied electric field, whereas for blue detuning $\omega_L > \omega_0$ it oscillates out of phase. This results in an attractive (repulsive) induced dipole potential for the case of red (blue) detuning (Grimm *et al.* 2000).

2.3.1 Optical lattices

Periodic potentials based on dipole forces can be formed by overlapping two counter-propagating laser beams. Owing to the interference between the two light fields, an optical standing wave with period $\lambda/2$ is formed, in which the atoms can be trapped. More generally, by choosing the two laser beams to interfere under an angle smaller than 180°, one can also realize periodic potentials with a larger period (Peil *et al.* 2003; Hadzibabic *et al.* 2004). For counterpropagating Gaussian laser beams, the resulting trapping potential is of the form

$$V(r,z) = -V_0 \cdot \mathrm{e}^{-2r^2/w^2(z)} \cdot \sin^2(kz), \tag{2.1}$$

where $k = 2\pi/\lambda$ is the wavevector of the laser light and V_0 denotes the depth of the lattice potential.

Periodic potentials in two or three dimensions can be formed by overlapping two or three optical standing waves along different, usually orthogonal, directions. For orthogonal polarization vectors or different frequencies of the individual standing waves, the resulting optical potential in the centre of the trap is then a simple sum of purely sinusoidal potentials in two or three directions.

In a two-dimensional optical lattice potential, the atoms are confined to arrays of tightly confining one-dimensional tubes (see Figure 2.1a). For typical experimental parameters, the harmonic trapping frequencies along the tube are very weak and of the order of 10–200 Hz, while in the radial direction, the trapping frequencies can become as high as 100 kHz. For sufficiently deep lattice depths, atoms can thus move only axially along the tube. In this manner, it is possible to create quantum wires with neutral atoms, which allow the study of strongly correlated gases in one dimension. Arrays of such quantum wires have been realized by several groups (Greiner *et al.* 2001; Moritz *et al.* 2003; Kinoshita *et al.* 2004; Paredes *et al.* 2004; Tolra *et al.* 2004).

In the case of three-dimensional lattice potentials, the confinement on a single lattice site is approximately harmonic. The atoms are then tightly confined at a single lattice site, with trapping frequencies ω_0 of up to $2\pi \times 100\,\mathrm{kHz}$. The energy $\hbar\omega_0 = 2E_r\,(V_0/E_r)^{1/2}$ of local oscillations in the well is of the order of several recoil energies, $E_r = \hbar^2k^2/2m$, which is a natural measure of energy scales in optical lattice potentials. Typical values of E_r are in the range of several kHz for ^{87}Rb.

(a)

(b)

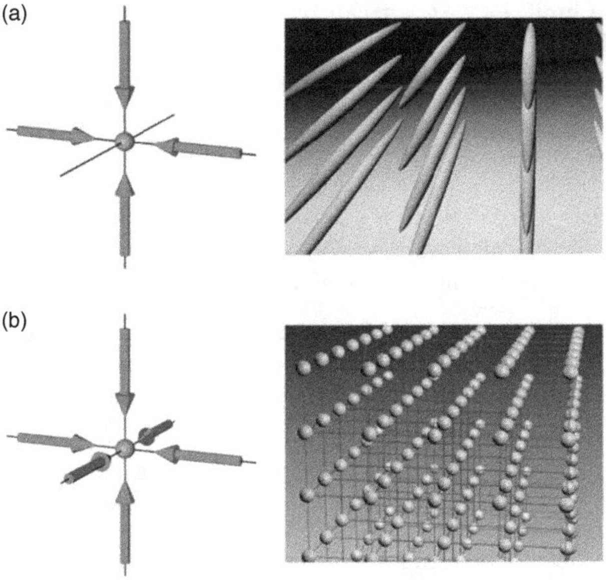

Figure 2.1 Two-dimensional (a) and three-dimensional (b) optical lattice potentials formed by superimposing two or three orthogonal standing waves. For a two-dimensional optical lattice, the atoms are confined to an array of tightly confining one-dimensional potential tubes, whereas in the three-dimensional case the optical lattice an be approximated by a three-dimensional simple cubic array of tightly confining harmonic oscillator potentials.

2.3.2 Tight binding regime

For sufficiently deep lattices above $V_0 \gtrsim 5E_{\rm r}$, quantum gases on a lattice are well described within a tight binding approximation. If the interaction energy between particles is, in addition, smaller than the separation between the lowest- and higher-energy bands, then we may restrict our discussion to particles only occupying a single energy band in the lattice—the lowest Bloch band. For neutral atoms interacting via short-ranged molecular potentials, this is the typical situation encountered in experiments and the one we shall focus on in the following discussion.

Kinetic energy The movement of particles on a lattice under such conditions is simply given by the hopping between neighbouring lattice sites,

$$H = -J \sum_{\langle \mathbf{R},\mathbf{R}' \rangle} \hat{a}^\dagger_{\mathbf{R},\sigma} \hat{a}_{\mathbf{R}',\sigma}, \tag{2.2}$$

where $\hat{a}_{\mathbf{R}',\sigma}$ denotes the fermionic (bosonic) particle destruction operator on lattice site $\mathbf{R}'$. The parameter $J > 0$ is the gain in kinetic energy due to nearest-neighbour tunnelling. In the limit $V_0 \gg E_r$, it can be obtained from the width $W \to 4J$ of the lowest band in the 1D Mathieu equation:

$$J = \frac{4}{\sqrt{\pi}} E_r \left(\frac{V_0}{E_r}\right)^{3/4} \exp\left[-2\left(\frac{V_0}{E_r}\right)^{1/2}\right]. \tag{2.3}$$

For static lattice potentials $J > 0$, however, it has recently been shown that by time modulation of the lattice potential position, one can also access regimes where effectively, after time averaging, $J < 0$ (Eckardt *et al.* 2005; Lignier *et al.* 2007).

Interaction energy For typical short-ranged collisional interactions between the particles, the interaction energy between two atoms can be described by an on-site interaction energy, typical of the form seen in Hubbard models:

$$H_{\text{int}}^{\text{B}} = \frac{1}{2} U \sum_{\mathbf{R}} \hat{n}_{\mathbf{R}}(\hat{n}_{\mathbf{R}} - 1) \quad \text{and} \quad H_{\text{int}}^{\text{F}} = U \sum_{\mathbf{R}} \hat{n}_{\mathbf{R},\sigma} \hat{n}_{\mathbf{R},\sigma'}, \tag{2.4}$$

for single species bosons or fermions in different spin states σ and σ', respectively. Within the pseudo-potential approximation for the interactions between the particles, the on-site interaction energy U is given by

$$U = g \int \mathrm{d}^3 r \, |w(\mathbf{r})|^4 = \sqrt{\frac{8}{\pi}} ka \, E_r \left(\frac{V_0}{E_r}\right)^{3/4}. \tag{2.5}$$

Here, a denotes the s-wave scattering length between two atoms and $w(\mathbf{r})$ is the Wannier function on a lattice site. Longer-ranged interactions could be generated by making use of dipolar interactions in polar molecules (Carr *et al.* 2009) or ground-state atoms dressed with a small Rydberg state admixture (Henkel *et al.* 2010; Honer *et al.* 2010; Pupillo *et al.* 2010). For collisional interactions alone, the description of the interactions as local on-site interactions is, however, an excellent approximation.

Trapping potential Typically, an overall harmonic trapping potential is used to confine the quantum gases to a trapping region. In a lattice, this gives rise to an additional term in the Hamiltonian of the form (in 1D):

$$H_{\text{trap}} = V_t \sum_{\mathbf{R},\sigma} \mathbf{R}^2 \hat{n}_{\mathbf{R},\sigma}, \tag{2.6}$$

where $\mathbf{R}$ denotes the position of the $\mathbf{R}$th lattice site. While the trapping potential is typically weak compared with the lattice potential and the corresponding characteristic length scale is long compared with the lattice period, the overall harmonic confinement has profound implications for resulting quantum phases in the trap. This can best be understood by introducing a local chemical potential $\mu_{\text{loc}}(\mathbf{R}) = \mu - V_t\mathbf{R}^2$

(for $\mu_{\text{loc}}(\mathbf{R}) > 0$, else $\mu_{\text{loc}}(\mathbf{R}) = 0$). Within a local density approximation (LDA), the inhomogeneous trapped quantum gas behaves locally as a homogeneous system with chemical potential μ_{loc}. As one moves from the centre of the trapped gas radially outwards to its border, one thus samples many-body phases at different chemical potentials $0 < \mu_{\text{loc}} < \mu$. In contrast with a homogeneous sample, where the system exhibits a single quantum phase, trapped quantum gases, therefore, typically exhibit several coexisting quantum phases in the trap. Although there are cases where such coexisting phases can complicate the analysis of the system, the harmonic trapping potential generally facilitates the generation of many-body quantum phases with ultracold atoms, as we will show in the case of Hubbard-type model systems. The approximation of the trapped quantum gas via LDA is expected to fail close to a quantum critical point, where the length scale of the fluctuating region tends to diverge (Pollet *et al.* 2010).

2.3.3 Non-standard lattice configurations

We have seen that by overlapping optical standing waves orthogonally to each other, one can create lattice potentials of simple cubic symmetry. Almost any other symmetry may be created via Fourier synthesis, i.e., superimposing standing waves of different wavelengths under different angles. Next to simple cubic type potentials, triangular, hexagonal (Becker *et al.* 2010), and Kagomé (Damski *et al.* 2005) lattice potentials have also been realized or proposed for ultracold atoms.

Spin-dependent lattices When using laser light, detuned not too far from a set of excited sublevels (for example, for alkalis, the detuning should be close to the fine-structure splitting), one may also realize lattice potentials, which can be vastly different for different spin states of an atom or molecule. Prominent examples in this respect are spin-dependent potentials that can be realized using counterpropagating laser beams with an angle θ between the two linear polarizations of the counterpropagating light fields (Jaksch *et al.* 1999; Mandel *et al.* 2003; Gadway *et al.* 2010). Such a configuration can be described by two σ^+ and σ^- polarized standing waves, whose relative position to each other, $\Delta x = \theta/\pi \times (\lambda/2)$, can be set by the angle θ. Shifting the two lattices relative to each other allows one to tune the interaction matrix element $U_{\uparrow\downarrow}$ between two spin components without the use of Feshbach resonances, but, instead, by changing the wavefunction overlap of the spin-components on a lattice site.

Optical superlattices When superimposing two standing wave light fields with a difference of exactly a factor of two between the two underlying lasers, optical superlattices can be created. Such optical superlattices mimic arrays of double wells, where the interwell and intrawell barrier of the double wells can be controlled dynamically and *in situ* by changing the frequency and amplitude of the two light fields relative to each other. Such optical superlattices have been used successfully to detect on-site exchange (Anderlini *et al.* 2007) and next-neighbour superexchange interactions (Trotzky *et al.* 2008) that will be discussed in Section 2.6. Extending such superlattices along two orthogonal directions enables one to realize systems of coupled plaquettes (Paredes and Bloch 2008) that can be useful to investigate plaquette superfluidity or access low-entropy d-wave superfluids in an adiabatic manner (Trebst *et al.* 2006; Rey *et al.* 2009).

2.4 Detection methods

The standard method of probing quantum gases has been connected to time-of-flight methods. Atoms are suddenly released from the trapping potential and subsequently imaged after a variable expansion time in free space. In the context of optical lattice quantum gases, such a time-of-flight image yields useful information about the momentum distribution and thereby coherence properties of the quantum gas. By adiabatically turning off the lattice, one may image the quasi-momentum distribution in the lattice rather than the momentum distribution. Both methods are discussed here in detail, as well as the possibility of using a noise-correlation analysis of time-of-flight images to learn about the correlation properties of the gas. More recently developed *in-situ* and single-site resolved imaging techniques are discussed in Section 2.7.

2.4.1 Time-of-flight imaging and adiabatic mapping

Sudden release When releasing ultracold quantum gases from an optical lattice, one of two possible release methods can be chosen. If the lattice potential is turned off abruptly and interaction effects can be neglected, a given Bloch state with quasi-momentum q will expand according to its momentum distribution as a superposition of plane waves with momenta $p_n = \hbar q \pm n \times 2\hbar k$. This is a direct consequence of the fact that Bloch waves can be expressed as a superposition of plane wave states $\exp \mathrm{i}(\mathbf{q}+\mathbf{G})\cdot\mathbf{r}$ with momenta $\mathbf{q}+\mathbf{G}$, which include arbitrary reciprocal lattice vectors $\mathbf{G}$. In a simple cubic lattice with lattice spacing $d = \pi/k$, the vectors $\mathbf{G}$ are integer multiples of the fundamental reciprocal lattice vector $2k$. After a certain time of flight, this momentum distribution can be imaged using standard absorption imaging methods. If only a single Bloch state is populated, as is the case for a Bose–Einstein condensate with quasi-momentum $q = 0$, this results in a series of interference maxima that can be observed after a time-of-flight period t. The density distribution observed after a fixed time of flight at position $\mathbf{x}$ is nothing but the momentum distribution of the particles trapped in the lattice,

$$n(\mathbf{x}) = \left(\frac{m}{\hbar t}\right)^3 |\tilde{w}(\mathbf{k})|^2 \mathcal{G}(\mathbf{k}). \tag{2.7}$$

Here $\mathbf{k}$ is related to $\mathbf{x}$ by $\mathbf{k} = m\mathbf{x}/\hbar t$, owing to the assumption of ballistic expansion, while $\tilde{w}(\mathbf{k})$ is the Fourier transform of the Wannier function. The coherence properties of the many-body state are characterized by the Fourier transform

$$\mathcal{G}(\mathbf{k}) = \sum_{\mathbf{R},\mathbf{R}'} \mathrm{e}^{\mathrm{i}\mathbf{k}\cdot(\mathbf{R}-\mathbf{R}')} G^{(1)}(\mathbf{R},\mathbf{R}') \tag{2.8}$$

of the one-particle density matrix $G^{(1)}(\mathbf{R},\mathbf{R}') = \langle \hat{a}^\dagger_{\mathbf{R}} \hat{a}_{\mathbf{R}'} \rangle$.

In a BEC, the long-range order in the amplitudes leads to a constant value of the first-order coherence function $G^{(1)}(\mathbf{R}, \mathbf{R}')$ at large separations $|\mathbf{R} - \mathbf{R}'|$. The resulting momentum distribution coincides with the standard multiple wave interference pattern obtained with light diffracting off a material grating. The atomic density distribution observed after a long enough time-of-flight time, thus yields information on the coherence properties of the many-body system (Gerbier *et al.* 2005; 2008).[1]

Adiabatic mapping One of the advantages of using optical lattice potentials is that the lattice depth can be dynamically controlled by simply tuning the laser power. This opens another possibility for releasing the atoms from the lattice potential, e.g., by adiabatically converting a deep optical lattice into a shallow one and eventually completely turning off the lattice potential. Under adiabatic transformation of the lattice depth, the quasi-momentum $\mathbf{q}$ is preserved and during the turn-off process a Bloch wave in the nth energy band is mapped onto a corresponding free particle momentum $\mathbf{p}$ in the nth Brillouin zone (see Figure 2.2) (Kastberg *et al.* 1995; Greiner *et al.* 2001; Köhl *et al.* 2005).

The adiabatic mapping technique has been applied to the case of bosonic (Greiner *et al.* 2001) and fermionic (Köhl *et al.* 2005) atoms. For the situation of a homogeneously filled lowest energy band, an adiabatic ramp-down of the lattice potential leaves the central Brillouin zone—a square of width $2\hbar k$—fully occupied (see Figure 2.3b). If, on the other hand, higher energy bands are populated, one also observes populations in higher Brillouin zones (see Fig. 2.3c). As in this method each Bloch wave is mapped onto a specific free-particle momentum state, it can be used efficiently to probe the distribution of the particles over Bloch states in different energy bands (Kastberg *et al.* 1995; Greiner *et al.* 2001; Köhl *et al.* 2005; Wirth *et al.* 2010).

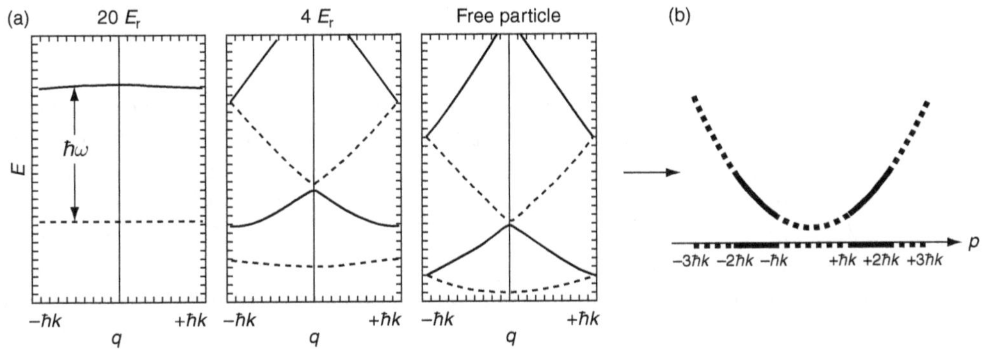

Figure 2.2 (a) Bloch bands for different potential depths. During an adiabatic ramp down, the quasi-momentum is conserved and (b) a Bloch wave with quasi-momentum q in the nth energy band is mapped onto a free particle with momentum p in the nth Brillouin zone of the lattice. From (Bloch, 2005).

[1]If interaction effects occur during time-of-flight expansion, the observed density distribution can deviate from the in-trap momentum distribution.

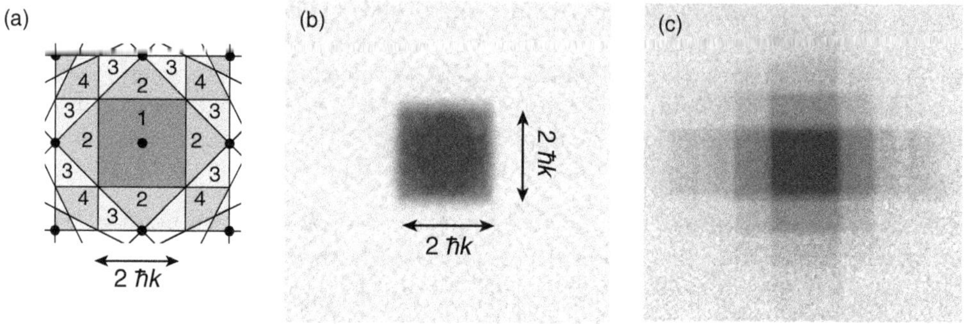

Figure 2.3 (a) Brillouin zones of a 2D simple cubic optical lattice. For a homogeneously filled lowest Bloch band, an adiabatic shut-off of the lattice potential leads to a homogeneously populated first Brillouin zone, which can be observed through absorption imaging after a time-of-flight expansion (b). If, in addition, higher Bloch bands are populated, higher Brillouin zones become populated as well (c).

2.4.2 Time-of-flight versus noise correlations

Let us begin by considering a quantum gas released from a trapping potential. After a finite time-of-flight time t, the resulting density distribution yields a three-dimensional density distribution $n_{3\mathrm{D}}(\mathbf{x})$. If interactions can be neglected during time of flight, the average density distribution is related to the in-trap quantum state via

$$\begin{aligned} \langle \hat{n}_{3\mathrm{D}}(\mathbf{x})\rangle_{\mathrm{tof}} &= \langle \hat{a}^{\dagger}_{\mathrm{tof}}(\mathbf{x})\hat{a}_{\mathrm{tof}}(\mathbf{x})\rangle_{\mathrm{tof}} \\ &\approx \langle \hat{a}^{\dagger}(\mathbf{k})\hat{a}(\mathbf{k})\rangle_{\mathrm{trap}} = \langle \hat{n}_{3\mathrm{D}}(\mathbf{k})\rangle_{\mathrm{trap}}, \end{aligned} \tag{2.9}$$

where $\mathbf{k}$ and $\mathbf{x}$ are related by the ballistic expansion condition $\mathbf{k} = M\mathbf{x}/\hbar t$ (a factor $(M/\hbar t)^3$ from the transformation of the volume elements $\mathrm{d}^3x \to \mathrm{d}^3k$ is omitted). Here we have used the fact that for long time of flight, the initial size of the atom cloud in the trap can be neglected. It is important to realize that, in each experimental image, a single realization of the density is observed, not an average. Moreover, each pixel in the image records on average a substantial number N_σ of atoms. For each of those pixels, however, the number of atoms recorded in a *single realization* of an experiment will exhibit shot noise fluctuations of relative order $1/\sqrt{N_\sigma}$ which will be discussed next. As shown in Eq. (2.3), the density distribution after the time of flight represents a momentum distribution reflecting the first-order coherence properties of the in-trap quantum state. This assumption is, however, only correct, if interactions between the atoms during the expansion process do not modify the initial momentum distribution, which we will assume throughout the text. When the interactions between the atoms have been enhanced, e.g., by a Feshbach resonance, or a high-density sample is prepared, such an assumption is not always valid. Near Feshbach resonances, one therefore often ramps back to the zero crossing of the scattering length before expansion.

Density–density correlations in time-of-flight images Let us now turn to the observation of density–density correlations in the expanding atom clouds (Altman *et al.* 2004). These are characterized by the density–density correlation function

$$\langle \hat{n}(\mathbf{x})\hat{n}(\mathbf{x}')\rangle = \langle \hat{n}(\mathbf{x})\rangle\langle \hat{n}(\mathbf{x}')\rangle g^{(2)}(\mathbf{x},\mathbf{x}') + \delta(\mathbf{x}-\mathbf{x}')\langle \hat{n}(\mathbf{x})\rangle, \tag{2.10}$$

which contains the normalized pair distribution $g^{(2)}(\mathbf{x},\mathbf{x}')$ and a self-correlation term. Relating the operators after time-of-flight expansion to the in-trap momentum operators, using Eq. (2.3), one obtains:

$$\begin{aligned}\langle \hat{n}_{3D}(\mathbf{x})\hat{n}_{3D}(\mathbf{x}')\rangle_{\mathrm{tof}} &\approx \langle \hat{a}^\dagger(\mathbf{k})\hat{a}(\mathbf{k})\hat{a}^\dagger(\mathbf{k}')\hat{a}(\mathbf{k}')\rangle_{\mathrm{trap}}\\ &= \langle \hat{a}^\dagger(\mathbf{k})\hat{a}^\dagger(\mathbf{k}')\hat{a}(\mathbf{k}')\hat{a}(\mathbf{k})\rangle_{\mathrm{trap}} + \delta_{\mathbf{kk}'}\langle \hat{a}^\dagger(\mathbf{k})\hat{a}(\mathbf{k})\rangle_{\mathrm{trap}}.\end{aligned} \tag{2.11}$$

The last term on the r.h.s. of the above equation is the autocorrelation term. It will be dropped in the subsequent discussion, as it only contributes to the signal for $\mathbf{x} = \mathbf{x}'$ and contains no more information about the initial quantum state than the momentum distribution itself. The first term, however, shows that for $\mathbf{x} \neq \mathbf{x}'$, subtle momentum–momentum correlations of the in-trap quantum states are present in the noise-correlation signal of the expanding atom clouds.

Let us discuss the obtained results for two cases that have been analysed in the experiment: (1) ultracold atoms in a Mott insulating state or a fermionic band insulating state released from a 3D optical lattice and (2) two interfering one-dimensional quantum gases separated by a distance **d**.

2.4.2.1 Noise correlations in bosonic Mott and fermionic band insulators

Consider a bosonic Mott insulating state or a fermionic band insulator in a three-dimensional simple cubic lattice. In both cases, each lattice site **R** is occupied by a fixed atom number $n_{\mathbf{R}}$. Such a quantum gas is released from the lattice potential and the resulting density distribution is detected after a time of flight t. In a deep optical lattice, the (in-trap) field operator $\hat{\psi}(\mathbf{r})$ can be expressed as a sum over destruction operators $\hat{a}_{\mathbf{R}}$ of localized Wannier states, neglecting all but the lowest band. The field operator for destroying a particle with momentum **k** is, therefore, given by

$$\hat{a}(\mathbf{k}) = \int \mathrm{e}^{-\mathrm{i}\mathbf{kr}}\hat{\psi}(\mathbf{r})\mathrm{d}^3r \simeq \tilde{w}(\mathbf{k})\sum_{\mathbf{R}} \mathrm{e}^{-\mathrm{i}\mathbf{kR}}\hat{a}_{\mathbf{R}}, \tag{2.12}$$

where $\tilde{w}(\mathbf{k})$ denotes the Wannier function in momentum space.

For the two states considered here, the expectation value in Eq. (2.5) factorizes into one-particle density matrices $\langle \hat{a}^\dagger_{\mathbf{R}}\hat{a}_{\mathbf{R}'}\rangle = n_{\mathbf{R}}\,\delta_{\mathbf{R},\mathbf{R}'}$ with vanishing off-diagonal order. The density–density correlation function after a time-of-flight is then given by (omitting the autocorrelation term of order $1/N$)

$$\langle \hat{n}_{3D}(\mathbf{x})\hat{n}_{3D}(\mathbf{x}')\rangle = |\tilde{w}(M\mathbf{x}/\hbar t)|^2|\tilde{w}(M\mathbf{x}'/\hbar t)|^2 N^2 \times \left[1 \pm \frac{1}{N^2}\left|\sum_{\mathbf{R}} \mathrm{e}^{\mathrm{i}(\mathbf{x}-\mathbf{x}')\cdot\mathbf{R}(M/\hbar t)} n_{\mathbf{R}}\right|^2\right]. \tag{2.13}$$

The plus sign in this equation corresponds to the case of bosonic particles and the minus sign to the case of fermionic particles in a lattice. Both in a Mott state of bosons and in a filled band of fermions, the local occupation numbers $n_{\mathbf{R}}$ are fixed integers. Equation (2.7) then shows that correlations or anticorrelations in the density–density expectation value appear for bosons or fermions, whenever the difference $\mathbf{k} - \mathbf{k}'$ is equal to a reciprocal lattice vector $\mathbf{G}$ of the underlying lattice. In real space, where the images are actually taken, this corresponds to spatial separations for which

$$|\mathbf{x} - \mathbf{x}'| = \ell = \frac{2ht}{\lambda M}. \tag{2.14}$$

Such spatial correlations or anticorrelations in the quantum noise of the density distribution of expanding atom clouds can, in fact, be traced back to the famous Hanbury Brown & Twiss effect (Hanbury Brown and Twiss 1956a; 1956b; Baym 1998) and its analogue for fermionic particles (Henny *et al.* 1999; Oliver *et al.* 1999; Kiesel *et al.* 2002; Iannuzzi *et al.* 2006; Rom *et al.* 2006; Jeltes *et al.* 2007). For the case of two atoms localized at two lattice sites this can be readily understood in the following manner: there are two possible ways for the particles to reach two detectors at positions $\mathbf{x}$ and $\mathbf{x}'$, which differ by exchange. A constructive interference for the case of bosons or a destructive interference for the case of fermions then leads to correlated or anticorrelated quantum fluctuations that are registered in the density–density correlation function (Baym 1998; Altman *et al.* 2004).

The correlations for the case of a bosonic Mott insulating state and anticorrelations for the case of a fermionic band insulating state have recently been observed experimentally (Fölling *et al.* 2005; Rom *et al.* 2006; Spielman *et al.* 2007). In these experiments, several single images of the desired quantum state are recorded after releasing the atoms from the optical trapping potential and observing them after a finite time of flight (for a single one of these images see, e.g., Figure 2.4a or Figure 2.5a). These individually recorded images only differ from each other in the atomic shot noise. A set of such absorption images is then processed to yield the spatially averaged second-order correlation function, $g^{(2)}_{\mathrm{exp}}(\mathbf{b})$:

$$g^{(2)}_{\mathrm{exp}}(\mathbf{b}) = \frac{\int \langle n(\mathbf{x}+\mathbf{b}/2)\cdot n(\mathbf{x}-\mathbf{b}/2)\rangle\, \mathrm{d}^2\mathbf{x}}{\int \langle n(\mathbf{x}+\mathbf{b}/2)\rangle\langle n(\mathbf{x}-\mathbf{b}/2)\rangle\, \mathrm{d}^2\mathbf{x}}. \tag{2.15}$$

As shown in Figure 2.4, the Mott insulating state exhibits long-range order in the pair correlation function $g^{(2)}(\mathbf{b})$. This order is not connected with the trivial

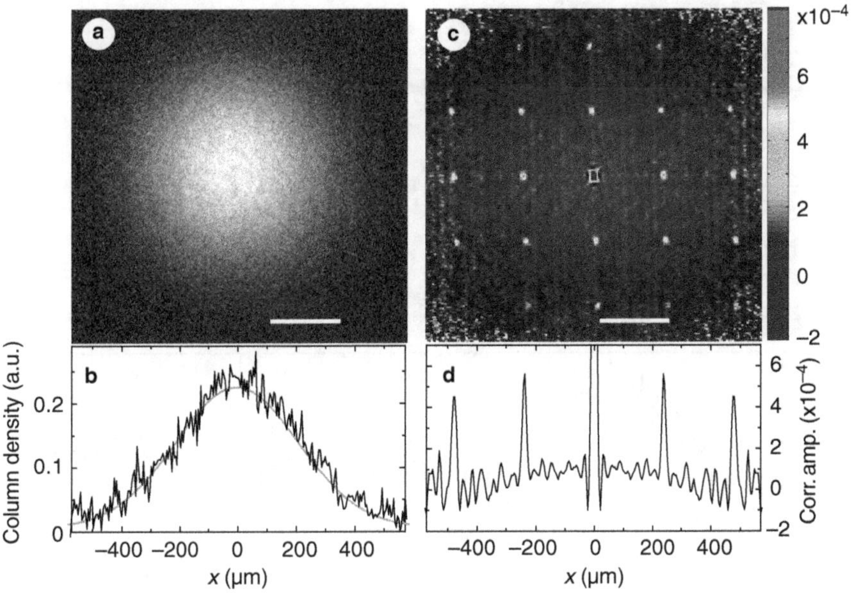

Figure 2.4 Noise correlations of a Mott insulator released from a 3D optical lattice. (a) Single shot absorption image of a Mott insulator released from an optical lattice and (b) associated cut through the image. A statistical correlation analysis over several independent images such as the one in (a) yields the correlation function (c). A cut through this two-dimensional correlation function reveals a Hanbury Brown & Twiss type bunching of the bosonic atoms (d). Reprinted with permission from (Fölling *et al.* 2005).

periodic modulation of the average density imposed by the optical lattice after the time of flight, which is factored out in $g^{(2)}(\mathbf{x}, \mathbf{x}')$ (see Eq. (2.4)). Therefore, in the superfluid regime, one expects $g^{(2)}(\mathbf{x}, \mathbf{x}') \equiv 1$, despite the periodic density modulation in the interference pattern after the time of flight. It is interesting to note that the correlations or anticorrelations can also be traced back to the enhanced fluctuations in the population of the Bloch waves with quasi-momentum q for the case of the bosonic particles, and the vanishing fluctuations in the population of Bloch waves with quasi-momentum q for the case of fermionic particles (Rom *et al.* 2006).

Note that, in general, the signal amplitude obtained in the experiments for the correlation function deviates significantly from the theoretically expected value of one. In fact, one typically observes signal levels of 10^{-4}–10^{-3} (see Figures 2.4, 2.5). This can be explained by the finite optical resolution when imaging the expanding atomic clouds, thus leading to a broadening of the detected correlation peaks and thereby a decreased amplitude, as the signal weight in each correlation peak is preserved in the detection process. Using single atom detectors with higher spatial and temporal resolution, such as the ones used in (Schellekens *et al.* 2005) and (Jeltes *et al.* 2007), one can overcome such limitations and thereby also evaluate higher-order correlation functions.

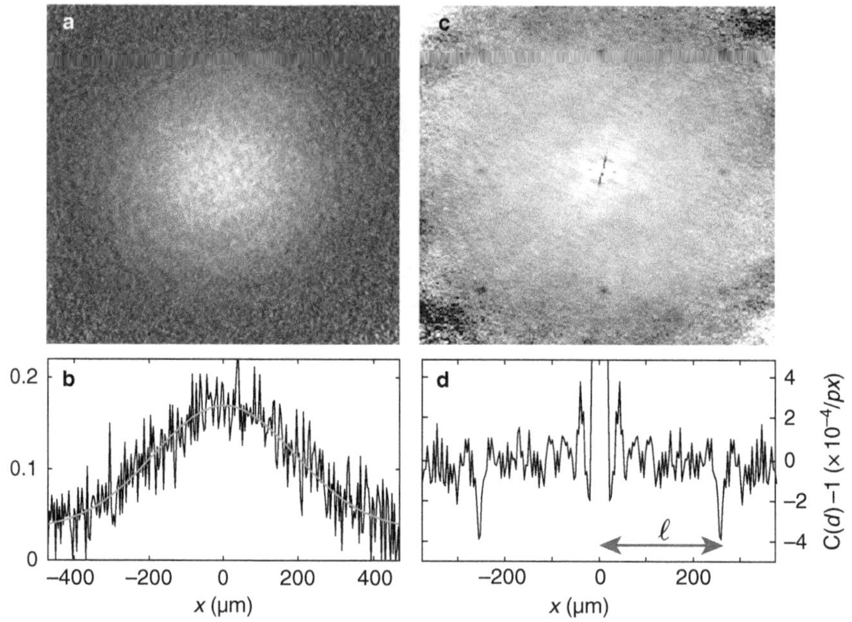

Figure 2.5 Noise correlations of a band insulating Fermi gas. Instead of the correlation bunching peaks observed in Figure 2.4, the fermionic quantum gas shows a HBT-type antibunching effect, with dips in the observed correlation function. From (Rom *et al.* 2006).

2.5 Bose– and Fermi–Hubbard models

The single-band Hubbard models play a paramount role in the context of condensed matter physics. Although they are among the simplest models used to describe interacting particles on a lattice, in several cases not even the phase diagram of the system is known and analytical solutions of the Hubbard model have not yet been found. It is also widely believed that the fermionic Hubbard model contains the essential physics for the explanation of high-temperature superconductivity (Lee *et al.* 2006; Le Hur and Maurice Rice 2009). The facts that both models can be realized efficiently with ultracold atoms (Jaksch *et al.* 1998; Hofstetter *et al.* 2002) and that all the underlying parameters of the Hubbard model may be tuned and controlled in cold atom and molecule experiments have led to widespread interest of ultracold gases as efficient quantum simulators of these foundational Hamiltonians (Jaksch and Zoller 2005; Lewenstein *et al.* 2007; Bloch *et al.* 2008). Current research is driven by the quest to explore the low-temperature (entropy) phases of these models, but has also opened a new path to studying non-equilibrium phenomena in strongly correlated quantum systems beyond a linear response (Polkovnikov *et al.* 2011).

As one of the most striking phenomena in both bosons and fermions, for strong repulsive interactions between the particles compared with their kinetic energy $U \gg J$ and integer fillings, the many-body system forms a Mott insulating state, with strongly suppressed density fluctuations. Mott insulators form the basis for states with magnetic order, when the temperature of the system becomes sufficiently lower than the superexchange coupling between two spin states on neighbouring lattice sites. The quest to realize such magnetically ordered states and the novel possibilities offered by ultracold atoms and molecules are discussed in subsequent sections.

Before we proceed, we will need to discuss a fundamental difference between a typical condensed matter and a cold quantum gas experiment. Typical condensed matter experiments are carried out under conditions where a probe sample is held at a constant temperature through a connection to a reservoir. Lowering the temperature then allows one to access novel phases of the many-body system. In the context of ultracold quantum gases, we are dealing with quantum systems that are completely isolated from their environment. A change of the underlying trapping or lattice parameters typically also leads to a change in the temperature of the isolated gas. A temperature reduction caused by a trap deformation, for example, might not bring one any closer to the transition point of a new phase, as the associated transition temperature is typically lowered as well. A much more useful system variable that is invariant to such adiabatic changes of the Hamiltonian parameters is the total entropy per particle of the quantum gas $S/(Nk_{\mathrm{B}})$. Whether or not a specific many-body phase can thus be reached via an adiabatic change of system parameters is only a question of whether the initial entropy of the initial system is low enough. For example, experiments typically begin with a fermionic or bosonic quantum gas cooled via evaporative cooling to a certain temperature in the harmonic trapping potential of a magnetic or an optical dipole trap. This temperature determines the entropy of the quantum gas. Then optical lattices are turned on by increasing the intensity of the corresponding light fields. Under the assumption of adiabatic changes of the lattice potential, the entropy of the systems remains constant. Typically, experiments are carried out under this assumption of conserved entropy; however it is not always fulfilled, as timescales for reaching equilibrium in interacting many-body systems have been shown to increase with interaction strength (Hung *et al.* 2010; Schneider *et al.* 2012; Strohmaier *et al.* 2010).

2.5.1 Bose–Hubbard model

The theory proposal (Fisher *et al.* 1989; Jaksch *et al.* 1998) and the subsequent realization of the Bose–Hubbard model with ultracold atoms (Greiner *et al.* 2002) marks the starting point for strong correlation physics with ultracold quantum gases. Within this model, a gas of interacting bosons occupying the lowest Bloch band of a periodic potential forms a superfluid below a critical ratio of interaction to kinetic energy ($U/J < (U/J)_{\mathrm{c}}$). For integer filling of the lattice and for $U/J > (U/J)_{\mathrm{c}}$, the system turns into a strongly correlated Mott insulator. By increasing the optical lattice depth, the ratio of interaction to kinetic energy of the system U/J can be tuned to increasingly large values. Even without employing scattering resonances to tune U, it is thus possible to bring the bosonic quantum gas into a strongly correlated regime of a Mott insulator simply by quenching the kinetic energy of the system. The transition from a superfluid to a Mott insulator has by now been the focus of numerous theoretical and experimental investigations and it is beyond the scope of this book to give a complete survey of these. The interested reader may find more detailed reviews on the topic in (Jaksch and Zoller 2005; Lewenstein *et al.* 2007; Bloch *et al.* 2008).

In the following, I will instead try to highlight a few characteristic properties that exemplify the dramatic changes occurring when a quantum gas is converted from a superfluid into a Mott insulator. Among one of the most prominent features is the measurement on the change of coherence properties when the transition is crossed. For a superfluid state, the underlying condensate exhibits long-range phase coherence and thus shows sharp matter-wave interference peaks when the quantum gas is released from the optical lattice. Deep in the Mott insulating phase $U/J \gg (U/J)_c$, the system is comprised of pure Fock states of integer on-site density and thus no interference pattern is observed (Greiner *et al.* 2002). For finite tunnelling, particle-hole fluctuations, induced by the kinetic energy of the quantum gas, can allow even a Mott insulator to exhibit finite coherence properties (Gerbier *et al.* 2005). However, as the particle-hole pairs are created only on nearest-neighbour lattice sites, this coherence is rather short-ranged in nature and distinctly different from the long-range phase coherence of a BEC. As the transition point to the superfluid is approached, the particle-hole pairs begin to extend over larger and larger distances, eventually inducing the formation of a superfluid at the transition point. It is thus natural to view the Mott insulator as a gas of bound particle-hole pairs on top of a fixed density background. As the transition to the superfluid is approached, these particle-hole pairs become more abundant and essentially deconfined at the transition point.

The harmonic trapping potential, which is typical in cold-atom experiments and is present in addition to the periodic potential, plays a crucial role for the realization of strongly interacting quantum phases on a lattice. Whereas for a homogeneous system one would need to tune the filling to integer values to observe the transition to a Mott insulator, the density distribution in a trap can change and the formation of a Mott insulator can be induced even for an initially non-commensurate lattice density. In fact, as the lattice is raised, Mott insulating regions will automatically form in the trap for sufficiently low temperatures. These Mott insulating regions become visible as constant-integer density plateaus, with jumps in density at characteristic trap radii. The resulting density distribution of a Mott insulator therefore resembles a 'wedding cake'-type structure in the trap (see Figure 2.6). The regions of constant density are characteristic for the incompressible nature of the Mott state with vanishing compressibility $\kappa = \mathrm{d}n/\mathrm{d}\mu = 0$. The compressibility of the quantum gas can also be related to the density fluctuations, for which one finds (Zhou and Ho 2011):

$$k_B T \frac{\partial n}{\partial \mu} = \int \langle \hat{n}(\mathbf{r})\hat{n}(\mathbf{r}') - \langle \hat{n}(\mathbf{r})\rangle\langle \hat{n}(\mathbf{r}')\rangle \mathrm{d}^3 r'. \tag{2.16}$$

For a system with short-ranged density fluctuations (such as a Mott insulator), the right-hand side of the equation takes essentially only non-zero values for $\mathbf{r} = \mathbf{r}'$ and thus corresponds to the on-site density fluctuations. The on-site number fluctuations of the quantum gas thus closely track the compressibility of the many-body state and may be used as an alternative means of identifying the incompressible regions of a Mott insulator. The wedding cake structure of the density distribution, including Mott

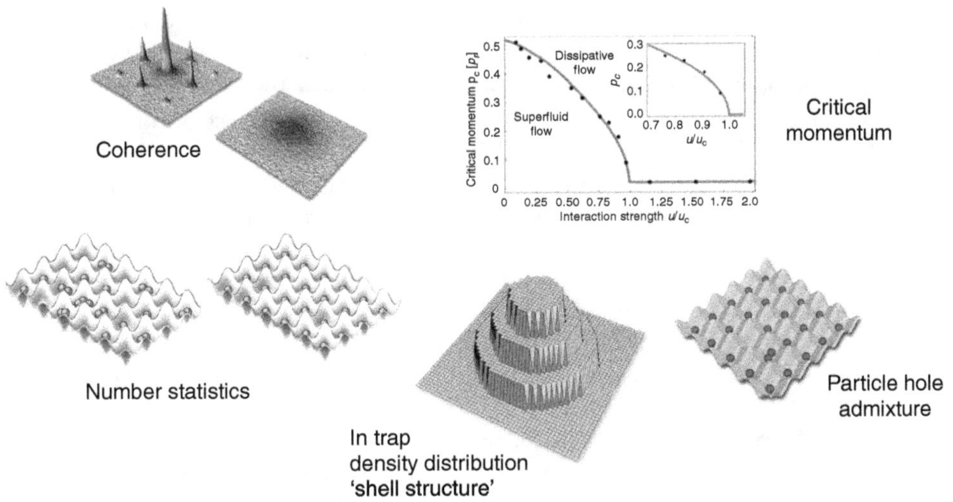

Figure 2.6 Summary of phenomena that have been measured in the context of the Bose–Hubbard model. Several characteristic properties in the transition from a superfluid to a Mott insulator have been investigated experimentally. Among these are the different coherence properties, number statistics, density distributions, and transport properties. Adapted from (Mun *et al.* 2007).

plateaus, has been revealed in several experiments (Fölling *et al.* 2006; Gemelke *et al.* 2009; Bakr *et al.* 2010; Sherson *et al.* 2010). In the most recent measurements, using single-atom and single-site resolved detection of atoms in an optical lattice, which are discussed in Section 2.7, the vanishing number fluctuations were also directly observed at the single-site level (Bakr *et al.* 2010; Sherson *et al.* 2010).

2.5.2 Fermi–Hubbard model

Restricting our discussion to the lowest energy band of a simple cubic 3D optical lattice, a fermionic quantum gas mixture can be modelled via the Hubbard–Hamiltonian (Hubbard 1963) with an additional term describing the underlying harmonic potential:

$$\begin{aligned}\hat{H} = &-J \sum_{\langle \mathbf{R},\mathbf{R}' \rangle,\sigma} \hat{a}^{\dagger}_{\mathbf{R},\sigma} \hat{a}_{\mathbf{R}',\sigma} + U \sum_{\mathbf{R}} \hat{n}_{\mathbf{R},\downarrow} \hat{n}_{\mathbf{R},\uparrow} \\ &+ V_t \sum_{\mathbf{R},\sigma} \mathbf{R}^2 \hat{n}_{\mathbf{R},\sigma}.\end{aligned} \tag{2.17}$$

The quantum phases of the fermionic Hubbard model with harmonic confinement are governed by the interplay between three energy scales: kinetic energy, whose scale is given by the lattice bandwidth $12J$, interaction energy U, and the strength of the harmonic confinement, which can conveniently be expressed by the characteristic trap energy $E_t = V_t(N_\sigma/(4\pi/3))^{2/3}$, denoting the Fermi energy of a non-interacting cloud in the zero-tunnelling limit, with N_σ being the number of atoms per spin state ($N_\downarrow = N_\uparrow$). The characteristic trap energy depends both on atom number and trap frequency via $E_t \propto \omega_\perp^2 N_\sigma^{2/3}$ and describes the effective compression of the quantum gas, controlled by the trapping potential in the experiment.

Depending on which term in the Hamiltonian dominates, different kinds of many-body ground states can occur in the trap centre. For weak interactions in a shallow trap, $U \ll E_t \ll 12J$, the Fermi energy is smaller than the lattice bandwidth ($E_F < 12J$) and the atoms are delocalized, to minimize their kinetic energy. This leads to compressible metallic states with central filling $n_{0,\sigma} < 1$, where the local filling factor $n_{\mathbf{R},\sigma} = \langle \hat{n}_{\mathbf{R},\sigma} \rangle$ denotes the average occupation per spin state of a given lattice site. A dominating repulsive interaction, $U \gg 12J$ and $U \gg E_t$, suppresses the double occupation of lattice sites and can lead to Fermi-liquid ($n_{0,\sigma} < 1/2$) or Mott-insulating ($n_{0,\sigma} = 1/2$) states at the trap centre, depending on the ratio of kinetic to characteristic trap energy. Stronger compressions lead to higher filling factors, ultimately ($E_t \gg 12J$, $E_t \gg U$) resulting in an incompressible band insulator with unity central filling at $T = 0$.

A finite temperature reduces all filling factors and enlarges the cloud size, as the system needs to accommodate the corresponding entropy. Furthermore, in the trap, the filling always varies smoothly from a maximum at the centre to zero at the edges of the cloud. For a dominating trap and strong repulsive interaction at low temperature ($E_t > U > 12J$), the interplay between the different terms in the Hamiltonian gives rise to a wedding-cake like structure consisting of a band-insulating core ($n_{0,\sigma} \approx 1$) surrounded by a metallic shell ($1/2 < n_{\mathbf{R},\sigma} < 1$), a Mott-insulating shell ($n_{\mathbf{R},\sigma} = 1/2$), and a further metallic shell ($n_{\mathbf{R},\sigma} < 1/2$) (Helmes *et al.* 2008). The outermost shell always remains metallic, independent of interaction and confinement; only its thickness varies.

Recent experiments (Figure 2.7) on ultracold fermionic spin mixtures of ^{40}K atoms have been able to produce a paramagnetic Mott insulating phase for increasing interactions in the quantum gases in the range of $U/(12J) \simeq 1.5 - 4$ (Jördens *et al.* 2008; Schneider *et al.* 2008). In the experiments, the suppression of double occupancy (doublons) or the incompressible nature of an insulating phase have been used to identify the Mott insulating phase. For example, for $k_B T < U$ and strong interactions, one expects doubly occupied sites $D = \langle \hat{n}_{\mathbf{R},\uparrow} \hat{n}_{\mathbf{R},\downarrow} \rangle / (\langle n_{\mathbf{R},\uparrow} \rangle + \langle n_{\mathbf{R},\downarrow} \rangle)$ to be strongly suppressed compared with the non-interacting case (DeLeo *et al.* 2008). Furthermore, when the system is in an insulating phase, the compressibility of the system will drop to a minimum. Both these quantities can be compared *ab-initio* dynamical mean field theory (DMFT) calculations (Schneider *et al.* 2008; Jördens *et al.* 2010). As the lowest achieved temperatures of the quantum gases are still above the single-particle hopping $k_B T \gtrsim J$, a high-temperature series expansion of the partition function has also been shown to be useful for comparison with the experimental results (Scarola *et al.* 2009; Jördens *et al.* 2010). The detailed comparison with theory has allowed one to determine the entropies per particle (those of current experiments being in the range of $S/N = (1 - 1.8)k_B$) for which Mott insulating behaviour could be observed.

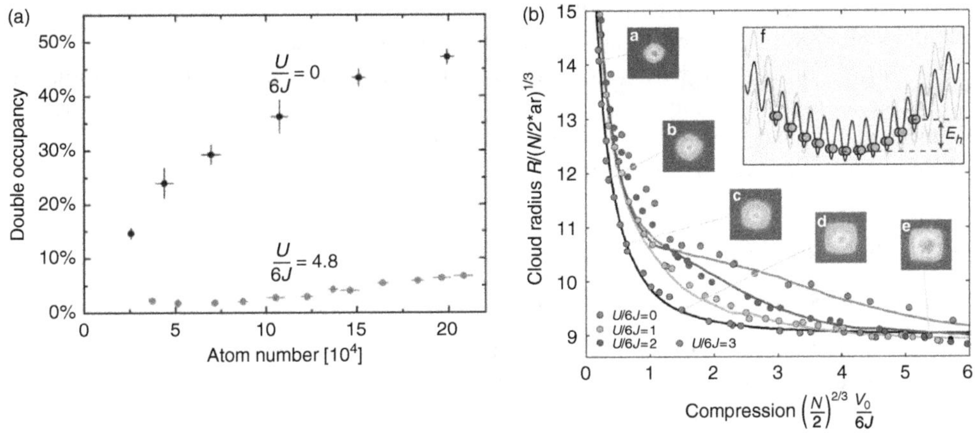

Figure 2.7 Signatures of Fermionic Mott insulators. (a) For strong interactions, a suppression of doubly occupied sites is observed compared to the non-interacting fermionic spin mixture. (b) The size behaviour of the quantum gas as a function of external trap compression can also be used as a signature to identify an insulating phase. The solid lines indicate *ab-initio* simulations using dynamical mean field theory (DMFT). Adapted from (Jördens *et al.* 2008) and (Schneider *et al.* 2008).

Interestingly, for the upper-limit values, these entropies are larger than the maximum entropy that can be stored in a homogeneous single-band Hubbard model system of $S_{\max}/N = k_B 2 \log 2$. For a homogeneous system, one would thus not expect the system to show Mott insulating behaviour; however, for the trapped quantum gas, the entropy per particle is distributed inhomogeneously throughout the system, such that in the metallic wings the excess entropy above $S/N = k_B \log 2$ can be efficiently stored as configurational entropy of the particles (see Figure 2.12). This inhomogeneous entropy distribution is, in fact, key to novel cooling (or better entropy reduction) methods that have been proposed for ultracold atoms in optical lattices (Bernier *et al.* 2009; Ho and Zhou 2009).

Simulating the repulsive model with a system of attractively interacting particles ($U < 0$) has recently been shown to exhibit advantages, especially when one wants to probe the effects of a controlled doping of, e.g., an antiferromagnetic Mott insulator (Ho *et al.* 2009). For fermions with attractive interactions, a preformed paired phase has been detected (Hackermüller *et al.* 2010); however, the generation of a superfluid state within a single band has not yet been achieved. The temperatures (and entropies) required to realize such a state are, in fact, comparable to the ones required for an antiferromagnet in the repulsive Hubbard model (Paiva *et al.* 2010).

2.6 Quantum magnetism with ultracold atoms in optical lattices

Magnetically ordered quantum phases play an important role in the low-temperature regime of the Hubbard model. The underlying spin–spin interactions responsible for the magnetically ordered phases arise from superexchange mediated coupling of neighbouring spins; our primary goal will be to understand how such superexchange couplings emerge in two-component quantum gases on a lattice in the regime of strong interactions. For this, it will be useful to introduce a 'toy model' of two spins in a double well—a system that can, in effect, be realized in the lab using superlattice potentials and that plays an important role in the context of realizing solid state qubits in electronic double-well quantum dots.

2.6.1 Superexchange spin interactions

2.6.1.1 Superexchange interactions in a double well

Spin–spin interactions between neighbouring atoms are mediated via so called superexchange processes. They arise directly from within the Hubbard model in the regime of strong interactions, leading to an effective Hamiltonian that couples the spin of neighbouring atoms in a lattice. Let us first discuss how such superexchange interactions can be derived theoretically (see also (Auerbach 1994)). As a starting point, we discuss the case of an atom with spin up $|\uparrow\rangle$ and another atom with spin down $|\downarrow\rangle$ loaded into a double-well potential. In the regime of strong repulsive interactions $U \gg J$, doubly occupied sites are energetically suppressed and our system can be described by the following basis states of the left and right well:

$$\mathcal{S} = \{|\downarrow,\uparrow\rangle, |\uparrow,\downarrow\rangle, |\uparrow,\uparrow\rangle, |\downarrow,\downarrow\rangle\}.$$

The action of the tunnelling operator of the Hubbard Hamiltonian can be evaluated in the strongly interacting regime via perturbation theory. First-order tunnelling processes lead out of the energetically allowed subspace and are therefore forbidden. However, second-order tunnelling processes that leave the system within $\mathcal{S}$ lead to an effective coupling between the different spin states. We can describe such processes via an effective Hamiltonian of the system, whose matrix elements within $\mathcal{S}$ can be evaluated via second-order perturbation theory:

$$H^{\text{eff}}_{a,b} = -\langle a|\hat{H}_J \frac{1-\hat{P}_{\mathcal{S}}}{U} \hat{H}_J|b\rangle = -\sum_{n\notin\mathcal{S}} \langle a|\hat{H}_J|n\rangle \frac{1}{\langle n|\hat{H}^{\text{int}}|n\rangle} \langle n|\hat{H}_J|b\rangle. \tag{2.18}$$

Here $\hat{H}^J$ denotes the tunnelling and $\hat{H}^{\text{int}}$ the interaction part of the Hubbard Hamiltonian and $\hat{P}_{\mathcal{S}}$ represents the projector into the subspace $\mathcal{S}$.

Let us consider, for example, the process that can lead to to an exchange of two spins (see also Figure 2.8):

$$\begin{aligned} |\downarrow,\uparrow\rangle \xrightarrow{J} \overbrace{|0,\uparrow\downarrow\rangle}^{-1/U} &\xrightarrow{J} |\uparrow,\downarrow\rangle \\ \text{or}\ \underbrace{|\uparrow\downarrow,0\rangle}_{-1/U} &\xrightarrow{J} |\uparrow,\downarrow\rangle\,. \end{aligned} \tag{2.19}$$

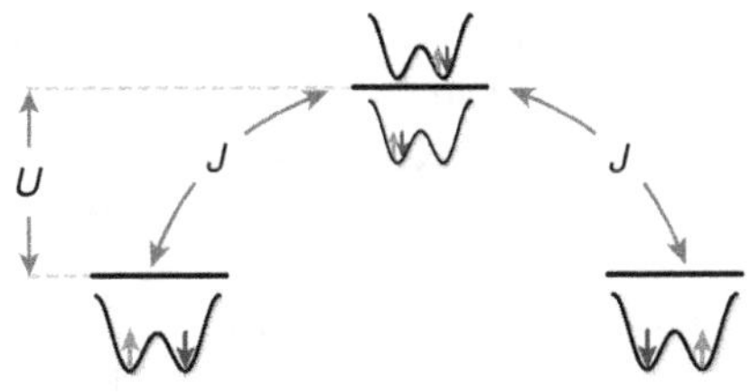

Figure 2.8 Schematic of superexchange interactions in a double well. For the case of strong repulsive interactions between the particles, atoms can mediate interactions between neighbouring sites via second-order tunnelling processes, depicted in the image. The second-order hopping of the particles via an intermediate state can lead to an exchange of the spins in the system.

The two possible pathways thus sum up to an effective coupling strength $-J_{\mathrm{ex}} = -2J^2/U$. Evaluating the other matrix elements in the same way, we obtain for the case of bosonic atoms the following matrix form of H^{eff} in the basis $\mathcal{S}$:

$$H^{\mathrm{eff}}_{\mathrm{bosons}} = J_{\mathrm{ex}} \begin{pmatrix} -1 & -1 & 0 & 0 \\ -1 & -1 & 0 & 0 \\ 0 & 0 & -1 & 0 \\ 0 & 0 & 0 & -1 \end{pmatrix}. \tag{2.20}$$

Diagonalizing the effective Hamiltonian yields the new eigenstates and eigenenergies:

$$\{|t_+\rangle, |t_0\rangle, |t_-\rangle\} \quad \text{with} \quad E = -2J_{\mathrm{ex}} \tag{2.21}$$

$$|s_0\rangle \quad \text{with} \quad E = 0. \tag{2.22}$$

Here, $|t_+\rangle = |\uparrow, \uparrow\rangle, |t_-\rangle = |\downarrow, \downarrow\rangle$, and $|t_0\rangle = 1/\sqrt{2}(|\downarrow, \uparrow\rangle + |\uparrow, \downarrow\rangle)$ are the spin-triplet eigenstates, whereas $|s_0\rangle = 1/\sqrt{2}(|\downarrow, \uparrow\rangle - |\uparrow, \downarrow\rangle)$ denotes the spin-singlet eigenstate of the two atoms. We may thus write $\hat{H}^{\mathrm{eff}}$ via a projector into the spin-triplet subspace $\hat{P}_{\mathrm{T}}$, as:

$$\hat{H}^{\mathrm{eff}} = -2J_{\mathrm{ex}}\hat{P}_{\mathrm{T}} = -J_{\mathrm{ex}}\left(\hat{1} + \hat{X}_{\mathrm{LR}}\right), \tag{2.23}$$

where $\hat{X}_{\mathrm{LR}}$ denotes the exchange operator between the left and right wells. The projection operator into the singlet and triplet subspace can be expressed via Dirac notation as:

$$P_{\mathrm{T}} = \frac{3}{4} + \hat{S}_{\mathrm{L}} \cdot \hat{S}_{\mathrm{R}}. \tag{2.24}$$

We can thus write the effective Hamiltonian as an interaction term between spins on the neighbouring wells (omitting the constant offset term):

$$H^{\text{eff}} = -2J_{\text{ex}}\hat{S}_{\text{L}} \cdot \hat{S}_{\text{R}}. \tag{2.25}$$

The minus sign for the case of bosons indicates ferromagnetic interactions, as the energy of the two spins is lowered if they align along the same direction.

For the case of fermionic spin mixtures, we can essentially follow the same derivation; however, for equal spin on neighbouring sites, second-order hopping processes are forbidden, owing to Pauli blocking in the intermediate state, where both particles with identical spin would occupy the same spin state. Also, when two particles are exchanged, we acquire an additional minus sign in the coupling, owing to the odd exchange symmetry of fermionic particles. The effective Hamiltonian for fermions thus reads:

$$H^{\text{eff}}_{\text{fermions}} = J_{\text{ex}} \begin{pmatrix} -1 & 1 & 0 & 0 \\ 1 & -1 & 0 & 0 \\ 0 & 0 & 0 & 0 \\ 0 & 0 & 0 & 0 \end{pmatrix}. \tag{2.26}$$

Diagonalizing the effective Hamiltonian in the fermionic case yields the eigenstates and eigenenergies:

$$\{|t_+\rangle, |t_0\rangle, |t_-\rangle\} \quad \text{with} \quad E = 0 \tag{2.27}$$

$$|s_0\rangle \quad \text{with} \quad E = -2J_{\text{ex}}. \tag{2.28}$$

For fermions, the spin-singlet state $|s_0\rangle$ is thus lower in energy and antiparallel orientation of the spins is thus favoured. For the effective Hamiltonian we thus obtain

$$\hat{H}^{\text{eff}} = -2J_{\text{ex}}\hat{P}_S = -J_{\text{ex}}\left(\hat{1} - \hat{X}_{\text{LR}}\right). \tag{2.29}$$

Making use of

$$P_S = \frac{1}{4} - \hat{S}_{\text{L}} \cdot \hat{S}_{\text{R}}. \tag{2.30}$$

we thus obtain for the fermionic spin Hamiltonian (up to a constant)

$$H^{\text{eff}} = +2J_{\text{ex}}\hat{S}_{\text{L}} \cdot \hat{S}_{\text{R}}, \tag{2.31}$$

favouring antiferromagnetic ordering between the particles. It is interesting to remind ourselves that the interactions between particles on neighbouring sites purely arise as a result of the strong repulsive contact interaction between the particles, virtual hopping of the spins and the exchange symmetry of the resulting quantum states. No long-ranged interactions are required to yield these interactions on neighbouring lattice sites.

2.6.1.2 Superexchange interactions on a lattice

This derivation can be extended to the case of a lattice system in a straightforward manner. For a bosonic or fermionic quantum system consisting of an equal mixture of two spin components in a Mott insulating regime with $\langle \hat{n}_{i,\uparrow} \rangle + \langle \hat{n}_{i,\downarrow} \rangle = 1$ and sufficiently low temperatures, one expects magnetically ordered quantum phases due to such superexchange spin–spin interactions. In the simplest case, such spin interactions take the form of an isotropic Heisenberg model:

$$H = \pm J_{\text{ex}} \sum_{\langle \mathbf{R},\mathbf{R}' \rangle} \hat{S}_{\mathbf{R}} \cdot \hat{S}_{\mathbf{R}'}, \tag{2.32}$$

with effective spin-1/2 operators $\hat{S}^x_{\mathbf{R}} = (\hat{a}^\dagger_{\mathbf{R},\uparrow} \hat{a}_{\mathbf{R},\downarrow} + \hat{a}^\dagger_{\mathbf{R},\downarrow} \hat{a}_{\mathbf{R},\uparrow})/2$, $\hat{S}^y_{\mathbf{R}} = (\hat{a}^\dagger_{\mathbf{R},\uparrow} \hat{a}_{\mathbf{R},\downarrow} - \hat{a}^\dagger_{\mathbf{R},\downarrow} \hat{a}_{\mathbf{R},\uparrow})/2\mathrm{i}$ and $\hat{S}^z_{\mathbf{R}} = (\hat{n}_{\mathbf{R},\uparrow} - \hat{n}_{\mathbf{R},\downarrow})/2$ and exchange coupling constant $J_{\text{ex}} = 2J^2/U$.

It is instructive to rewrite the Heisenberg Hamiltonian using the spin raising and lowering operators $\hat{S}^+_{\mathbf{R}} = (\hat{S}^x_{\mathbf{R}} + \mathrm{i}\hat{S}^y_{\mathbf{R}}) = \hat{a}^\dagger_{\mathbf{R},\uparrow} \hat{a}_{\mathbf{R},\downarrow}$ and $\hat{S}^-_{\mathbf{R}} = (\hat{S}^x_{\mathbf{R}} - \mathrm{i}\hat{S}^y_{\mathbf{R}}) = \hat{a}^\dagger_{\mathbf{R},\downarrow} \hat{a}_{\mathbf{R},\uparrow}$. We find:

$$H = \pm \frac{J_{\text{ex}}}{2} \sum_{\langle \mathbf{R},\mathbf{R}' \rangle} \left(\hat{S}^+_{\mathbf{R}} \hat{S}^-_{\mathbf{R}'} + \hat{S}^-_{\mathbf{R}} \hat{S}^+_{\mathbf{R}'} \right) \pm \frac{J_{\text{ex}}}{2} \sum_{\langle \mathbf{R},\mathbf{R}' \rangle} \hat{S}^z_{\mathbf{R}} \hat{S}^z_{\mathbf{R}'}. \tag{2.33}$$

Note that in this form it becomes especially apparent that the first part of the spin Hamiltonian has exactly the same structure as the tunnelling operator in the Hubbard Hamiltonian, while the second term acts like a nearest-neighbour interaction term.

2.6.2 Tunable superexchange interactions

Let us turn to the more general case of spin–spin interactions in two-component Bose–Bose or Fermi–Fermi mixtures in the strong coupling regime and half-filling, but in general unequal hopping matrix elements $J_\sigma \neq J_{\sigma'}$ and tunable interspecies and intraspecies interactions $U_{\uparrow\downarrow}$ and $U_{\sigma\sigma}$ (for fermions in a single-band model $U_{\sigma\sigma} = 0$). In this case one finds an xxz-type spin Hamiltonian of the form (Duan *et al.* 2003; Kuklov and Svistunov 2003),

$$H = \sum_{\langle \mathbf{R},\mathbf{R}' \rangle} \left[J^z_{\text{ex}} \hat{S}^z_{\mathbf{R}} \hat{S}^z_{\mathbf{R}'} \pm J^\perp_{\text{ex}} \left(\hat{S}^x_{\mathbf{R}} \hat{S}^x_{\mathbf{R}'} + S^x_{\mathbf{R}} \hat{S}^x_{\mathbf{R}'} \right) \right], \tag{2.34}$$

with tunable exchange coefficients for bosonic particles,

$$J^z_{\text{ex}} = \frac{J^2_\uparrow + J^2_\downarrow}{U_{\uparrow\downarrow}} - \frac{2J^2_\uparrow}{U_{\uparrow\uparrow}} - \frac{2J^2_\uparrow}{U_{\downarrow\downarrow}} \quad \text{and} \quad J^\perp_{\text{ex}} = \frac{2J_\uparrow J_\downarrow}{U_{\uparrow\downarrow}}. \tag{2.35}$$

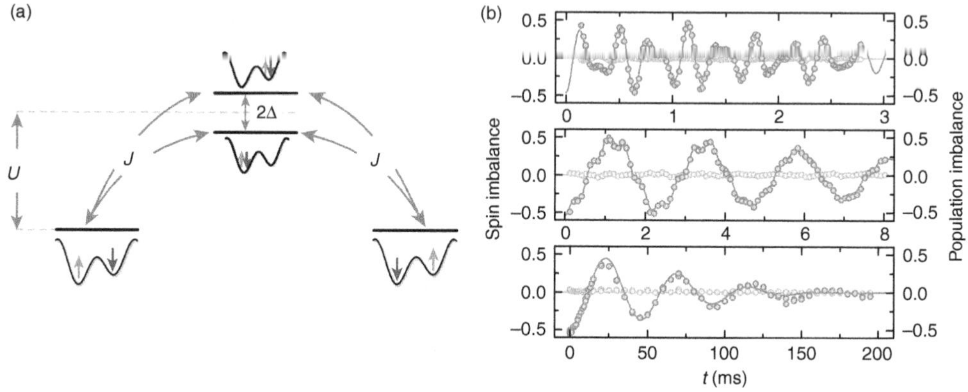

Figure 2.9 Detecting and controlling superexchange interactions: (a) superexchange interactions are mediated via second-order hopping processes. By introducing an energy offset Δ between neighbouring lattice sites, one may tune the exchange coupling J_{ex}. (b) Dynamical observation of superexchange interaction in double wells, initially prepared in a z-Néel order (see (a)). For increasing interactions (top row to bottom row) one observes how Heisenberg-type superexchange spin–spin interactions emerge and single-particle hopping becomes increasingly suppressed, owing to the increased repulsive interactions between the particles. Adapted from (Trotzky *et al.* 2008).

For fermionic atoms, the expression for $J_{\text{ex}}^{\perp}$ is identical; however, in the expression for J_{ex}^{z} the last two terms vanish. By tuning the different interaction matrix elements, e.g., via Feshbach resonances, or by tuning the different hopping matrix element, one may thus easily tune the system from an Ising-type interaction through a isotropic Heisenberg interaction to a dominating transverse spin–spin interaction.

Time-resolved superexchange interactions between ultracold atoms have been observed with ultracold bosons in optical double-well set-ups using optical superlattices (see Figure 2.9). By introducing an energy offset Δ between neighbouring lattice sites, Trotzky *et al.* (2008) showed that the superexchange coupling can be tuned via $J_{\text{ex}} = J^2/(U+\Delta) + J^2/(U-\Delta)$. When $\Delta > U$, the sign of the superexchange coupling is thus reversed, allowing one to tune between ferromagnetic and antiferromagnetic spin interactions of the particles. It is thus also conceivable that one might generate quantum systems for which ferromagnetic and antiferromagnetic spin interactions are simultaneously present along different lattice axes.

2.6.2.1 Magnetic phases of bosonic mixtures on a lattice

For the case of bosonic spin mixtures at half-filling per spin state and strong coupling ($U \gg J$), two distinct magnetically ordered quantum phases are expected at $T = 0$. When $J_{\text{ex}}^{z} > J_{\text{ex}}^{\perp} > 0$, an Ising antiferromagnet with z-Néel order represents the ground state of the system, whereas for dominating transverse spin interactions $J_{\text{ex}}^{\perp} > J_{\text{ex}}^{z} > 0$, a ferromagnet oriented in the $x-y$ plane is formed. Such a state has also been identified as a counterflow superfluid, for which the order parameter in each of the spin components vanishes $\langle\hat{\psi}_{\uparrow}\rangle = \langle\hat{\psi}_{\downarrow}\rangle = 0$; however, the easy-plane spin order of the ferromagnet

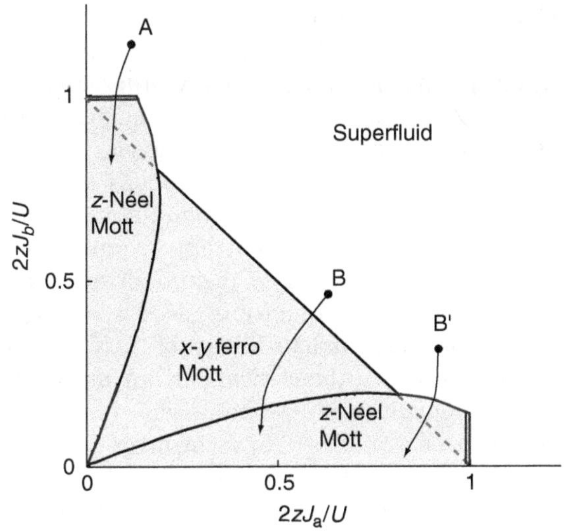

Figure 2.10 Phase diagram of two-component bosons at half-filling of each species, including quantum fluctuation corrections to mean-field theory. In the filled area above the broken lines, the superfluid state is metastable. Hysteresis behaviour is expected when the system is driven across the double lines. From (Altman *et al.* 2003).

corresponds to non-vanishing order in $\langle \hat{\psi}_\uparrow^\dagger \hat{\psi}_\downarrow \rangle \neq 0$. The counter superfluid is thus a result of the pairing of particles of spin component $|\uparrow\rangle$ with holes of spin component $|\downarrow\rangle$. Whereas the individual subsystems display no superfluid properties, owing to their vanishing order parameter, the *relative motion* between the two spin components shows superfluidity. One possibility of detecting such a counterflow superfluid would thus be to induce a sudden relative displacement of the two spin components and subsequently observe undamped oscillations in the relative oscillations of the two components (Kuklov and Svistunov 2003).

While these arguments were derived in the strong coupling regime of both spin components (i.e., deep within the Mott phase), work by several groups (Altman *et al.* 2003; Kuklov *et al.* 2004; Capogrosso-Sansone *et al.* 2010) has shown that the corresponding magnetically ordered phases should persist even into the regime up to the transition point to a superfluid (see also Figure 2.10). The corresponding maximum critical entropies required to observe such phases have recently been analysed in 3D using QMC methods and found to be $S_c/N \simeq 0.5 \pm 20\%$ and $S_c/N \simeq 0.35 \pm 10\%$ for the case of an Ising antiferromagnet and the $x-y$ ferromagnet, respectively (Capogrosso-Sansone *et al.* 2010). As we will see, these entropies are remarkably close to those expected to be needed to achieve a Heisenberg antiferromagnet in fermionic spin mixtures.

2.6.2.2 Magnetic phases of fermionic mixtures on a lattice

At half-filling, strong coupling $U/J \gg 1$ and temperatures below the exchange coupling $k_B T < J_{ex}$, the ground state of the fermionic Hubbard model is a Heisenberg antiferromagnet. Such an antiferromagnetic ordering of the spins is believed to be connected to a d-wave superconducting phase that is experimentally observed to occur under hole doping of initially antiferromagnetically ordered cuprate compounds (Lee *et al.* 2006). The goal of achieving such initial antiferromagnetic ordering with ultracold fermionic quantum gases and subsequently doping the gas has thus become one of the major research goals in the field. Ultimately, the hope is to reveal whether a d-wave superconducting phase is indeed present as a ground state of the repulsive Hubbard model or not. Let us, therefore, recall the entropy requirements for the creation of an antiferromagnetically ordered state. The maximum entropy that can be stored per lattice site as spin-disorder of the quantum gas is $s = k_{\rm B} \log 2$, corresponding to the equivalent entropy per particle (at half-filling) of $S/N = k_{\rm B} \log 2$. In (dynamical) mean field theory this coincides with the critical entropy needed to reach an antiferromagnetically ordered state. Quantum fluctuations will, in general, tend to lower this critical entropy. This has been confirmed by Schwinger boson calculations (Werner *et al.* 2005) and QMC simulations (Paiva *et al.* 2010; Wessel 2010), where the critical entropy needed to reach a Néel ordered phase was found to be of the order of $S/N = k_{\rm B}\, 0.5 \log 2$. Using DMFT calculations, Werner *et al.* (2005) have shown that this critical entropy undergoes a maximum close to the point where the Néel temperature is also maximal around $U/J \simeq 10{-}15$ in the intermediate coupling regime. For weak coupling (small U/J) one expects the quantum gas to show only a weak modulation of the sublattice magnetization, corresponding to a spin-density instability of the system, with a transition temperature that is exponentially small in U/J. These resulting quantum phases and critical entropies calculated via DMFT are summarized in Figure 2.11 (see (Werner *et al.* 2005)).

For current state-of-the-art experiments, where entropies per particle of $S/N \simeq (1 - 1.5)k_{\rm B}$ are reached, it is clear that novel entropy removal techniques will be required to lower the entropies by a factor of 3–5, to bring the system reliably into a magnetically ordered quantum phase. One should note that for larger exchange coupling that could be achieved, e.g., with polar molecules or Rydberg atoms, the same low critical entropies have to be realized, to bring the spin mixture into a magnetically ordered state. In that respect, reaching a magnetically ordered state is equally difficult for weak or strong exchange interactions. An important advantage of larger exchange coupling should, however, be that the timescales for the formation of the spin-ordered state would be much lower, and the effect of adverse technical heating sources should thereby be greatly ameliorated (Gerbier and Castin 2010; Pichler *et al.* 2010).

2.6.2.3 Novel cooling approaches

Current experiments are faced with the challenge of lowering the entropy of the quantum gas mixtures, to reach a magnetically ordered ground state. As cooling the quantum gases initially to even lower temperatures proves difficult, novel cooling (or better entropy removal) techniques will be required to reach this goal. One fact that can be used to advantage in this respect is that because of the inhomogeneous overall trapping potential, the entropy of the quantum gas is, in general, not distributed equally throughout the system. In fact, for a Mott insulator or a band insulating

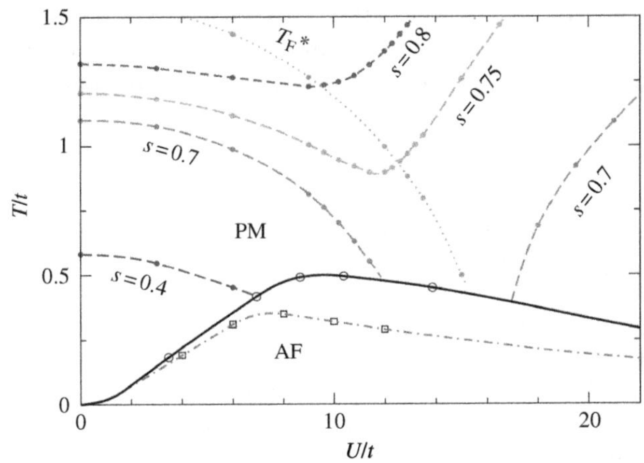

Figure 2.11 Phase diagram of the half-filled fermionic Hubbard model for a simple cubic lattice. For high temperatures, a paramagnetic phase is formed (PM), whereas for low temperatures, an antiferromagnetically (AF) ordered phase emerges. The solid line marks the transition temperature calculated in DMFT, whereas the dash-dotted line shows that expected using QMC. Dashed lines indicate isentropic curves calculated within DMFT. Within DMFT, the AF-ordered phase would be reached for entropies below $s < k_B \log 2$, however, quantum fluctuations tend to reduce the required maximum entropy. From (Werner *et al.* 2005).

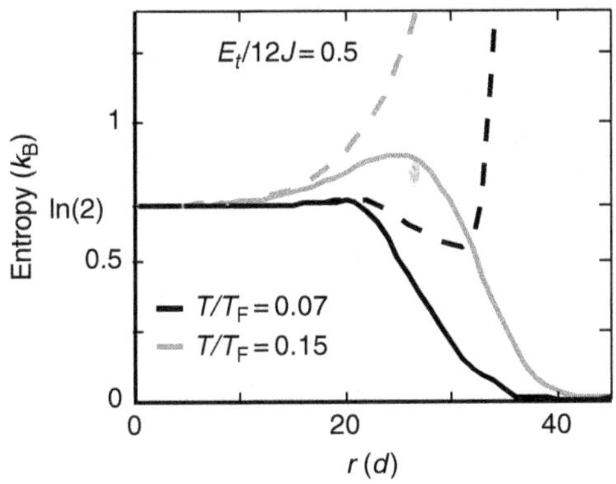

Figure 2.12 Entropy distribution of a Mott insulator in a trap for two initial temperatures of $T/T_F = 0.07$ (black) and $T/T_F = 0.15$ (grey). Solid (dashed) lines indicate entropy per lattice site (per particle). From (Schneider *et al.* 2008).

core, it is dominantly located in the metallic wings of the system (see Figure 2.12). By removing such highly entropic wings, one should thus be able to dramatically lower the overall entropy and thereby achieve a significant cooling of the system. The ability to shape the trapping potentials with high-resolution has been shown to be crucial in this respect, such that entropy can first be confined to spatial regions and then efficiently removed in a second step. Of course, care will have to be taken not to heat the gas during such an entropy removal procedure, but several schemes have been analysed theoretically that render this approach promising (Bernier *et al.* 2009; Ho *et al.* 2009).

2.7 Single-site and single-atom resolved imaging of quantum gases in optical lattices

Imaging single quantum particles has revolutionized the field of quantum optics. For several years, researchers have aspired to employ such single-particle detection methods for the analysis of ultracold quantum gases. Only recently has it become possible to realize such imaging techniques, marking a milestone for the characterization and manipulation of ultracold quantum gases (Gericke *et al.* 2008; Bakr *et al.* 2009, Bakr *et al.* 2010; Sherson *et al.* 2010). In this discussion, we will focus on a novel detection technique: high-resolution fluorescence imaging of strongly interacting quantum gases. The imaging of a bosonic Mott insulator will serve as an example of the novel analysis possibilities available with this technique.

2.7.1 High resolution and single-atom sensitive fluorescence imaging

One of the standard imaging techniques in ultracold quantum gases—absorption imaging—cannot be easily extended to the regime of single-atom sensitivity. This is mainly due to the limited absorption a laser beam experiences when interacting with a single atom. For typical experimental conditions, this absorption is smaller than the accompanying photon shot noise. While high-resolution images of down to 1 μm resolution have been successfully used to record *in-situ* absorption images of trapped quantum gases (Gemelke *et al.* 2009), they have not reached the single-atom sensitive detection regime. Fluorescence imaging (Figure 2.13) can, however, overcome this limited signal-to-noise ratio and, therefore, provides a viable route for combining high-resolution imaging with single-atom sensitivity. By using laser-induced fluorescence and by trapping the atoms in a very deep potential, several hundred thousand photons can be scattered from a single atom, of which a few thousand are ultimately detected. An excellent signal-to-noise ratio in the detection of a single atom can, therefore, be achieved.

This idea was first pioneered for the case of optical lattices by the group of D. Weiss (Nelson *et al.* 2007), who loaded atoms from a magneto-optical trap into a three-dimensional lattice with a lattice constant of 6 μm (Nelson *et al.* 2007). For typical condensed matter oriented experiments, such large spaced lattices are, however, of limited use, owing to their almost vanishing tunnel coupling between neighbouring wells. Extending fluorescence imaging, to a regime where the resolution can be comparable to typical lattice spacings of $d = 500\,\text{nm}$, thus requires microscope objectives working in the regime of large numerical apertures (NAs), as the smallest resolvable distances in classical optics are determined by $\sigma = \lambda/(2\text{NA})$.

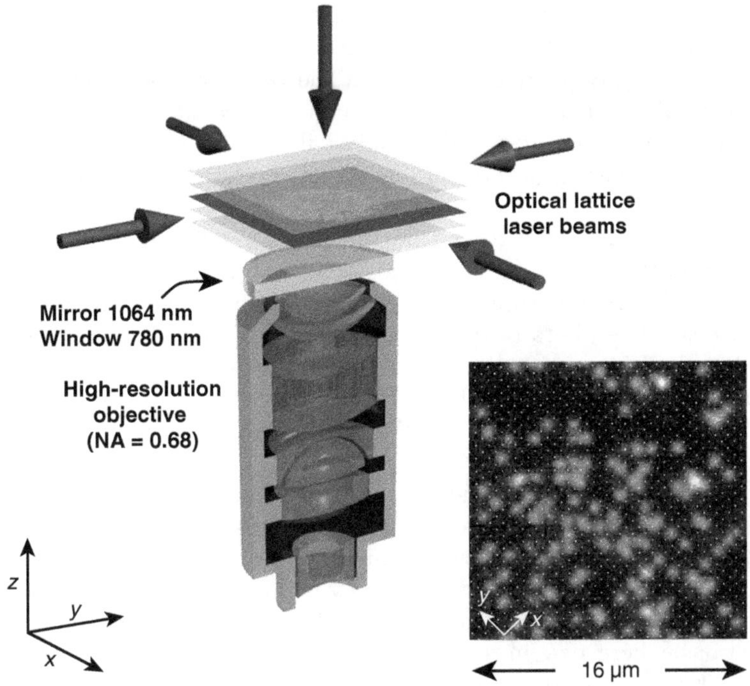

Figure 2.13 Schematic set-up for high resolution fluorescence imaging of a two-dimensional quantum gas. Two-dimensional bosonic quantum gases are prepared in a single 2D plane of an optical standing wave along the z-direction, which is created by retroreflecting a laser beam ($\lambda = 1064\,\text{nm}$) on the coated vacuum window. Additional lattice beams along the x- and y-directions are used to bring the system into the strongly correlated regime of a Mott insulator. The atoms are detected using fluorescence imaging via a high resolution microscope objective. Fluorescence of the atoms was induced by illuminating the quantum gas with an optical molasses that simultaneously laser cools the atoms. The inset shows a section from a fluorescence picture of a dilute thermal cloud (points mark the lattice sites).

In recent publications, Bakr *et al.* (2009; 2010) and Sherson *et al.* (2010) have demonstrated such high-resolution imaging and applied it to image the transition of a superfluid to a Mott insulator in 2D. In the experiments, 2D Bose–Einstein condensates were first created in tightly confining potential planes. Subsequently, the depth of a two-dimensional simple-cubic type lattice was increased, leaving the system in either a superfluid or Mott insulating regime. The lattice depths were then suddenly increased to very deep values of $300\,\mu\text{K}$, essentially freezing out the density distribution of the atoms in the lattice. A near-resonant optical molasses was then used to induce fluorescence of the atoms in the deep lattice and also provided laser cooling, such that atoms remained on lattice sites while fluorescing. High-resolution microscope objectives with numerical apertures of $\text{NA} \approx 0.7–0.8$ were used to record the fluorescence and image the atomic density distribution on CCD cameras. A limitation of the detection method is that inelastic light induced collisions occurring during the illumination period only allow one to record the parity of the on-site atom number. Whenever pairs of atoms are present on a single lattice site, both atoms of the pair are rapidly lost within the first millisecond of illumination, owing to a large energy release caused by radiative escape and fine-structure changing collisions (DePue *et al.* 1999).

In both experiments, high-resolution imaging has allowed reconstruction of the atom distribution (modulo 2) on the lattice down to a single-site level. Results for the case of Mott insulators of such a digital particle number reconstruction are displayed in Figure 2.14c,d. The digitally reconstructed images have the significant advantage of essentially exhibiting an infinite signal-to-noise ratio that is of great advantage in further correlation analyses. Non-local order parameters that are characteristic for topologically ordered quantum phases (Wen 2004) can also easily be revealed with the detection method.

Being able to resolve single-lattice sites in addition allows one to manipulate atoms with single-site resolution. This was achieved by focusing a manipulation laser beam through the high-resolution objective, which enables one to perform arbitrary spin rotations on an atom at a given lattice site (Weitenberg *et al.* 2011). High-resolution imaging and addressing will, therefore, also be highly beneficial for quantum information applications, where, e.g., in the case of a one-way quantum computer (Raussendorf and Briegel 2001), it is essential to measure the spin state of an atom at a specified lattice site.

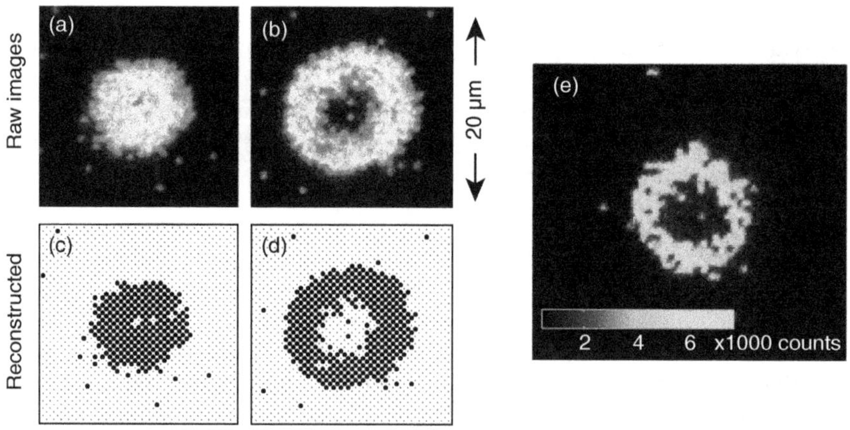

Figure 2.14 High-resolution fluorescence images of Mott insulators. (a) $n = 1$ and (b,e) wedding cake structure of $n = 1$ and $n = 2$ Mott insulators. Using a numerical algorithm, the corresponding atom distribution on a lattice can be reconstructed. In (c,d) such a reconstructed atom distribution is displayed (small points mark lattice sites, large points mark position of a single atom). Figures adapted from (Sherson *et al.* 2010) and (Bakr *et al.* 2010).

2.7.2 Thermodynamics of the Bose–Hubbard model in the atomic limit

Modelling the density distribution and the fluctuations of the strongly interacting bosonic quantum gas becomes especially simple in the so-called atomic limit of the Bose–Hubbard model, for which $J = 0$ or $U/J \to \infty$. In this limit, the grand canonical partition function of the trapped quantum gas $Z^{(0)}$ can be written as a product of on-site partition functions:

$$Z^{(0)} = \prod_{\mathbf{R}} Z_{\mathbf{R}}^{(0)}, \tag{2.36}$$

where the on-site partition function is given by

$$Z^{(0)} = \sum_{n} \mathrm{e}^{-\beta(E_n - \mu(\mathbf{R})n)}. \tag{2.37}$$

The local chemical potential at lattice site $\mathbf{R}$ is denoted by $\mu(\mathbf{R})$ and the eigenenergy of n atoms on this lattice site is given by the standard single-band Bose–Hubbard interaction term $E_n = 1/2\,Un(n-1)$. In particular, we can use these functions to calculate the on-site probability of finding n atoms per lattice site as

$$P_{\mathbf{R}}(n) = \frac{\mathrm{e}^{-\beta(E_n - \mu(\mathbf{R})n)}}{Z^{(0)}}. \tag{2.38}$$

We can thus think of our system as consisting of disconnected sites, whose thermodynamics is determined only by the ratio of $U/(k_{\mathrm{B}}T)$ and the local chemical potential. In this limit, the problem becomes analytically tractable and simple to analyse.

2.7.2.1 Radial density and fluctuation profiles

As a simple application of our result, let us calculate the density profile and its fluctuations for a two-dimensional radially symmetric trapping potential. All sites with the same distance r from the trapping centre exhibit the same chemical potential $\mu(r)$. The average density at this radial distance is thus given by:

$$\bar{n} = \frac{1}{Z(r)} \sum_{n} n\mathrm{e}^{-\beta(E_n - \mu(r))}. \tag{2.39}$$

To evaluate this, we would need to sum over all possible occupation states in our on-site partition function. In practice, we may truncate our sum around values of $\mathrm{int}[\mu/U] \pm 1$ for temperatures $k_{\mathrm{B}}T \ll U$, as thermal fluctuations become exponentially suppressed in this regime. This is the so called *particle-hole* approximation.

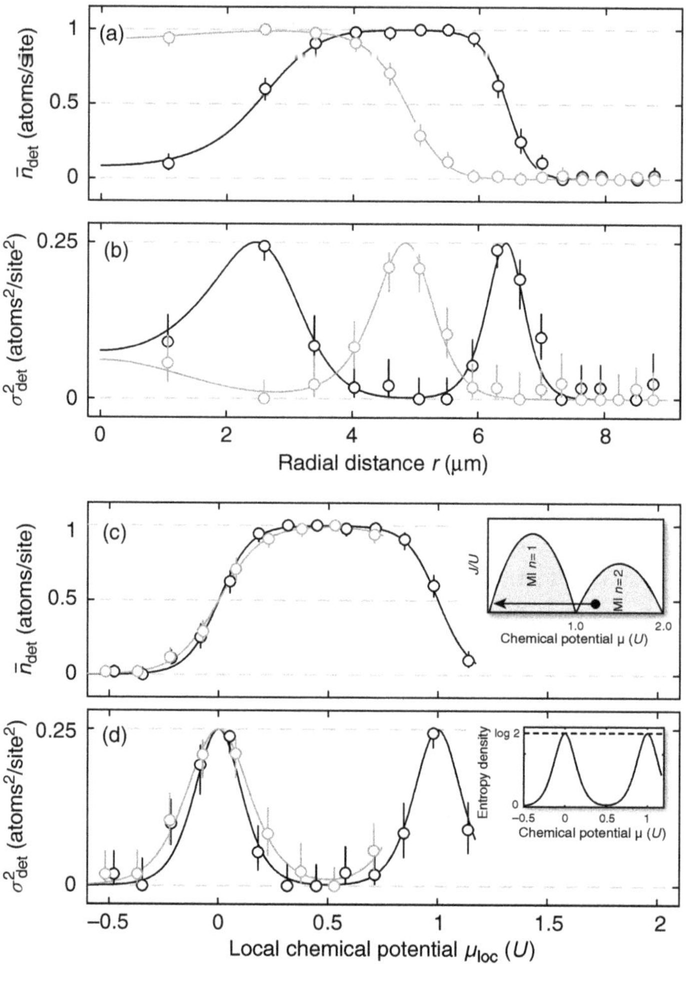

Figure 2.15 Radial atom density and variance profiles. Radial profiles were obtained from the digitized reconstructed images by azimuthal averaging. (a, b), Grey and black points correspond to the $n = 1$ and $n = 2$ MI images of Figure 2.14 d,e. For the two curves, the fits yielded temperatures $T = 0.090(5)\,U/k_{\mathrm{B}}$ and $T = 0.074(5)\,U/k_{\mathrm{B}}$, chemical potentials $\mu = 0.73(3)\,U$ and $\mu = 1.17(1)\,U$, and radii $r_0 = 5.7(1)$ μm and $r_0 = 5.95(4)$ μm, respectively. From the fitted values of T, μ, and r_0, we determined the atom numbers of the system to be $N = 300(20)$ and $N = 610(20)$. (c, d), The same data plotted versus the local chemical potential using local-density approximation. The inset of (c) is a Bose–Hubbard phase diagram ($T = 0$) showing the transition between the characteristic MI lobes and the superfluid region. The line starting at the maximum chemical potential μ shows the part of the phase diagram existing simultaneously at different radii in the trap due to the external harmonic confinement. The inset of (d) is the entropy density calculated for the displayed $n = 2$ MI. From (Sherson *et al.* 2010).

In the case of fluorescence imaging, light-assisted collisions only allow us to detect the parity of the on-site atom number, i.e., the on-site particle number modulo 2. We thus find for the detected average density:

$$\bar{n}_{\text{det}} = \frac{1}{Z(r)} \sum_n \quad \text{mod}_2(n) e^{-\beta(E_n - \mu(r))}. \tag{2.40}$$

For the fluctuations of the quantum gas, we find that, as a result of the parity projection in the detection process, $\overline{n^2(r)} = \bar{n}_{\text{det}}$, resulting in

$$\sigma^2_{\text{det}}(r) = \bar{n}_{\text{det}}(r) - \bar{n}^2_{\text{det}}(r) \tag{2.41}$$

for the detected fluctuations. Both the detected fluctuations and the average density profile are functions of three parameters $\mu/U, k_B T/U$ and the trapping frequency ω^2 of the overall harmonic confinement. While the trap frequency can be independently measured, the chemical potential and temperature of the quantum gas can be extracted via a fit to azimuthally averaged radial density profiles of single images of the quantum gas. This is shown in Figure 2.15 for the two images of an $n = 1$ and $n = 2$ Mott insulator in the core of the gas.

In the atomic limit, these fit functions thus allow an efficient determination of temperature and chemical potential of the quantum gas. Note that apart from the particle-hole and local density approximation (LDA), no further approximations have to be made in this case, giving exact numerical model functions for the trapped Bose gas. The radial density and fluctuation profiles can be converted to density and fluctuation profiles versus chemical potential by using, again, LDA $\mu_{\text{loc}}(r) = \mu - 1/2m\omega^2 r^2$. We see that the data for the two distinct data sets of the $n = 1$ and $n = 2$ Mott insulators fall on top of each other when plotted in this way, emphasizing the fact that radial profiles correspond to cuts through the phase diagram (see inset in Figure 2.15c) of the Bose–Hubbard model. Residual small differences between the two curves can be attributed to the slightly different temperatures of the atom clouds. Interestingly, as in the case of fermionic atoms, one notes that fluctuations are concentrated at the border of the Mott insulating regions, where the system is superfluid for $T = 0$, and it becomes thermodynamically easiest to introduce fluctuations in this spatial region, where the energy needed to create an excitation is lowest.

References

Altman, E, Hofstetter, W, Demler, E, and Lukin, MD (2003). Phase diagram of two-component bosons on an optical lattice. *New J. Phys.*, **5**, 113.

Altman, E, Demler, E, and Lukin, MD (2004). Probing many-body states of ultracold atoms via noise correlations. *Phys. Rev. A*, **70**, 13603.

Anderlini, M, Lee, PJ, Brown, BL, Sebby-Strabley, J, Phillips, WD, and Porto, JV (2007). Controlled exchange interaction between pairs of neutral atoms in an optical lattice. *Nature*, **448**(7152), 452.

Auerbach, A (1994). *Interacting Electrons and Quantum Magnetism*. Springer-Verlag, New York.

Bakr, WS, Gillen, JI, Peng, A, Fölling, S, and Greiner, M (2009). A quantum gas microscope for detecting single atoms in a Hubbard-regime optical lattice. *Nature*, **462**(7269), 74.

Bakr, WS, Peng, A, Tai, ME, *et al.* (2010). Probing the superfluid-to-Mott-insulator transition at the single-atom level. *Science*, **547**, 547.

Baym, G (1998). The physics of Hanbury Brown–Twiss intensity interferometry: from stars to nuclear collisions. *Act. Phys. Pol. B*, **29**, 1839.

Becker, C, Soltan-Panahi, P, Kronjäger, J, Dörscher, S, Bongs, K, and Sengstock, K (2010). Ultracold quantum gases in triangular optical lattices. *New J. Phys.*, **12**(6), 065025.

Bernier, J-S, Kollath, C, Georges, A, *et al.* (2009). Cooling fermionic atoms in optical lattices by shaping the confinement. *Phys. Rev. A*, **79**(6), 061601.

Bloch, I (2005). Ultracold quantum gases in optical lattices. *Nat. Phys.*, **1**, 23.

Bloch, I, Dalibard, J, and Zwerger, W (2008). Many-body physics with ultracold gases. *Rev. Mod. Phys.*, **80**(3), 885.

Capogrosso-Sansone, B, Söyler, SG, Prokofev, NV, and Svistunov, BV (2010). Critical entropies for magnetic ordering in bosonic mixtures on a lattice. *Phys. Rev. A*, **81**(5), 053622.

Carr, LD, DeMille, D, Krems, RV, and Ye, J (2009). Cold and ultracold molecules: science, technology and applications. *New. J. Phys.*, **11**(5), 055049.

Chin, C, Grimm, R, Julienne, P, and Tiesinga, E (2010). Feshbach resonances in ultracold gases. *Rev. Mod. Phys.*, **82**(2), 1225.

Damski, B, Fehrmann, H, Everts, H-U, Baranov, M, Santos, L, and Lewenstein, M (2005). Quantum gases in trimerized Kagomé lattices. *Phys. Rev. A*, **72**(5), 053612.

DeLeo, L, Kollath, C, Georges, A, Ferrero, M, and Parcollet, O (2008). Trapping and cooling fermionic atoms into Mott and Néel states. *Phys. Rev. Lett.*, **101**(21), 210403.

DePue, MT, McCormick, C, Winoto, SL, Oliver, S, and Weiss, DS (1999). Unity occupation of sites in a 3D optical lattice. *Phys. Rev. Lett.*, **82**, 2262.

Duan, L-M, Demler, E, and Lukin, MD (2003). Controlling spin exchange interactions of ultracold atoms in an optical lattice. *Phys. Rev. Lett.*, **91**, 90402.

Eckardt, A, Weiss, C, and Holthaus, M (2005). Superfluid-insulator transition in a periodically driven optical lattice. *Phys. Rev. Lett.*, **95**, 260404.

Fisher, MPA, Weichman, PB, Grinstein, G, and Fisher, DS (1989). Boson localization and the superfluid-insulator transition. *Phys. Rev. B*, **40**, 546.

Fölling, S, Gerbier, F, Widera, A, Mandel, O, Gericke, T, and Bloch, I (2005). Spatial quantum noise interferometry in expanding ultracold atom clouds. *Nature*, **434**(7032), 481.

Fölling, S, Widera, A, Müller, T, Gerbier, F, and Bloch, I (2006). Formation of spatial shell structure in the superfluid to Mott insulator transition. *Phys. Rev. Lett.*, **97**(6), 060403.

Gadway, B, Pertot, D, Reimann, R, and Schneble, D (2010). Superfluidity of interacting Bosonic mixtures in optical lattices. *Phys. Rev. Lett.*, **105**(4), 045303.

Gemelke, N, Zhang, X, Hung, C-L, and Chin, C (2009). *In situ* observation of incompressible Mott-insulating domains in ultracold atomic gases. *Nature*, **460**(7258), 995.

Gerbier, F and Castin, Y (2010). Heating rates for an atom in a far-detuned optical lattice. *Phys. Rev. A*, **82**(1), 013615.

Gerbier, F, Widera, A, Fölling, S, Mandel, O, Gericke, T, and Bloch, I (2005). Phase coherence of an atomic Mott insulator. *Phys. Rev. Lett.*, **95**(5), 050404.

Gerbier, F, Trotzky, S, Fölling, S, *et al.* (2008). Expansion of a quantum gas released from an optical lattice. *Phys. Rev. Lett.*, **101**(15), 155303.

Gericke, T, Würtz, P, Reitz, D, Langen, T, and Ott, H (2008). High-resolution scanning electron microscopy of an ultracold quantum gas. *Nat. Phys.*, **4**(12), 949.

Greiner, M, Bloch, I, Mandel, MO, Hänsch, TW, and Esslinger, T (2001). Exploring phase coherence in a 2D lattice of Bose–Einstein condensates. *Phys. Rev. Lett.*, **87**, 160405.

Greiner, M, Mandel, O, Esslinger, T, Hänsch, TW, and Bloch, I (2002). Quantum phase transition from a superfluid to a Mott insulator in a gas of ultracold atoms. *Nature*, **415**, 39.

Grimm, R, Weidemüller, M, and Ovchinnikov, YB (2000). Optical dipole traps for neutral atoms. *Adv. At. Mol. Opt. Phys.*, **42**, 95.

Hackermüller, L, Schneider, U, Moreno-Cardoner, M, *et al.* (2010). Anomalous expansion of attractively interacting fermionic atoms in an optical lattice. *Science*, **327**(5973), 1621.

Hadzibabic, Z, Stock, S, Battelier, B, Bretin, V, and Dalibard, J (2004). Interference of an array of independent Bose–Einstein condensates. *Phys. Rev. Lett.*, **93**, 180403.

Hanbury Brown, R and Twiss, RQ (1956a). Correlation between photons in two coherent beams of light. *Nature*, **177**, 27.

Hanbury Brown, R and Twiss, RQ (1956b). The question of correlation between photons in coherent light rays. *Nature*, **178**, 1447.

Helmes, R, Costi, T, and Rosch, A (2008). Mott transition of Fermionic atoms in a three-dimensional optical trap. *Phys. Rev. Lett.*, **100**(5), 056403.

Henkel, N, Nath, R, and Pohl, T (2010). Three-dimensional roton excitations and supersolid formation in Rydberg-excited Bose–Einstein condensates. *Phys. Rev. Lett.*, **104**(19), 4.

Henny, M, Oberholzer, S, Strunk, C, *et al.* (1999). The fermionic Hanbury Brown and Twiss experiment. *Science*, **284**, 296.

Ho, T-L and Zhou, Q (2009). Universal cooling scheme for quantum simulation. arXiv:0911.5506.

Ho, A, Cazalilla, M, and Giamarchi, T (2009). Quantum simulation of the Hubbard model: the attractive route. *Phys. Rev. A*, **79**(3), 033620.

Hofstetter, W, Cirac, JI, Zoller, P, Demler, E, and Lukin, MD (2002). High-temperature superfluidity of fermionic atoms in optical lattices. *Phys. Rev. Lett.*, **89**, 220407.

Honer, J, Weimer, H, Pfau, T, and Büchler, HP (2010). Collective many-body interaction in Rydberg dressed atoms. *Phys. Rev. Lett.*, **105**, 160404.

Hubbard, J (1963). Electron correlations in narrow energy bands. *Proc. R. Soc. Lond. A*, **276**, 238.

Hung, C-L, Zhang, X, Gemelke, N, and Chin, C (2010). Slow mass transport and statistical evolution of an atomic gas across the superfluid Mott-insulator transition. *Phys. Rev. Lett.*, **104**(16), 160403.

Iannuzzi, M, Orecchini, A, Sacchetti, F, Facchi, P, and Pascazio, S (2006). Direct experimental evidence of free-fermion antibunching. *Phys. Rev. Lett.*, **96**, 80402.

Jaksch, D and Zoller, P (2005). The cold atoms Hubbard toolbox. *Ann. Phys.*, **315**, 52.

Jaksch, D, Bruder, C, Cirac, JI, Gardiner, CW, and Zoller, P (1998). Cold bosonic atoms in optical lattices. *Phys. Rev. Lett*, **81**, 3108.

Jaksch, D, Briegel, HJ, Cirac, JI, Gardiner, CW, and Zoller, P (1999). Entanglement of atoms via cold controlled collisions. *Phys. Rev. Lett.*, **82**, 1975.

Jeltes, T, McNamara, JM, Hogervorst, W, *et al.* (2007). Hanbury Brown Twiss effect for bosons versus fermions. *Nature*, **445**, 402.

Jördens, R, Strohmaier, N, Günter, K, Moritz, H, and Esslinger, T (2008). A Mott insulator of fermionic atoms in an optical lattice. *Nature*, **455**, 204.

Jördens, R, Tarruell, L, Greif, D, *et al.* (2010). Quantitative determination of temperature in the approach to magnetic order of ultracold fermions in an optical lattice. *Phys. Rev. Lett.*, **104**(18), 180401.

Kastberg, A, Phillips, WD, Rolston, SL, Spreeuw, RJC, and Jessen, PS (1995). Adiabatic cooling of cesium in an optical lattice. *Phys. Rev. Lett.*, **74**, 1542.

Kiesel, H, Renz, A, and Hasselbach, F (2002). Observation of Hanbury Brown–Twiss anticorrelations for free electrons. *Nature*, **418**, 392.

Kinoshita, T, Wenger, T, and Weiss, DS (2004). Observation of a one-dimensional Tonks–Girardeau gas. *Science*, **305**, 1125.

Köhl, M, Moritz, H, Stöferle, T, Günter, K, and Esslinger, T (2005). Fermionic atoms in a three dimensional optical lattice: observing Fermi surfaces, dynamics, and interactions. *Phys. Rev. Lett.*, **94**, 80403.

Kuklov, AB and Svistunov, BV (2003). Counterflow superfluidity of two-species ultracold atoms in a commensurate optical lattice. *Phys. Rev. Lett.*, **90**, 100401.

Kuklov, A, Prokofev, N, and Svistunov, B (2004). Commensurate two-component bosons in an optical lattice: ground state phase diagram. *Phys. Rev. Lett.*, **92**(5), 050402.

Le Hur, K and Maurice Rice, T (2009). Superconductivity close to the Mott state: from condensed-matter systems to superfluidity in optical lattices. *Ann. Phys.*, **324**(7), 1452.

Lee, PA, Nagaosa, N, and Wen, X-G (2006). Doping a Mott insulator: physics of high-temperature superconductivity. *Rev. Mod. Phys.*, **78**, 17.

Lewenstein, M, Sanpera, A, Ahufinger, V, Damski, B, Sen (De) A., and Sen, U (2007). Ultracold atomic gases in optical lattices: mimicking condensed matter physics and beyond. *Adv. Phys.*, **56**, 243.

Lignier, H, Sias, C, Ciampini, D, *et al.* (2007). Dynamical control of matter-wave tunneling in periodic potentials. *Phys. Rev. Lett.*, **99**(22), 220403.

Mandel, O, Greiner, M, Widera, A, Rom, T, Hänsch, TW, and Bloch, I (2003). Coherent transport of neutral atoms in spin-dependent optical lattice potentials. *Phys. Rev. Lett.*, **91**, 10407.

Moritz, H, Stöferle, T, Köhl, M, and Esslinger, T (2003). Exciting collective oscillations in a trapped 1D gas. *Phys. Rev. Lett.*, **91**, 250402.

Mun, J, Medley, P, Campbell, GK, Marcassa, L, Pritchard, D, Ketterle, W (2007). Phase diagram for a Bose-Einstein condensate moving in an optical lattice. *Phys. Rev. Lett.*, **99**, 150604.

Nelson, KD, Li, X, and Weiss, DS (2007). Imaging single atoms in a three-dimensional array. *Nat. Phys.*, **3**(8), 556.

Oliver, WD, Kim, J, Liu, RC, and Yamamoto, Y (1999). Hanbury Brown and Twiss-type experiment with electrons. *Science*, **284**, 299.

Paiva, T, Scalettar, R, Randeria, M, and Trivedi, N (2010). Fermions in 2D optical lattices: temperature and entropy scales for observing antiferromagnetism and superfluidity. *Phys. Rev. Lett.*, **104**(6), 066406.

Paredes, B and Bloch, I (2008). Minimum instances of topological matter in an optical plaquette. *Phys. Rev. A*, **77**, 23603.

Paredes, B, Widera, A, Murg, V, *et al.* (2004). Tonks-Girardeau gas of ultracold atoms in an optical lattice. *Nature*, **429**, 277.

Peil, S, Porto, JV, Laburthe-Tolra, B, *et al.* (2003). Patterned loading of a Bose–Einstein condensate into an optical lattice. *Phys. Rev. A*, **67**, 51603.

Pichler, H, Daley, A, and Zoller, P (2010). Nonequilibrium dynamics of bosonic atoms in optical lattices: decoherence of many-body states due to spontaneous emission. *Phys. Rev. A*, **82**(6), 063605.

Polkovnikov, A, Sengupta, K, Silva, A, and Vengalattore, M (2011). Colloquium: nonequilibrium dynamics of closed interacting quantum systems. *Rev. Mod. Phys.*, **83**(3), 863.

Pollet, L, Prokofev, N, and Svistunov, B (2010). Criticality in trapped atomic systems. *Phys. Rev. Lett.*, **104**(24), 245705.

Pupillo, G, Micheli, A, Boninsegni, M, Lesanovsky, I, and Zoller, P (2010). Strongly correlated gases of Rydberg-dressed atoms: quantum and classical dynamics. *Phys. Rev. Lett.*, **104**(22), 223002.

Raussendorf, R and Briegel, HJ (2001). A one-way quantum computer. *Phys. Rev. Lett.*, **86**, 5188.

Rey, AM, Sensarma, R, Fölling, S, Greiner, M, Demler, E, and Lukin, MD (2009). Controlled preparation and detection of *d*-wave superfluidity in two-dimensional optical superlattices. *Europhys. Lett.*, **87**(6), 60001.

Rom, T, Best, T, van Oosten, D, *et al.* (2006). Free fermion antibunching in a degenerate Fermi gas released from an optical lattice. *Nature*, **444**, 733.

Scarola, VW, Pollet, L, Oitmaa, J, and Troyer, M (2009). Discerning incompressible and compressible phases of cold atoms in optical lattices. *Phys. Rev. Lett.*, **102**(13), 135302.

Schellekens, M, Hoppeler, R, Perrin, A, *et al.* (2005). Hanbury Brown Twiss effect for ultracold quantum gases. *Science*, **310**, 648.

Schneider, U, Hackermüller, L, Will, S, *et al.* (2008). Metallic and insulating phases of repulsively interacting fermions in a 3D optical lattice. *Science*, **322**, 1520.

Schneider, U, Hackermüller, L, Ronzheimer, JP, *et al.* (2012). Fermionic transport and out-of-equilibrium dynamics in a homogeneous Hubbard model with ultracold atoms. *Nature Physics*, **8**, 213–218.

Sherson, JF, Weitenberg, C, Endres, M, Cheneau, M, Bloch, I, and Kuhr, S (2010). Single-atom-resolved fluorescence imaging of an atomic Mott insulator. *Nature*, **467**, 68.

Spielman, IB, Phillips, W, and Porto, J (2007). Mott-insulator transition in a two-dimensional atomic Bose gas. *Phys. Rev. Lett.*, **98**(8), 080404.

Strohmaier, N, Greif, D, Jördens, R, *et al.* (2010). Observation of elastic doublon decay in the Fermi–Hubbard model. *Phys. Rev. Lett.*, **104**(8), 080401.

Tolra, B, O'Hara, K, Huckans, J, Phillips, W, Rolston, S, and Porto, J (2004). Observation of reduced three-body recombination in a correlated 1D degenerate Bose gas. *Phys. Rev. Lett.*, **92**(19), 190401.

Trebst, S, Schollwöck, U, Troyer, M, and Zoller, P (2006). d-wave resonating valence bond states of fermionic atoms in optical lattices. *Phys. Rev. Lett.*, **96**(25), 250402.

Trotzky, S, Cheinet, P, Fölling, S, *et al.* (2008). Time-resolved observation and control of superexchange interactions with ultracold atoms in optical lattices. *Science*, **319**(5861), 295.

Weitenberg, C, Endres, M, Sherson, JF, *et al.* (2011). Single-spin addressing in an atomic Mott insulator. *Nature*, **471**(7338), 319.

Wen, XG (2004). *Quantum Field Theory of Many-Body Systems*. Oxford Graduate Texts. Oxford University Press, Oxford.

Werner, F, Parcollet, O, Georges, A, and Hassan, SR (2005). Interaction-induced adiabatic cooling and antiferromagnetism of cold fermions in optical lattices. *Phys. Rev. Lett.*, **95**, 56401.

Wessel, S (2010). Critical entropy of quantum Heisenberg magnets on simple-cubic lattices. *Phys. Rev. B*, **81**(5), 052405.

Wirth, G, Ölschläger, M, and Hemmerich, A (2010). Evidence for orbital superfluidity in the P-band of a bipartite optical square lattice. *Nat. Physics*, **7**(2), 147.

Zhou, Q and Ho, T-L (2011). Universal thermometry for quantum simulation. *Phys. Rev. Lett.*, **106**, 225301.

3
The few-atom problem

Dmitry S. Petrov

Laboratoire de Physique Théorique et Modèles Statistiques, CNRS,
Université Paris Sud, 91405 Orsay, France;
Russian Research Center Kurchatov Institute, Kurchatov Square,
123182 Moscow, Russia

Chapter Contents

3.1 Overview

The few-body problem plays a very important role in the physics of ultracold gases. It describes the structure of molecules and their formation in three-body collisions (three-body recombination), atom–molecule and molecule–molecule collisional properties, structure of trimers and larger clusters, numerous phenomena that go under the name of Efimov physics, and many other problems.

In ultracold gases, we benefit from a remarkable separation of scales. The atomic de Broglie wavelengths are much larger than the range, R_e, of the van der Waals interatomic potential. The ultracold regime is defined by the inequality $kR_e \ll 1$, where k is the typical atomic momentum. In this regime, very few terms in the effective range expansion for the scattering amplitude are required, and most of the time the interaction is characterized by a single parameter—the s-wave scattering length a. From the two-body viewpoint, all short-range potentials are equivalent, as long as they have the same scattering length, and, therefore, one can use an idealized zero-range potential (pseudo-potential) with the same a. The zero-range model aptly describes the weakly interacting BEC (Huang 1963) as well as the BCS–BEC crossover in fermionic mixtures (Giorgini *et al.* 2008).

There is a class of few-body problems where both long and short length scales are important. For example, knowledge of a is not sufficient for calculating the spectrum of Efimov trimers or the rates of recombination and relaxation to deeply bound molecular states. Such problems can be solved in the *universal* limit, $|a| \gg R_e$, by introducing the so-called three-body parameter, which absorbs all the short-range three-body physics in the same manner that the scattering length absorbs the short-range two-body physics. Universality (Braaten and Hammer 2006; 2007) in this context reflects the amazing fact that different systems with the same scattering length and three-body parameter exhibit the same physics. The possibility of modifying a in atomic gases using Feshbach resonances makes ultracold gases an ideal playground for checking the universal theory. The Efimov effect, predicted 40 years ago (Efimov 1970), was observed first in a cold gas of ^{133}Cs in Innsbruck (Kraemer *et al.* 2006; see also Ferlaino *et al.* 2011 for review) and subsequently in other alkali atoms and mixtures (Ottenstein *et al.* 2008; Barontini *et al.* 2009; Huckans *et al.* 2009; Pollack *et al.* 2009; Zaccanti *et al.* 2009; Gross *et al.* 2009, 2010; Lompe *et al.* 2010; Nakajima *et al.* 2010, 2011).

The few-body analysis, apart from being interesting in its own right, can be used in many-body problems to integrate out few-body degrees of freedom, thus making the many-body problem more tractable. For example, a two-component fermionic mixture on the BEC side of the BCS–BEC crossover is actually a Bose gas of molecules (or an atom-molecule mixture in the density-imbalanced case). If the molecules are sufficiently small (a is much smaller than the interparticle distance), we can forget about their composite nature and treat them as elementary objects as long as the atom–molecule and molecule–molecule scattering parameters are known, the latter being given by the solution of the corresponding three- and four-atom problems. Here, we profit from the fact that the few- and many-body processes separate along the length and energy scales. Then the equation of state and the stability of the system with respect to collapse or phase separation are much easier to analyse. We can also consider lattice problems where, to describe the tight binding limit, one has to solve the few body problem on a single site and use the result as an input for the Hubbard model.

Another path from 'few' to 'many' is the high-temperature virial expansion, which involves solving few-body problems with more and more particles (Huang 1963; Liu *et al.* 2010*a*). The results are then used for consecutive approximations of thermodynamic quantities. The procedure rapidly converges, which emphasizes the importance of the few-body aspect of many-body problems and leads to the conjecture that even at low temperatures and in strongly interacting regimes, where we lack a small parameter, the natural sorting criterion for different sets of Feynman diagrams is the number of particles (particle–hole pairs) involved.

3.2 The two-body problem and resonance width

This section discusses the effective range expansion for the two-body scattering amplitude and puts a special emphasis on the role of the resonance width. Near a scattering resonance, the scattering length can be modified and, in particular, can take anomalously large values (i.e., $|a| \gg R_e$). Another very important parameter characterizing the resonance is its width, which is determined by the strength of the coupling between the closed and open channels. The narrower the resonance, the stronger the collision energy dependence of the scattering amplitude. In the effective range expansion, this leads to an increased value of the effective range.

To familiarize ourselves with the notions of scattering length, resonance width, and effective range, we will consider a simple model of a single-channel potential with a barrier. It mimics the Feshbach resonance picture and clearly illustrates the role of the barrier. Then we will derive a general formula for the effective range and discuss the relation between the effective range and the probability of finding two atoms in the closed channel. This quantity plays a very important role in many problems involving Feshbach resonances with finite width. As a particular example, we will consider the structure of a Feshbach molecule.

3.2.1 Potential well with delta-function barrier

Consider two atoms with reduced mass μ interacting via an isotropic interaction potential $V(r)$. The radial Schrödinger equation for the relative motion with zero orbital angular momentum reads

$$\left[-\frac{1}{2\mu}\frac{\partial^2}{\partial r^2} + V(r) - \frac{k^2}{2\mu}\right]\chi(r) = 0, \tag{3.1}$$

where $E = k^2/2\mu$ is the collision energy and we use $\chi(r) = r\psi(r)$.

Let us take $V(r)$ in the form

$$V(r) = g\delta(r - R_e) - \begin{cases} \kappa_0^2/2\mu, & r < R_e, \text{ region I}, \\ 0, & r \geq R_e, \text{ region II}. \end{cases} \tag{3.2}$$

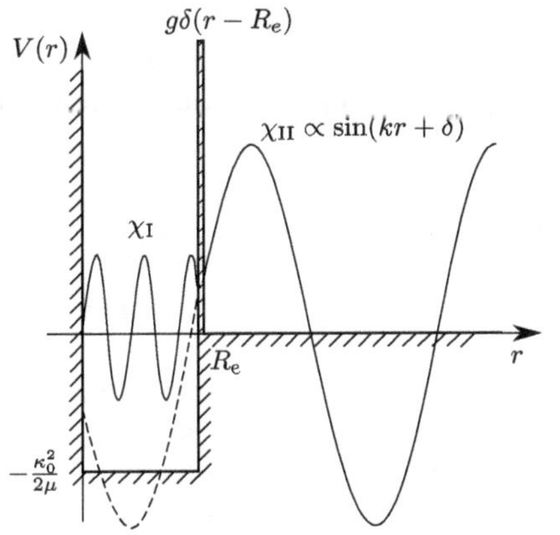

Figure 3.1 The square-well potential with a delta-function barrier, Eq. (3.2), and the wavefunction in the inner, χ_I, and the outer, χ_{II}, regions at the collision energy $k^2/2\mu$ (solid line). The spike of the wavefunction at $r = R_e$ is governed by Eq. (3.3). The dashed line shows the wavefunction χ_{II} extrapolated to the inner region I.

The potential of Eq. (3.2) is plotted in Figure 3.1. It allows one to vary the resonance width at will by varying the barrier strength g. The solutions of Eq. (3.1) in regions I and II are, respectively, $\chi_I \propto \sin(\sqrt{\kappa_0^2 + k^2}r)$ and $\chi_{II} \propto \sin(kr + \delta)$. The matching condition at $r = R_e$ reads

$$\chi'_{II}(R_e)/\chi_{II}(R_e) - \chi'_I(R_e)/\chi_I(R_e) = g, \tag{3.3}$$

which gives the phase shift δ:

$$\cot\delta = \frac{\sqrt{\kappa_0^2 + k^2} + \tan(\sqrt{\kappa_0^2 + k^2}R_e)[g + k\tan(kR_e)]}{k - \tan(kR_e)[g + \sqrt{\kappa_0^2 + k^2}\tan(\sqrt{\kappa_0^2 + k^2}R_e)]}. \tag{3.4}$$

The s-wave scattering amplitude is given in terms of the phase shift as (Landau and Lifshitz 1987)

$$f(k) = 1/(k\cot\delta(k) - ik). \tag{3.5}$$

The idea of the zero-range approximation is to extrapolate the solution χ_{II} into region I (the dashed line in Figure 3.1) and use its logarithmic derivative at the origin to construct the zero-range boundary condition, which then replaces the potential V:

$$\chi'(0)/\chi(0) = k\cot\delta(k). \tag{3.6}$$

By construction, the s-wave scattering amplitudes obtained by solving Eq. (3.1) on the one hand and the free-motion Schrödinger equation with the boundary condition (3.6) on the other hand are identical. In general, substituting the potential by the boundary condition (3.6) does not make the scattering problem easier, since we do have to calculate the phase shift. However, the zero-range approximation becomes valuable for small momenta, when one needs only a few parameters to describe the scattering. Namely, one writes down the effective range expansion

$$k \cot \delta(k) = -1/a + (r_0/2)k^2 + \ldots, \tag{3.7}$$

where a is the scattering length, r_0 is the effective range, and the terms denoted by ... contain higher powers of k.

Typically (i.e., for most interatomic interactions provided by nature), the length parameters a and r_0 are of the order of the physical range of the potential R_e. Then, in the ultracold limit, the first term on the right-hand side of Eq. (3.7) suffices and one arrives at the so-called Bethe–Peierls boundary condition for the wavefunction

$$\chi'(0)/\chi(0) = (r\psi(r))'/r\psi(r)|_{r=0} = -1/a, \tag{3.8}$$

or, equivalently,

$$\psi(r) \propto 1/r - 1/a, \ r \to 0. \tag{3.9}$$

Equation (3.9) sets the relation between the coefficient in front of the singular term and the regular term of the wavefunction. Moreover, it points to the physical meaning of the scattering length—for positive a, the node of the zero-energy wavefunction is found at $r = a$.[1] In this sense, at sufficiently low energies, any potential characterized by $a > 0$ is equivalent to a hard-core one of radius a.

In the case of the potential of Eq. (3.2) the scattering length is given by

$$a = R_e - 1/[g + \kappa_0 \cot(\kappa_0 R_e)]. \tag{3.10}$$

Figure 3.2, shows the dependence $a(\kappa_0)$ for $g = 0$ (left panel) and $gR_e = 10$ (right panel). We see that by changing the depth of the well we can tune the scattering length to a resonance ($a = \infty$). This happens each time a new bound state crosses the zero energy threshold.

In the limit $g = \infty$, the regions I and II completely decouple; we can call the former a closed channel, and the latter an open channel. Obviously, in this case stationary states in the closed channel live independently and the potential of Eq. (3.2) is equivalent to a hard-core one characterized by $a = R_e$. For finite but large g the stationary states at positive energies become quasi-stationary and when the collision energy coincides with the position of one of them we encounter a narrow scattering resonance. This type of resonant scattering is called the Breit–Wigner scattering or the scattering on a quasi-stationary level.

Expanding Eq. (3.4) at small momenta, one obtains the effective range r_0 for our model potential (3.2). In fact, it makes sense to discuss r_0 only for very large values

[1] Here, we mean the wavefunction obtained by extrapolating its large-distance asymptote to the region $r \sim R_e$.

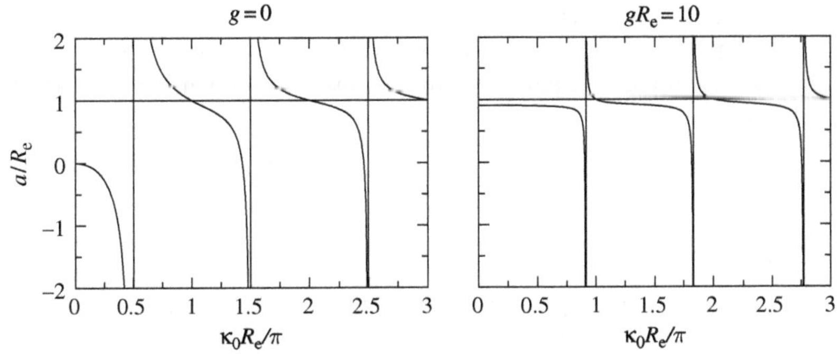

Figure 3.2 The scattering length versus κ_0 for $g = 0$ (left panel) and $gR_e = 10$ (right panel).

of a when $1/ak$ is comparable to r_0k (see Eq. 3.7). In such a narrow vicinity of the resonance, r_0 can be considered equal to its value at $a = \infty$:

$$r_0(a = \infty) = R_e - g(1 + gR_e)/\kappa_0^2 \xrightarrow[g\to\infty]{} -(g/\kappa_0)^2 R_e. \tag{3.11}$$

We see that in the limit of a narrow resonance ($g \to \infty$) the effective range r_0 is negative and can be much larger than the physical range of the potential. Next, we will see that this statement holds generally for narrow resonances.

Let us introduce a positive length parameter

$$R^* = -r_0/2 > 0. \tag{3.12}$$

We call a resonance narrow if $R^* \gg R_e$. Keeping only the first two terms in the effective range expansion, Eq. (3.7), we write the scattering amplitude, Eq. (3.5), in the form

$$f(k) = -\frac{1}{1/a + R^*k^2 + ik}. \tag{3.13}$$

Equation (3.13) has the well-known low-energy Breit–Wigner resonance shape, which becomes apparent if we rewrite it in energy units:

$$f(E) = -\frac{\hbar\gamma/\sqrt{2\mu}}{E - E_{res} + i\gamma\sqrt{E}}, \tag{3.14}$$

where the position of the quasi-stationary state E_{res} and the tunnelling amplitude γ are related to the lengths scales a and R^* as $a = -\hbar\gamma/\sqrt{2\mu}E_{res}$ and $R^* = \hbar/\sqrt{2\mu}\gamma$. The parameter $\gamma\sqrt{E}$ (multiplied by two) is nothing else than the decay rate of the quasi-stationary state, and therefore, both γ and R^* are necessarily positive.

As with resonances in the model potential with barrier of Eq. (3.2), magnetic Feshbach resonances occur when the collision energy of two atoms is close to the energy of a quasi-discrete molecular state in another hyperfine domain, in this case a closed channel. The tuning of the scattering length is achieved by shifting the open and closed channels with respect to each other in an external magnetic field (hyperfine states corresponding to the open and closed channels have different magnetic moments). For magnetic resonances

$$R^* = \frac{1}{2\mu a_{\rm bg}\mu_{\rm rel}\Delta B}, \tag{3.15}$$

where $a_{\rm bg}$ is the background scattering length, $\mu_{\rm rel}$ is the difference in the magnetic moments of the closed and open channels, and ΔB is the magnetic width of the Feshbach resonance. Usually, there are many Feshbach resonances for a given atomic gas or mixture in the realistic magnetic field range and, most of the time, one has the chance of working with a couple of 'nice' wide resonances. However, there are systems where all available resonances are narrow. For example, parameters of the widest resonance in ^{87}Rb (at 1007.4 G) are $\Delta_{\rm B} \approx 0.17$ G, $R_{\rm e} \approx 4.4$ nm, $R^* \approx 32$ nm (Volz *et al.* 2003). Similarly, all ^{6}Li–^{40}K interspecies resonances discussed so far are characterized by $R^* \gtrsim 100\,$nm (Wille *et al.* 2008; Tiecke *et al.* 2010), whereas the van der Waals range for this alkali pair is $R_{\rm e} \approx 2.2$ nm. The review of Chin *et al.* (2010) broadly covers the phenomenon of Feshbach resonances and related physics in ultracold gases and, in particular, presents parameters for many resonances.

As already mentioned, the situation where a is much larger than the physical range of the potential is rare and it usually requires a fine tuning of the potential. Large R^* arises for narrow resonances when the scattering amplitude is characterized by a strong dependence on the collision energy. One can imagine an interatomic potential for which the higher-order terms in the effective range expansion of Eq. (3.7) are also anomalously large. For example, we can introduce one or several additional closed channels with quasi-stationary states very close to the open channel threshold resulting in a rather exotic scattering amplitude. However, in all practically relevant cases, these terms can be neglected as the typical distance between neighbouring molecular states is rather large—of order $1/\mu R_{\rm e}^2$. In particular, for the model potential of Eq. (3.2), the next term in the expansion of Eq. (3.7) at $a = \infty$ and in the limit of large g is proportional to $(g/\kappa_0)^4 R_{\rm e}^3 k^4 \propto R^{*2} R_{\rm e} k^4$. This means that the potential of Eq. (3.2) has a well-defined zero-range limit: tending $R_{\rm e}$ to zero, one can accordingly modify κ_0 and g in such a way that a and R^* remain unchanged, whereas all other terms in the effective range expansion vanish. In this limit, Eqs (3.13) and (3.14) become exact and the Bethe–Peierls boundary condition reads

$$\chi'(0)/\chi(0) = (r\psi(r))'/r\psi(r)|_{r=0} = -1/a - 2\mu R^* E, \tag{3.16}$$

or, in analogy with Eq. (3.9),

$$\psi(r) \propto 1/r - 1/a - 2\mu R^* E, \; r \to 0, \tag{3.17}$$

where E is the collision energy.

Clearly, a is not always enough to characterize the interaction strength. For example, a can be infinite, but typical energies in the problem can be so high that the interactions are off-resonant. Thus, as a more appropriate quantity, it is reasonable to introduce the so-called effective energy-dependent scattering length $\tilde{a}(E)$ defined by the equation

$$1/\tilde{a}(E) = 1/a + 2\mu R^* E. \tag{3.18}$$

Comparison of $\tilde{a}(E)$ with other length scales in the problem gives us an idea of how strong the interaction is at a given (or typical for this problem) collision energy.

3.2.2 Effective range and population of closed channel

Let us now demonstrate the relation between the effective range and the probability of finding two atoms within the physical range of the potential. We will then use this relation to understand the structure of a weakly bound molecular state near a narrow resonance.

Consider the Schrödinger equation, Eq. (3.1), with an arbitrary short-range potential $V(r)$. We require that $V(r)$ can be neglected at $r > R_e$ and that the wavefunction χ vanishes at the origin. Assume that χ_0 and χ_k are two real solutions of Eq. (3.1) at zero and finite energy, respectively:

$$[-\partial^2/\partial r^2 + 2\mu V(r)]\chi_0(r) = 0, \tag{3.19}$$

$$[-\partial^2/\partial r^2 + 2\mu V(r) - k^2]\chi_k(r) = 0. \tag{3.20}$$

We multiply Eq. (3.19) by χ_k and Eq. (3.20) by χ_0. Subtracting the results we obtain

$$(-\chi_0\chi_k' + \chi_k\chi_0')' = k^2\chi_0\chi_k, \tag{3.21}$$

Let us now introduce the functions $\tilde{\chi}_0$ and $\tilde{\chi}_k$, which satisfy, respectively, Eqs (3.19) and (3.20), with $V(r) \equiv 0$, and which equal χ_0 and χ_k at distances $r > R_e$. These functions satisfy

$$(-\tilde{\chi}_0\tilde{\chi}_k' + \tilde{\chi}_k\tilde{\chi}_0')' = k^2\tilde{\chi}_0\tilde{\chi}_k. \tag{3.22}$$

We subtract Eq. (3.22) from Eq. (3.21) and integrating the result from 0 to R_e, we obtain

$$-\tilde{\chi}_0(0)\tilde{\chi}_k'(0) + \tilde{\chi}_k(0)\tilde{\chi}_0'(0) = k^2 \int_0^{R_e} [\chi_0(r)\chi_k(r) - \tilde{\chi}_0(r)\tilde{\chi}_k(r)]\mathrm{d}r. \tag{3.23}$$

In the limit $k \to 0$, we can use the boundary condition of Eq. (3.6) for $\tilde{\chi}_0$ and $\tilde{\chi}_k$ and the effective range expansion of Eq. (3.7) up to the effective range term. For the effective range, we obtain

$$R^* = -r_0/2 = \tilde{\chi}_0^{-2}(0)\left(\int_0^{R_e}\chi_0^2(r)\mathrm{d}r - \int_0^{R_e}\tilde{\chi}_0^2(r)\mathrm{d}r\right). \tag{3.24}$$

The first integral on the right-hand side of Eq. (3.24) gives the probability of finding the two atoms in the region $r < R_e$:

$$P_{r<R_e} = 4\pi\int_0^{R_e}\chi_0^2(r)\mathrm{d}r. \tag{3.25}$$

Near a narrow resonance, $P_{r<R_e}$ is dominated by the population of the quasi-stationary state in the closed channel. The second integral in Eq. (3.24) equals

$$\int_0^{R_e}\tilde{\chi}_0^2(r)\mathrm{d}r = \tilde{\chi}_0^2(0)R_e[1 - R_e/a + (R_e/a)^2/3], \tag{3.26}$$

where we have used the equality $\tilde{\chi}_0(r) = \tilde{\chi}_0(0)(1 - r/a)$. In the case $R^* \gg R_e$ the second integral can be neglected and we get

$$P_{r<R_e} \approx 4\pi\tilde{\chi}_0^2(0)R^*. \tag{3.27}$$

Equation (3.27) allows one to find the occupation of the closed channel (sometimes called the bare molecular state) without solving coupled-channel equations. One just needs to know the effective range parameter R^*, which can be found from Eq. (3.15), and the short distance asymptote of the open-channel wavefunction, $\psi(r) \approx \tilde{\chi}_0(0)(1/r - 1/a)$. Note that $\tilde{\chi}_0(0)$ is none other than the coefficient in front of the singular part of the wavefunction of the two-particle relative motion just outside of the potential support.

Equation (3.27) becomes exact in the zero-range limit and, together with Eqs (3.13), (3.14), and (3.16), it generalizes the usual zero-range theory to the case of narrow resonances. Let us now demonstrate how this modified zero-range theory works by calculating parameters of a weakly bound molecular state near a narrow resonance.

3.2.3 Weakly bound molecule near a narrow resonance

The wavefunction and the energy of a weakly bound molecular state can be found in the zero-range approximation, the applicability of which, in this case, requires that the size of the bound state be much larger than the physical range of the potential R_e. The solution of the free-motion Schrödinger equation

$$[-\partial^2/\partial r^2 + \kappa^2]\chi(r) = 0, \tag{3.28}$$

decaying at large distances reads

$$\chi(r) = \alpha_{\text{norm}} \exp(-\kappa r), \tag{3.29}$$

where α_{norm} is a real normalization coefficient and κ is determined from the Bethe–Peierls boundary condition (3.16):

$$\chi'(0)/\chi(0) = -\kappa = -1/a + R^* \kappa^2. \tag{3.30}$$

This quadratic equation can also be obtained by looking for the poles of the scattering amplitude (3.14). One of its roots gives the energy of the bound state

$$\varepsilon_0 = -\frac{\kappa^2}{2\mu} = -\frac{1}{2\mu}\left(\frac{\sqrt{1+4R^*/a}-1}{2R^*}\right)^2. \tag{3.31}$$

The normalization constant α is calculated in the following way. The integral of the square of χ gives the probability of finding the atoms in the ‘open’ channel $P_{\text{open}} = P_{r>R_e}$ (in this case $R_e = 0$):

$$P_{\text{open}} = 4\pi \int_0^\infty \chi^2(r)\mathrm{d}r = 2\pi\alpha_{\text{norm}}^2/\kappa. \tag{3.32}$$

At the same time, the closed channel population is given by Eq. (3.27):

$$P_{\text{closed}} = 4\pi\chi^2(0)R^* = 4\pi\alpha_{\text{norm}}^2 R^*. \tag{3.33}$$

Using Eq. (3.31) and the fact that $P_{\text{open}} + P_{\text{closed}} = 1$, we get

$$P_{\text{open}} = 1/\sqrt{1+4R^*/a}, \tag{3.34}$$

and

$$\alpha_{\text{norm}} = \sqrt{\frac{\kappa}{2\pi(1+2R^*\kappa)}} = \sqrt{\frac{1-1/\sqrt{1+4R^*/a}}{4\pi R^*}}. \tag{3.35}$$

Depending on the value of the ratio R^*/a, we distinguish two regimes: the case $R^*/a \ll 1$ we call the regime of small detuning, and the opposite limit—the regime of intermediate detuning. To be more precise, in the latter case we require $R_e \ll a \ll R^*$, since we would like to stay within the region of validity of the zero-range approximation. In the regime of small detuning, the molecular energy equals $\varepsilon_0 \approx -1/2\mu a^2$ and the probability of finding the atoms in the closed channel is small. The physics in this case is similar to that near a wide resonance. In the regime of intermediate detuning $\varepsilon_0 \approx -1/2\mu R^* a \approx E_{\text{res}}$ and $P_{\text{closed}} \approx 1$, which means that the molecular state in this case is dominated by the closed-channel or bare molecular state.

3.3 Basics of the three-body problem with short-range interactions

In this section we apply the zero-range approach to the problem of three atoms. We consider the phenomenon of the three-body collapse, or Thomas collapse, and define the Efimov effect. Then, we introduce a model system of one light and two heavy atoms, where one can use the Born–Oppenheimer approximation. The Efimov physics in this model is extremely transparent and one can understand the role of the two-body interaction parameters (scattering length, resonance width), masses of the atoms, and their quantum statistics. We also discuss the question of inelastic three-body losses.

3.3.1 Thomas collapse

Consider three identical bosons interacting with each other via the square well potential of Eq. (3.2) with $g = 0$. Let us gradually decrease R_e, and at the same time increase the depth of the potential in such a way that the scattering length a is unchanged. In this way we arrive at a zero-range potential equivalent to the initial one in the ultracold limit. What happens with the spectrum of the three-body problem as we decrease R_e? Below this, we show that its ground-state energy is unbound from below in the limit $R_e \to 0$. This phenomenon was discovered by Thomas (1935), who theoretically addressed the mass defect of tritium and arrived at a number of important conclusions on the character of the proton–neutron and neutron–neutron interaction.

The Hamiltonian for three bosons in the centre-of-mass reference frame reads (we set $\hbar = m = 1$)

$$H = -\nabla_{\mathbf{x}}^2 - \nabla_{\mathbf{y}}^2 + V(y) + \sum_{\pm} V(|\sqrt{3}\mathbf{x} \pm \mathbf{y}|/2), \tag{3.36}$$

where $\mathbf{y}$ is the distance between two of them and $\sqrt{3}\mathbf{x}/2$ is the distance from their centre of mass to the third boson. Consider the variational wavefunction,

$$\Psi_{\text{var}}(\mathbf{x}, \mathbf{y}) = (\omega/2\pi)^{3/2} \exp[-\omega(x^2 + y^2)/4], \tag{3.37}$$

which is the normalized ground-state wavefunction of a six-dimensional harmonic oscillator of mass 1/2 and frequency ω—our variational parameter. The ground-state energy of the Hamiltonian, Eq. (3.36) satisfies

$$\begin{aligned} E_0 \le \langle \Psi_{\text{var}} | H | \Psi_{\text{var}} \rangle &= 3\omega/2 + 3(\omega/2\pi)^{3/2} \int V(y) \exp(-\omega y^2/2) \mathrm{d}^3 y \\ &= 3\omega/2 - 3\kappa_0^2 \left[\operatorname{erf}\sqrt{\omega R_e^2/2} - \sqrt{2\omega R_e^2/\pi} \exp(-\omega R_e^2/2) \right], \end{aligned} \tag{3.38}$$

where we use Eq. (3.2) and the fact that the three interaction terms in (3.36) contribute equally to the matrix element $\langle \Psi_{\text{var}} | H | \Psi_{\text{var}} \rangle$.

When we reach the condition $R_e \ll |a|$, the depth of the potential becomes practically independent of the exact value of a and is determined by the equation $\cot(\kappa_0 R_e) = 0$ (see Eq. (3.10)). The minimum depth of the potential is then given by $\kappa_0 = \pi/2R_e$. Substituting this value in Eq. (3.38) we find that its right-hand side is minimized for $\omega \approx 2.1/R_e^2$ and we get

$$E_0 \leq -0.167/R_e^2, \tag{3.39}$$

which confirms the statement that the ground-state energy is unbound from below as $R_e \to 0$.

In the case of three bosons this result is expected: the inequality $|a| \gg R_e$ means that there is a bound or quasi-bound two-body state. In 3D, this requires that the potential is of a certain finite attractive strength. If we assume, a priori, that two of the atoms are very close to each other, the third atom will experience twice the attractive potential and will then be tightly bound, which verifies the initial assumption. Interestingly, Thomas showed that the statement holds for the proton–neutron–neutron (tritium) system in which the neutron–neutron interaction is neglected. In this case, the above hand-waving argument does not work because the interaction potential in between a neutron and a deuteron is the same as the neutron–proton interaction. However, the Thomas effect can be explained by the larger effective mass of the neutron–deuteron system compared with the neutron–proton one. The strict variational analysis is rather bulky in this case (Thomas 1935).

The system of three ^{4}He atoms is one of the most spectacular examples in which this physics actually takes place. The ^{4}He–^{4}He potential is just attractive enough to accommodate a single bound state close to the continuum, while the scattering length is positive and is an order of magnitude larger than the range of the potential. However, the $^4He_3$ trimer is much smaller in size and significantly stronger bound than the dimer. In fact, even if the interatomic potential were a bit weaker, so that the dimer state were just pushed into the continuum, the trimer state would still be there. This phenomenon is called the Borromean binding—none of the two-body subsystems has a bound state, but there exists a trimer state. An example of the Borromean binding is the ^{6}He nucleus, which is stable in spite of the fact that neither the α–n nor the n–n system is bound.

We now know that the ground state of such systems is a trimer state with the binding energy $\sim 1/R_e^2$. Are there other trimer states and what does their spectrum look like?

3.3.2 Efimov effect and discrete scaling invariance

Consider again three identical bosons interacting with each other via a potential of the range R_e. This time, let us keep R_e constant, but increase the value of a, no matter whether positive or negative, by slightly tuning the depth of the potential. Efimov showed (Efimov 1970) that as $a \to \infty$ the number of trimer states is proportional to $(s_0/\pi)\log(|a|/R_e) \to \infty$, where s_0 is a dimensionless parameter. A new trimer state emerges from the continuum each time a is increased by a factor of $\exp(\pi/s_0)$. When $a = \infty$, we have an infinite number of trimer states with binding energies $|\varepsilon_n| \propto \exp(-2\pi n/s_0)$.

The effect is quite general—it can take place in a system of three particles with different masses and statistics, as long as at least two of the three interactions are resonant. The parameter s_0 depends on the masses and statistics of the atoms, for identical bosons it equals $s_0 \approx 1.00624$ so that the scaling parameter $\exp(\pi/s_0) \approx 22.7$.

In fact, the Efimov effect goes along with the phenomenon of discrete scaling invariance: a three-body observable of the dimension of length to the power α close to the resonance ($a \to \infty$) behaves as $a^\alpha f(a/R_e)$, where f is a dimensionless function periodic on the logarithmic scale, i.e., $f(x) = f(x \exp[\pi/s_0])$.

We are used to the idea of continuous scaling invariance: once we multiply all length scales in the Hamiltonian by a factor of λ, all eigenenergies get multiplied by $1/\lambda^2$ and all eigenfunctions trivially rescale to satisfy the normalization condition. It is important to keep in mind that in this discussed example we do change a, but we do not rescale R_e. Therefore, the discrete scaling invariance does not contradict our understanding of the continuous scaling. As we see next, when $R_e \to 0$ the short-range physics does not completely drop out of the (three-body) problem.

3.3.3 Born–Oppenheimer approximation

Let us consider a light atom of mass m interacting resonantly via a short-range potential with two heavy atoms of mass M. For $M/m \gg 1$ one can use the Born–Oppenheimer approximation to illustrate the essential physics leading to the Efimov effect (Fonseca *et al.* 1979). The Born–Oppenheimer approach takes advantage of the large mass ratio by assuming that the state of the light atom adiabatically adjusts itself to the distance $\mathbf{R}$ between the heavy atoms. The three-body problem thus splits into two 'one-body' pieces. First, we solve the problem of the light atom in the field of two fixed scatterers at positions $\mathbf{R}/2$ and $-\mathbf{R}/2$ (a version of the double-well potential problem). The binding energy of the light atom then serves as the interaction potential between the heavy ones. The second step is then to solve the Schrödinger equation for the relative motion of the heavy atoms.

If there were only one heavy atom at position $\mathbf{R}/2$, the wavefunction of the light atom could be written as

$$\psi(\mathbf{r}) \propto \frac{e^{-\kappa|\mathbf{r}-\mathbf{R}/2|}}{|\mathbf{r}-\mathbf{R}/2|}, \tag{3.40}$$

where κ satisfies Eq. (3.30). In the case of two heavy atoms, we look for the solution in the form

$$\psi_{\mathbf{R}}(\mathbf{r}) \propto C_1 \frac{e^{-\kappa(R)|\mathbf{r}-\mathbf{R}/2|}}{|\mathbf{r}-\mathbf{R}/2|} + C_2 \frac{e^{-\kappa(R)|\mathbf{r}+\mathbf{R}/2|}}{|\mathbf{r}+\mathbf{R}/2|}. \tag{3.41}$$

The wavefunction of Eq. (3.41) satisfies the Schrödinger equation

$$\left[-\frac{1}{2m}\nabla_{\mathbf{r}}^2 + \sum_{\pm} V(|\mathbf{r} \pm \mathbf{R}/2|)\right] \psi_{\mathbf{R}}(\mathbf{r}) = \epsilon(R)\psi_{\mathbf{R}}(\mathbf{r}), \tag{3.42}$$

with the energy

$$\epsilon(R) = -\kappa^2(R)/2m. \tag{3.43}$$

The parameters $\kappa(R)$, C_1, and C_2 are obtained by applying the Bethe–Peierls boundary condition, Eq. (3.16) to $\psi_{\mathbf{R}}(\mathbf{r})$ at vanishing $\mathbf{r} \pm \mathbf{R}/2$ and by using the fact that the collisional energy equals $\epsilon(R)$. In particular, for the boundary $\mathbf{r} - \mathbf{R}/2 \to 0$, the resulting equation reads

$$[\kappa(R) - 1/a + R^*\kappa^2(R)]\, C_1 - \{\exp[-\kappa(R)R]/R]\}\, C_2 = 0, \tag{3.44}$$

and for the other boundary we get the same equation with interchanged C_1 and C_2.

These equations have two solutions: $\{C_1, C_2\}_+ = \{1, 1\}$ and $\{C_1, C_2\}_- = \{1, -1\}$. Substituting them into Eq. (3.41) we obtain the wavefunctions $\psi_{\mathbf{R},+}(\mathbf{r})$ and $\psi_{\mathbf{R},-}(\mathbf{r})$, which are none other than the symmetric and antisymmetric bound states of our double-well potential problem. The equation for $\kappa_\pm(R)$ reads

$$\kappa_\pm(R) \mp \exp\left[-\kappa_\pm(R)R\right]/R = 1/a - R^*\kappa_\pm^2(R). \tag{3.45}$$

Solving Eq. (3.45), and using Eq. (3.43), we get the induced interaction potentials $\epsilon_\pm(R)$ for the heavy atoms. In Figure 3.3 we plot these potentials for different values of R^*/a. We observe that $\epsilon_+(R)$ is purely attractive and $\epsilon_-(R)$ is repulsive. In the

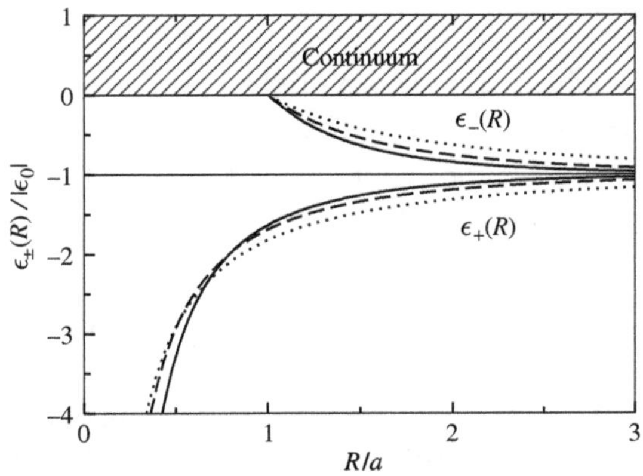

Figure 3.3 Born–Oppenheimer potentials ϵ_- and ϵ_+ in units of the molecular binding energy $|\epsilon_0|$ for different detunings: $R^*/a = 0$ (solid), $R^*/a = 1$ (dashed), and $R^*/a = 10$ (dotted).

limit of large R, both potentials tend to the molecular binding energy (note that the Born–Oppenheimer value of ϵ_0 differs from Eq. (3.31) by the factor μ/m). At distances $R < a$, the symmetric state (+) is the only bound state since the antisymmetric one (−) goes to the continuum at $R = a$.

The second step of the Born–Oppenheimer method consists of solving the Schrödinger equation for the heavy atoms by using $\epsilon_{\pm}(R)$ as the potential energy surface. Let us postpone the question of how this induced interaction depends on R^*/a and first consider the case $R^* = 0$. Then, in the limit $R \gg a$, the parameter $\kappa_{\pm}(R)$ can be expanded in a power series in the small parameter $\exp(-R/a)$. The first three terms read

$$\kappa_{\pm}(R) \approx a^{-1} \pm R^{-1} \exp(-R/a) - R^{-1} \exp(-2R/a), \tag{3.46}$$

which gives the long-distance asymptotes:

$$\epsilon_{\pm}(R) \approx -\frac{1}{2ma^2} \mp \frac{\exp(-R/a)}{maR} + \frac{\exp(-2R/a)}{maR}(1 - a/2R). \tag{3.47}$$

This means that at large distances the exchange of the light atom leads to a Yukawa-type force between the heavy atoms. In the limit $R \ll a$, only the symmetric state is bound, the right-hand side of Eq. (3.45) can be neglected, and we obtain $\kappa_+(R) \approx C/R^2$, where $C \approx 0.567$ is the solution of the transcendental equation $C = \exp(-C)$. Accordingly, the short-distance asymptote of $\epsilon_+(R)$ reads

$$\epsilon_+(R) \approx -\frac{C^2}{2mR^2} \approx -\frac{0.16}{mR^2}. \tag{3.48}$$

The $1/R^2$ potential does not have a characteristic length scale and is at the origin of the Efimov effect. Note that in the case $a = \infty$ Eq. (3.48) holds for all R.

3.3.4 Efimov effect in heavy–heavy–light system

Let us assume that the light atom is in the symmetric state and that it follows adiabatically the displacement of the heavy atoms ($M/m \gg 1$). Then, we can integrate out its motion and write down the Schrödinger equation for the heavy atoms in the form

$$[-\nabla^2_{\mathbf{R}}/M + \epsilon_+(R) - E]\phi(\mathbf{R}) = 0. \tag{3.49}$$

At small R, we can neglect E compared with other terms in Eq. (3.49) and use the short-distance asymptote of the potential of Eq. (3.48). The radial Schrödinger equation then reads

$$(-\partial^2/\partial R^2 + \beta/R^2)\chi(R) = 0, \tag{3.50}$$

where $\beta = l(l+1) - 0.16M/m$ and $\chi(R)$ is the radial part of the lth angular momentum component of $R\phi(\mathbf{R})$.

Equation (3.50) is satisfied by the function R^ν, where the power ν can take two values

$$\nu_{1,2} = 1/2 \pm \sqrt{\beta + 1/4}. \tag{3.51}$$

The case $\beta < -1/4$ corresponds to the well-known problem of the fall of a particle to the centre in an attractive R^{-2} potential (Landau and Lifshitz 1987). In this case, a real solution of Eq. (3.50) is a linear superposition of R^{ν_1} and R^{ν_2} with complex conjugate coefficients. It can be expressed in the form

$$\chi(R) \propto \sqrt{R}\cos[s_0 \log(R/R_0)], \tag{3.52}$$

where $s_0 = \sqrt{-\beta - 1/4}$ and R_0 is the three-body parameter, which fixes the phase in between the oscillations of R^{ν_1} and R^{ν_2}.

The three-body parameter is sensitive to the short-range (high-energy) details of the interatomic potentials, i.e., two systems with different interparticle potentials are, generally, characterized by different three-body parameters even if the corresponding scattering lengths are the same. The task of calculating R_0 for a realistic potential is very challenging, but, most of the time, we need to know only a few general properties of this parameter. For example, for one and the same system it is reasonable to assume that the three-body phase is a continuous function of the potential depth, and in a sufficiently narrow vicinity of a two-body resonance it can be approximated by a constant. It then enters the zero-range theory as a fixed external parameter, whereas the scattering length can vary over a very wide range.

In the case of finite R_e and a the wavefunction of Eq. (3.52) remains a good approximation for distances $R_e \lesssim R \lesssim |a|$. We see that it has approximately $N_b \approx (s_0/\pi)\log|a|/R_e$ nodes, which is an estimate of the number of bound states (trimer states). This number becomes infinite in the limit $a \to \infty$, exactly as predicted by Efimov.

Let us now look at the spectrum of trimers in the case $a = \infty$. Equation (3.49) in this case reads

$$(-\partial^2/\partial R^2 + \beta/R^2 - ME)\chi(R) = 0. \tag{3.53}$$

Let us assume that there is a bound state with energy $E = -E_0 < 0$. The wavefunction $\chi_0(R)$ corresponding to this state is a linear superposition of $\sqrt{R}J_{\mathrm{i}s_0}(\mathrm{i}\sqrt{ME_0}R)$ and its complex conjugate, where J is the Bessel function. The dependence of E_0 on the three-body parameter is established by requiring that $\chi_0(R)$ be finite for large $R \gg 1/\sqrt{ME_0}$ and take the form of Eq. (3.52) in the opposite limit. In principle, E_0 itself can be used as a three-body parameter.

Let us now consider a state given by the wavefunction $\chi_0(\lambda R)$. It obviously satisfies Eq. (3.53) with energy $E = -\lambda^2 E_0$ and decays at large R. The parameter λ is not arbitrary—Eq. (3.52) gives the condition $s_0 \log\lambda = \pi n$, where n is an integer. This is one of the consequences of the discrete scaling invariance. We understand now that all the Efimov trimers are self-similar and that their spectrum is geometric.

$$E_n = E_0 \exp(2\pi n/s_0). \tag{3.54}$$

Another manifestation of the discrete scaling invariance is the log-periodic dependence of three-body observables on the scattering length. This can be shown by rescaling the coordinates $R = a\tilde{R}$. In the new coordinates, we have to solve the three-body problem for the unit scattering length and use the small-$\tilde{R}$ three-body condition,

$$\chi(\tilde{R}) \propto \sqrt{\tilde{R}} \cos[s_0 \log(\tilde{R}/\tilde{R}_0)], \tag{3.55}$$

where the rescaled three-body parameter depends on the scattering length, $\tilde{R}_0 = R_0/a$. We then see that all the three-body observables should be log-periodic functions of a (apart from the 'trivial' dimensional scaling with a) with the discrete scaling factor $\exp(\pi/s_0)$.

As an example, let us consider the s-wave atom-molecule scattering. To calculate the corresponding scattering length, we take the $l = 0$ component of Eq. (3.49) with $E = -|\epsilon_0| = -1/ma^2$, i.e., with the atom–molecule collision energy equal to zero. In the rescaled coordinates, $\tilde{R} = R/a$, this equation reads

$$(-\partial^2/\partial\tilde{R}^2 + M\epsilon_{+,a=1}(\tilde{R}) + M/m)\chi(\tilde{R}) = 0. \tag{3.56}$$

Let us take the solution of Eq. (3.56) that at $\tilde{R} \ll 1$ behaves as $\tilde{R}^{1/2+\mathrm{i}s_0}$. Its long-range asymptote is a certain linear superposition of free solutions $\tilde{R}^0$ and $\tilde{R}^1$. We write it as $A + B\tilde{R}$, where A and B are complex coefficients, which could, in principle, be found numerically. Since Eq. (3.56) is real, the second linearly independent solution can be constructed as the complex conjugate of the first one. Then we should take their linear combination, satisfying Eq. (3.55), and compare its long-range asymptote with the expression $\propto (\tilde{R} - \tilde{a}_{\mathrm{ad}})$, where $\tilde{a}_{\mathrm{ad}}$ is the rescaled atom–dimer scattering length. Performing this operation we find that

$$\tilde{a}_{\mathrm{ad}}(\tilde{R}_0) = -\frac{A\tilde{R}_0^{-\mathrm{i}s_0} + A^*\tilde{R}_0^{\mathrm{i}s_0}}{B\tilde{R}_0^{-\mathrm{i}s_0} + B^*\tilde{R}_0^{\mathrm{i}s_0}} = \xi + \zeta \tan(s_0 \log \tilde{R}_0 + \delta), \tag{3.57}$$

where ξ, ζ, and δ are real numbers. We see that the non-rescaled atom-molecule scattering length has the form $a_{\mathrm{ad}} = a\tilde{a}_{\mathrm{ad}}(R_0/a)$, i.e., it is indeed a log-periodic function of a (up to the dimensional prefactor). The resonances in the atom–dimer scattering length denote the emergence of a new trimer state each time the scattering length is multiplied by $\exp(\pi/s_0)$.

We emphasize once again that the discrete scaling invariance is a consequence of the fact that we assume R_0 to be constant while we modify a. This is a good approximation close to a Feshbach resonance; however, theoretically we could rescale both the depth and the range of the potential so that R_0 is proportional to a. Then, we would observe the usual continuous scaling, i.e., the self-similarity for any value of the scaling factor.

3.3.5 Role of quantum statistics and masses of atoms

In the Born–Oppenheimer approach, we can also understand the role played by the quantum statistics of the heavy atoms. Indeed, the total three-body wavefunction is proportional to the product $\phi(\mathbf{R})\psi_{\mathbf{R},\pm}(\mathbf{r})$. It should be symmetric (antisymmetric) with respect to the permutation $\mathbf{R} \rightleftarrows -\mathbf{R}$ if the heavy particles are identical bosons (fermions). Then the symmetry of ϕ depends on the choice of the light atom wavefunction. Remember that $\psi_{\mathbf{R},+}(\mathbf{r})$ is symmetric and $\psi_{\mathbf{R},-}(\mathbf{r})$ is antisymmetric with respect to the permutation of the heavy particles. Since only the symmetric state leads to the induced attraction, the Efimov effect is expected for even angular momentum channels in the case of bosonic heavy atoms and for odd ones in the case of fermions. Of course, if the atoms are distinguishable, even if they are different hyperfine components of one and the same isotope, all angular momenta are allowed.

Although the Born–Oppenheimer approximation is supposed to work only for large mass ratios, exact calculations show that it captures the essentials of the Efimov physics quite well, even for moderate M/m. In particular, the formula (see Section 3.3.4)

$$s_0 = \sqrt{0.16M/m - (l + 1/2)^2} \tag{3.58}$$

correctly predicts that the scaling factor $\lambda = \exp(\pi/s_0)$ decreases with the mass ratio and increases with the angular momentum. The explanation of this phenomenon in the Born–Oppenheimer picture is straightforward: an increase in the mass ratio strengthens the exchange $1/R^2$ attraction compared with the kinetic energy of the heavy atoms. In turn, the centrifugal barrier competes with the exchange attraction and prevents the heavy particles from falling to the centre.

In Figure 3.4 we plot the scaling factor $\exp(\pi/s_0)$ and s_0 versus M/m for $l = 0$ and $l = 1$. The result of Eq. (3.58) (dashed lines) is compared with the exact one (solid lines), the derivation of which will be discussed in Section 3.4.2. In the case of heavy bosons ($l = 0$), the Efimov effect always takes place. However, large mass ratios are more favorable for its observation as the scaling factor λ rapidly decreases with M/m, which then allows for a larger number of log periods within a fixed window of a.

In the case of heavy fermions ($l = 1$), we see that the Efimov effect does not happen for $M/m < 13.6$ because the centrifugal barrier for the heavy fermions is stronger than the attraction induced by the light atom. The absence of the Efimov effect has advantages: first of all, the three-body problem in *non-Efimovian* cases is quite interesting in its own right. For example, there can be trimer states that are qualitatively different from the Efimov trimers (Kartavtsev and Malykh 2007). Another striking feature of non-Efimovian systems is their collisional stability—the reason why a two-component strongly interacting fermionic mixture is long-lived even at unitarity, $a = \infty$, whereas bosonic systems decay very rapidly near a Feshbach resonance. Let us now discuss the

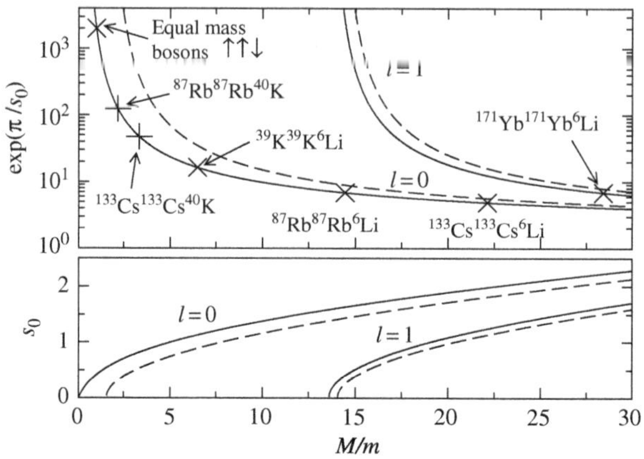

Figure 3.4 Scaling factor $\exp(\pi/s_0)$ (upper panel) and s_0 (lower panel) versus M/m. The dashed lines are calculated using the Born–Oppenheimer approximation and the solid lines are results of the exact approach of Section 3.6. Symbols show the scaling factors for several realistic three-atom systems.

three-body recombination into deeply bound states—one of the most important loss mechanisms in a strongly interacting gas.

3.3.6 Three-body recombination into deeply bound states: Efimovian case

In the absence of, or in the case of a strong suppression of, two-body inelastic channels, a third body is needed for a pair of atoms to form a bound state. The binding energy is then transferred into the kinetic energy of recombination products (an atom and a molecule). If the final bound state is deeply bound (typically of size R_e), the products are not ultracold and they escape from the trapped sample. The frequency of three-body recombination events in a unit volume is proportional to the number of all possible triples in this volume. For example, for a one-component Bose gas, it is proportional to n^3, where n is the density. The three-body recombination rate ν is the frequency per atom; it is thus proportional to n^2. The coefficient $\alpha_{\rm rec}$ in the equation $\nu = \alpha_{\rm rec} n^2$ is called the three-body recombination rate constant and has dimensions cm^{6}s^{-1}. The loss rate equals 3ν as three atoms are lost at each event, and the density evolves according to the equation $\dot{n} = -3\nu n^3$.

The recombination into deeply bound states can also happen in collisions of atoms and weakly bound molecules. In this case, it is called collisional relaxation. Its frequency per unit volume is proportional to the product of the atomic and molecular densities, $n_a n_m$. Thus, the relaxation rate per atom equals $\alpha_{\rm rel} n_m$, and per molecule, $\alpha_{\rm rel} n_a$. The coefficient $\alpha_{\rm rel}$ has dimensions cm^{3}s^{-1}.

Clearly, the task of calculating or, at least, estimating $\alpha_{\rm rec}$ and $\alpha_{\rm rel}$ is very important from a practical viewpoint. The knowledge of these constants allows one to tune experimental parameters and maximize the gas longevity. On the other hand, these parameters contain a lot of information on few-body correlations in the gas. In particular, the Efimov effect has so far been observed exclusively by measuring losses as a function of the scattering length.

We now discuss the zero-range approach to those types of inelastic process that happen at very short distances, $\sim R_e$. As an illustrative example, let us consider the relaxation in atom–dimer collisions and calculate the corresponding rate constant $\alpha_{\rm rel}$. It turns out that in the Efimovian case this problem can be solved rather elegantly by allowing the three-body parameter to be complex (Braaten and Hammer 2007). Indeed, the solution of the heavy-atom Schrödinger equation at distances $R_e \ll R \ll a$ is the sum of the incoming wave, $\sqrt{R}(R/R_0)^{-{\rm i}s_0}$, and the outgoing one, $\sqrt{R}(R/R_0)^{{\rm i}s_0}$. In the presence of the relaxation to deeply bound states, the outgoing flux should be a certain fraction of the incoming one. Theoretically, this is modelled by introducing the so-called elasticity parameter η_* and writing R_0 in the form $R_0 = |R_0|\exp(-{\rm i}\eta_*/s_0)$. The modulus of R_0 then fixes the relative phase of the incoming and outgoing waves, and $\eta_* > 0$ ensures that the incoming flux is larger than the outgoing one by a factor $\exp 4\eta_*$. Everything that we have said about the three-body parameter also holds for the elasticity parameter. Namely, it is very difficult to calculate η_* *ab initio*, but with in a sufficiently narrow vicinity of the resonance we can approximate it by a constant, which is inserted into the zero-range theory as a parameter.

Now, let us consider a weakly bound molecule and an atom in a unit volume. The interaction energy shift for this system equals $\Delta E = 2\pi a_{\rm ad}/\tilde{\mu}$, where $\tilde{\mu} \approx M/2$ is the atom–molecule reduced mass and $a_{\rm ad}$ is the atom–dimer scattering length given by Eq. (3.57). Once we introduce a finite elasticity parameter $[\tilde{R}_0 = |\tilde{R}_0|\exp(-{\rm i}\eta_*/s_0)$ into Eq. (3.57)], the atom–dimer scattering length acquires an imaginary part, which in turn leads to the shift of ΔE into the lower complex half-plane. We thus see that a single atom–dimer pair decays with the rate $-2{\rm Im}(\Delta E)$, which in this case equals the relaxation rate constant

$$\alpha_{\rm rel,bosons} = -\frac{4\pi {\rm Im}\, a_{\rm ad}}{\tilde{\mu}} = \frac{4\pi a}{\tilde{\mu}}\zeta\frac{\sinh\eta_*\cosh\eta_*}{\cos^2(s_0\log|R_0/a|+\delta)+\sinh^2\eta_*}, \tag{3.59}$$

where ζ and δ are the same as in Eq. (3.57). We see that $\alpha_{\rm rel,bosons}/a$ is a log-periodic function of a. Quite naturally it reaches its maximum at the atom–dimer scattering resonances when the cosine in the denominator in the right-hand, side of Eq. (3.59) vanishes.

In general, whenever we know the dependence of an arbitrary three-body observable on the real three-body parameter, we can analytically continue this function in the complex R_0 plane, thus obtaining the result for the situation when the relaxation is present. As an exercise, we suggest that the reader calculate the energies and lifetimes of the Efimov states in the case $a = \infty$.

The case $\eta_* = \infty$ is quite peculiar. It corresponds to complete absorption—only the incoming wave is present at short distances in the three-body wavefunction. In this case, we recover the continuous scaling invariance, but three-body observables acquire a significant imaginary part. For example, the atom–dimer scattering length is complex and is simply proportional to a without log-periodicity.

The real and imaginary parts of R_0 can be found by fitting experimental loss data. Existing experimental measurements for various isotopes and Feshbach resonances are consistent with values of η^* between 0.1 and 0.4.

3.3.7 Three-body recombination into deeply bound states: non-Efimovian case

The non-Efimovian case corresponds to $\beta > -1/4$ in Eq. (3.51). Then the three-body wavefunction at distances $R_e \ll R \ll a$ is the sum

$$\chi(R) \propto A_1(R/R_e)^{\nu_1} + A_2(R/R_e)^{\nu_2}, \tag{3.60}$$

where $\nu_1 > \nu_2$. Equation (3.61) should be matched with the solution of the three-body problem at $R \lesssim R_e$. To find the latter is challenging. However, in the absence of a three-body resonance, the matching procedure implies that both terms on the right-hand side of Eq. (3.60) are of the same order of magnitude, i.e., $A_1 \sim A_2$. Then, at distances $R \gg R_e$, the second term can be neglected compared with the first one. Note that we use exactly the same argument when we neglect the interaction between identical fermions in the ultracold limit: out of the two possible free solutions of the p-wave radial Schrödinger equation, R^1 and R^{-2}, we choose the former one (the one that increases with R).

The zero-range approach to calculating the relaxation into deeply bound states in non-Efimovian cases is perturbative. It uses the unperturbed three-body wavefunction to predict the probability of finding three atoms at distances $\sim R_e$, and gives the functional dependence of $\alpha_{\rm rel}$ on the scattering length. This means that if the relaxation rate constant is known (measured) for a certain a, one can predict its value for any other $a \gg R_e$.

As a specific example, let us estimate the atom–dimer relaxation rate in p-wave collisions of heavy fermions with heavy–light molecules in the Born–Oppenheimer approximation. In this case, the light atom is in the state $\psi_{\mathbf{R},+}(\mathbf{r})$, which means that at $R \ll a$ it is always close to the heavy fermions. Then we simply need to calculate the probability to find the heavy atoms at distances $R \sim R_e$. To do this, we take the short-distance ($R \lesssim a$) asymptote of the heavy-atom wavefunction, $\Phi(\mathbf{R}) \approx CR^{\nu_1 - 1}\cos\theta_{\mathbf{k},\mathbf{R}}$, and match it with the long-distance one, $\Phi(\mathbf{R}) \approx \sqrt{2}\sin(\mathbf{kR})$. The latter is properly normalized to a single atom–molecule pair in a unit volume (we assume that we are far from any atom–molecule p-wave resonance), $\mathbf{k}$ is the relative atom–dimer momentum, and $\theta_{\mathbf{k},\mathbf{R}}$ is the angle between $\mathbf{k}$ and $\mathbf{R}$. Matching the short- and long-distance asymptotes at $R \sim a$, we obtain $C \sim ka^{2-\nu_1}$, which gives the probability $P_{R\lesssim R_e} \sim (R_e/a)^{2\nu_1 - 2}(ka)^2 R_e^3$. To obtain a dimensional estimate of $\alpha_{\rm rel,fermions}$, we can multiply $P_{R\lesssim R_e}$ by the frequency of recombination processes that would take place if all three atoms were confined to distances $\sim R_e$. This frequency is of order $1/mR_e^2$. We thus obtain

$$\alpha_{\rm rel,fermions} = (R_e/m)(R_e/a)^{2\nu_1 - 2}(ka)^2. \tag{3.61}$$

The three factors in Eq. (3.61) are interpreted as follows: the factor R_e/m is of the order of the relaxation rate constant for s-wave collisions of atoms and deeply bound molecules (with the size $\sim R_e$). The factor $(R_e/a)^{2\nu_1-2}$ is the suppression factor that comes from the fact that the atoms have to tunnel under the effective repulsive potential, which is the sum of the centrifugal barrier and the exchange attraction. The power ν_1 depends on the mass ratio. The factor $(pa)^2$ corresponds to the low-energy Wigner law for reactions with unit angular momentum ($l = 1$ in this case).

Accordingly, the relaxation rate constant for s-wave atom-molecule collisions has the form (3.61) without the last factor. Unfortunately, in this case, the Born–Oppenheimer approximation does not work because the state $\psi_{\mathbf{R},-}(\mathbf{r})$ is unbound at $R < a$. The problem should then be solved without relying on the adiabatic approximation (see Section 3.4.5).

3.3.8 Role of resonance width

As explained in Section 3.3.4, the Efimov effect is related to the effective $1/R^2$ attraction emerging in a three-body system at distances $R \lesssim |a|$. Let us now discuss how this effective potential is modified in the case of a narrow resonance. First, consider the regime of small detuning, $R_e \ll R^* \ll |a|$. In this regime, the right-hand side of Eq. (3.45) can be neglected at distances $R^* \ll R \ll |a|$, and we recover the $1/R^2$ behaviour of the effective potential of Eq. (3.48). However, at distances $R_e \ll R \ll R^*$, the parameter $\kappa_+(R)$, is determined mostly by the last terms on each side of Eq. (3.45). We thus find that $\kappa_+^2(R) \approx 1/R^*R$ and

$$\epsilon_+(R) \approx -1/2mR^*R, \quad R_e \ll R \ll R^* \ll |a|. \tag{3.62}$$

The Coulomb potential of Eq. (3.62) is qualitatively different from the inverse-square one of Eq. (3.48). In particular, in this potential, heavy atoms do not fall to the centre. Moreover, close to the origin, $R \ll (m/M)R^*$, the heavy-atom wavefunction behaves similarly to the non-interacting case, i.e., at these distances, the kinetic energy operator in the heavy-atom Schrödinger equation is dominant. Therefore, we do not have to worry about the boundary condition at the origin (as in Section 3.3.7, we simply choose the solution that grows faster). The wavefunction is thus uniquely defined. Then, at distances $R \sim R^*$ it can be matched with the Efimov-like wavefunction of Eq. (3.52), the three-body parameter being determined by R^*. Remarkably, in this case three-body observables depend only on the two-body parameters a and R^*.

It is instructive to discuss more qualitatively what happens with the system as R decreases from $R \gg R^*$ to $R \ll R^*$. One can see that in the former case the effective energy-dependent scattering length $\tilde{a}(E)$ introduced in Eq. (3.18) is larger than R and, therefore, drops out of the problem, as in the wide-resonance case with $a = \infty$. However, for $R \sim R^*$, the energy of the light atom becomes sufficiently detuned from the resonance, so that $\tilde{a}(E)$ is comparable to R, and the effective potential acquires a characteristic length scale and is no longer that steep. Another important point to mention is a qualitative change of the wavefunction at $R \sim R^*$. By using Eqs (3.27) and (3.41), it is straightforward to show that for $R \gg R^*$ the light atom predominantly occupies the open channel, whereas at distances $R \ll R^*$ the open-channel occupation is negligible. Therefore, at these distances the wavefunction of the system to the leading order describes the free motion of a heavy atom and a closed-channel molecule. A small open-channel occupation can be treated perturbatively.

In the regime of intermediate detuning, $R_e \ll |a| \ll R^*$, the effective potential is nowhere proportional to $1/R^2$. At distances $R \lesssim |a|$ it is approximated by the Coulomb potential of Eq. (3.62). Then in the case $a < 0$ there are no bound states for the light atom if $R > |a|$, and the potential $\epsilon_+(R)$ terminates at the point $R = |a|$, where it reaches the three-atom continuum. For positive a, the effective potential changes its shape from the Coulomb one for $R \lesssim a$ to the constant $\epsilon_+(R) \approx -|\epsilon_0|$ for $R \gtrsim a$. Accordingly, in both cases, trimer states appear only when the characteristic Bohr radius, $\propto (m/M)R^*$, corresponding to the potential of Eq. (3.62) is smaller than $|a|$.

The last point to be touched upon in this section is the problem of inelastic losses in few-body systems near a narrow resonance R^*. The approach to this problem is similar to the one discussed in Section 3.3.7 for non-Efimovian systems. In fact, in the narrow-resonance case, the probability of relaxation or recombination to deeply bound states can always be related to the relaxation rate constant for collisions of atoms and closed-channel molecules. Indeed, the relaxation is a local process, it requires three atoms to approach each other to distances $\sim R_e \ll R^*$ and, as we have argued, at these distances the wavefunction describes the relative motion of an atom and a closed-channel molecule. We will return to this question in Section 3.4.5. More details on the few-body problem near a narrow resonance can be found in (Petrov 2004; Levinsen and Petrov 2011; Wang *et al.* 2011a).

3.4 The method of Skorniakov and Ter-Martirosian (STM) for few-body problems with resonant short-range interactions

In spite of numerous advantages and simplifications provided by the Born–Oppenheimer approximation, we still have to discuss methods suitable for a quantitative description of the case of comparable masses and even some highly mass-imbalanced ones (for example, the s-wave scattering of a heavy fermion and a weakly bound heavy–light molecule when the symmetry forces the light atom to be in the state $\psi_{\mathbf{R},-}(\mathbf{r})$, which does not exist at small R).

A natural generalization of the Born–Oppenheimer approximation is the adiabatic hyperspherical method, in which, after separating out the centre-of-mass motion, one introduces the coordinate system of hyperradius and hyperangles, $\{\rho, \hat{\Omega}\}$. The hyperradius ρ is the square root of the sum of squares of all interparticle distances with mass-dependent weights, and all other coordinates of the system can be written as a set of dimensionless hyperangles $\hat{\Omega}$. The adiabatic idea in this case is to consider ρ formally as a slow coordinate and $\hat{\Omega}$ as fast ones. Accordingly, one fixes ρ and diagonalizes the hyperangular part of the Schrödinger equation, thus obtaining the channel potentials. The second step is then to solve the hyperradial part, which is a set of second-order ordinary differential equations for each channel. The difference from the simple Born–Oppenheimer approach is that all these differential equations are now coupled by non-adiabatic matrix elements. In practice, one truncates the set of these coupled equations according to a given accuracy goal. Usually only a few channels suffice to obtain convergent results.

An advantage of the adiabatic hyperspherical approach is that it is quite general: it can be used for finite- and long-range interaction potentials; one can easily include three-body forces, etc. In fact, the complexity of the method is equivalent to the complexity of the initial few-body Schrödinger equation and is practically independent of the form of the interatomic potentials. Let us consider an N-body problem in three dimensions ($N \geq 3$). We can use the centre-of-mass reference frame and separate out the three global Euler angles by using the rotational invariance. We then have $3N - 6$ degrees of freedom, $3N - 7$ of which are hyperangular. Thus, for $N = 3$, the hyperangular space is two-dimensional, and each additional particle adds three degrees of freedom, which makes the hyperangular part of the calculation quite challenging. For details on the adiabatic hyperspherical approach see, for example, (Lin 1995; Nielsen *et al.* 2001)

Another approach, appropriate for systems with short-range interactions, was first introduced by Skorniakov and Ter-Martirosian (1957). For $N = 3$ it leads to a one-dimensional integral equation in momentum space (the STM equation), and its generalization to $N > 3$ gives an integral equation for a function of $3N - 9$ coordinates. The STM equation can be obtained using the effective field theory (EFT) (Bedaque *et al.* 1999) and other diagrammatic techniques (Brodsky *et al.* 2005; Levinsen and Gurarie 2006). We should mention that the famous Faddeev–Yakubovsky few-body method can be understood as a generalization of the STM approach to the case of finite-range (and long-range) potentials (Faddeev 1961; Yakubovskii 1967; Faddeev and Merkuriev 1993).

We will now derive the STM equations in coordinate space directly from the Schrödinger equation. This derivation clearly shows how one can integrate out all short-range physics and use the fact that, almost everywhere in space, the motion of the atoms is free.

3.4.1 STM equation

Let us first demonstrate the general idea behind this method. Assume that we are in two-dimensional space and we solve the problem of scattering by a curve S (see Figure 3.5). Namely, we solve the free Schrödinger (or Helmholtz) equation

$$(-\nabla^2_{\boldsymbol{\rho}} - E)\psi = 0, \tag{3.63}$$

with the boundary condition

$$[\partial\psi/\partial\mathbf{n}]/\psi = F(x), \tag{3.64}$$

where x is the natural (by the arc length) parametrization of S, F is a known function, and $[\partial\psi/\partial\mathbf{n}]$ is the sum of the normal outward derivatives at the point x of the curve.

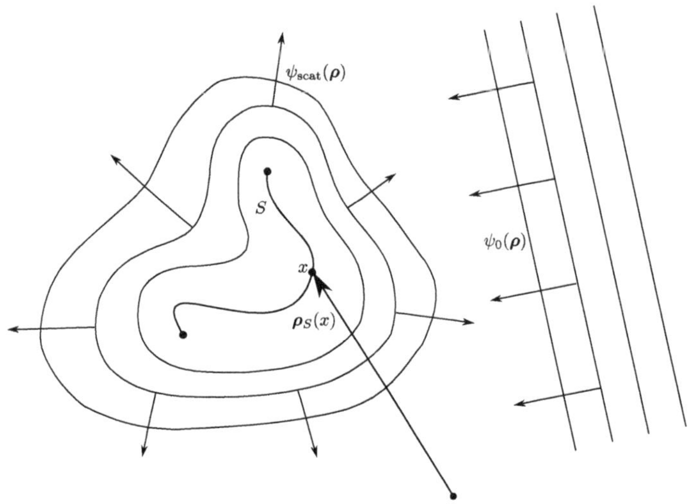

Figure 3.5 Two-dimensional scattering of a particle by a curve.

An efficient way of solving this scattering problem is to introduce an auxiliary function $f(x)$, defined on the boundary, and look for the solution in the form

$$\psi(\boldsymbol{\rho}) = \psi_0(\boldsymbol{\rho}) + \int_S G_E[\boldsymbol{\rho} - \boldsymbol{\rho}_S(x)]f(x)\mathrm{d}x, \tag{3.65}$$

where $G_E(\boldsymbol{\rho})$ is the Green's function of Eq. (3.63), $\boldsymbol{\rho}_S(x)$ is the coordinate on the curve, and $\psi_0(\boldsymbol{\rho})$ is the incoming wave, which is a free solution of Eq. (3.63).

Clearly, the function of Eq. (3.65) satisfies Eq. (3.63). We just have to ensure the boundary condition of Eq. (3.64). It can be shown (by choosing a proper contour around the curve element $\mathrm{d}x$ and applying the Gauss–Ostrogradsky theorem) that $[\partial\psi/\partial\mathbf{n}] = f(x)$. Then Eq. (3.64) gives the equation

$$\psi_0[\boldsymbol{\rho}_S(x)] + \int_S G_E[\boldsymbol{\rho}_S(x) - \boldsymbol{\rho}_S(x')]f(x')\mathrm{d}x' = 1/F(x). \tag{3.66}$$

Note that Eq. (3.66) is one-dimensional, in contrast with the original two-dimensional Schrödinger equation (3.63). In fact, it is much more suitable for analytical and especially numerical calculations as this reduction of the configurational space allows for rapid computational schemes.

The idea of using free Green functions is quite natural when one deals with boundary value problems. It can be encountered in electrostatics, hydrodynamics, problems of heat diffusion, etc. One can mention the so-called method of boundary elements used for calculating spectra of quantum billiards (Berry 1981; Bohigas *et al.* 1984) and based on the Korringa–Kohn–Rostoker method (Korringa 1947; Khon and Rostoker 1954) in

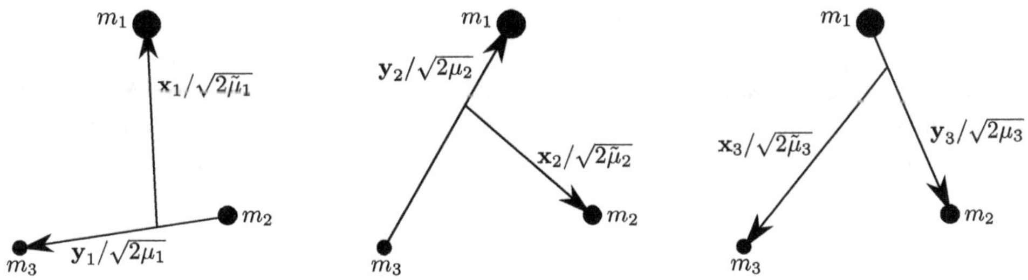

Figure 3.6 Three sets of coordinates for the three-body problem.

solid-state theory. Let us now discuss how one can do similar things for the few-atom problem.

Consider three atoms with coordinates $\mathbf{r}_i$ and masses m_i, where $i = 1, 2, 3$. We separate out the centre of mass motion and introduce the three sets of rescaled Jacobi coordinates (see Figure 3.6):

$$\begin{aligned} \mathbf{x}_i &= \sqrt{2\tilde{\mu}_i}[\mathbf{r}_i - (m_j\mathbf{r}_j + m_k\mathbf{r}_k)/(m_j + m_k)], \\ \mathbf{y}_i &= \sqrt{2\mu_i}(\mathbf{r}_k - \mathbf{r}_j), \end{aligned} \tag{3.67}$$

where $\{i, j, k\}$ are cyclic permutations of $\{1, 2, 3\}$, i.e. $\{1, 2, 3\}$, $\{2, 3, 1\}$, and $\{3, 1, 2\}$, $\mu_i = m_j m_k/(m_j + m_k)$ is the reduced mass for atoms j and k, and $\tilde{\mu}_i = m_i(m_j + m_k)/(m_i + m_j + m_k)$ is the reduced mass for the relative motion of atom i with respect to pair jk. The purpose of the mass-dependent rescaling is to express the kinetic energy operator in the symmetric form $-\nabla^2_{\mathbf{x}_i} - \nabla^2_{\mathbf{y}_i}$.

Let us choose to work in the first coordinate system ($i = 1$ in Eq. (3.67)). It is related to the other two by the equations:

$$\begin{aligned} \mathbf{x}_1 &= \mathbf{X}_{1\leftarrow 2}(\mathbf{x}_2, \mathbf{y}_2) = -(\sqrt{\mu_1\mu_2}/m_3)\mathbf{x}_2 + \sqrt{\mu_2/\tilde{\mu}_1}\mathbf{y}_2, \\ \mathbf{y}_1 &= \mathbf{Y}_{1\leftarrow 2}(\mathbf{x}_2, \mathbf{y}_2) = -\sqrt{\mu_1/\tilde{\mu}_2}\mathbf{x}_2 - (\sqrt{\mu_1\mu_2}/m_3)\mathbf{y}_2, \end{aligned} \tag{3.68}$$

and

$$\begin{aligned} \mathbf{x}_1 &= \mathbf{X}_{1\leftarrow 3}(\mathbf{x}_3, \mathbf{y}_3) = -(\sqrt{\mu_1\mu_3}/m_2)\mathbf{x}_3 - \sqrt{\mu_3/\tilde{\mu}_1}\mathbf{y}_3, \\ \mathbf{y}_1 &= \mathbf{Y}_{1\leftarrow 3}(\mathbf{x}_3, \mathbf{y}_3) = \sqrt{\mu_1/\tilde{\mu}_3}\mathbf{x}_3 - (\sqrt{\mu_1\mu_3}/m_2)\mathbf{y}_3. \end{aligned} \tag{3.69}$$

Now the original three-body problem can be represented as a single particle scattering in the six-dimensional space by a potential that is localized on the three-dimensional hyperplanes defined by the equations $\mathbf{y}_i = 0$, $i = 1, 2, 3$. Everywhere outside these hyperplanes the three-body wavefunction satisfies the free-motion six-dimensional Schrödinger equation

$$(-\nabla^2_{\mathbf{x}_1} - \nabla^2_{\mathbf{y}_1} - E)\Psi(\mathbf{x}_1, \mathbf{y}_1) = 0. \tag{3.70}$$

As in the usual scattering theory, we write $\Psi(\mathbf{x}_1,\mathbf{y}_1) = \Psi_0(\mathbf{x}_1,\mathbf{y}_1) + \Psi_{\text{scat}}(\mathbf{x}_1,\mathbf{y}_1)$, where $\Psi_0(\mathbf{x}_1,\mathbf{y}_1)$ is the incoming wave, a free solution of Eq. (3.70) without singularities (as if there were no interactions), and $\Psi_{\text{scat}}(\mathbf{x}_1,\mathbf{y}_1)$ is the scattered wave which contains $1/y$ singularities at each of the boundaries. We now introduce auxiliary functions defined at the boundaries, $f_i(\mathbf{x}_i)$, and write $\Psi(\mathbf{x}_1,\mathbf{y}_1)$ in the form

$$\begin{aligned}\Psi(\mathbf{x}_1,\mathbf{y}_1) = \Psi_0(\mathbf{x}_1,\mathbf{y}_1) &+ \int G_E\left(\sqrt{(\mathbf{x}_1-\mathbf{x}')^2+y_1^2}\right) f_1(\mathbf{x}')\mathrm{d}^3x' \\ &+ \sum_{i=2,3}\int G_E\left(\sqrt{[\mathbf{x}_1-\mathbf{X}_{1\leftarrow i}(\mathbf{x}',0)]^2+[\mathbf{y}_1-\mathbf{Y}_{1\leftarrow i}(\mathbf{x}',0)]^2}\right) f_i(\mathbf{x}')\mathrm{d}^3x',\end{aligned} \tag{3.71}$$

where G_E is the Green's function of Eq. (3.70):

$$G_E(X) = \begin{cases} \dfrac{-EK_2(\sqrt{-E}\,|X|)}{8\pi^3X^2}, & E<0 \\ \dfrac{iEH_2(\sqrt{E}\,|X|)}{16\pi^2X^2}, & E>0 \end{cases} \xrightarrow[E\to 0]{} \frac{1}{4\pi^3X^4}. \tag{3.72}$$

By construction, the wavefunction (3.71) satisfies the Schrödinger equation (3.70) for arbitrary boundary functions $f_i(\mathbf{x}_i)$. As in the previous example, this freedom is removed by boundary conditions. We first discuss the wide resonance case. Having in mind the Bethe–Peierls boundary condition of Eq. (3.9), let us study the singular and regular terms of $\Psi(\mathbf{x}_1,\mathbf{y}_1)$ close to the boundary $y_1 \to 0$. In this limit, $\Psi_0(\mathbf{x}_1,\mathbf{y}_1)$ and the two terms in the sum in the second line of Eq. (3.71) are not singular. We can thus set $y_1 = 0$ there. In the remaining term we make the $1/y_1$ singularity explicit by subtracting and adding the quantity

$$f_1(\mathbf{x}_1)\int G_E\left(\sqrt{(\mathbf{x}_1-\mathbf{x}')^2+y_1^2}\right)\mathrm{d}^3x' = f_1(\mathbf{x}_1)e^{-\sqrt{-E}y_1}/4\pi y_1, \tag{3.73}$$

where, for $E>0$ we use the convention $\sqrt{-E} = -\mathrm{i}\sqrt{E}$. We thus have the following asymptotic expression for $\Psi(\mathbf{x}_1,\mathbf{y}_1)$:

$$\Psi(\mathbf{x}_1,\mathbf{y}_1) \xrightarrow[y_1\to 0]{} (1/4\pi)[(1/y_1-\sqrt{-E})f_1(\mathbf{x}_1) + 4\pi\Psi_0(\mathbf{x}_1,0) - \hat{L}_{1,E}\{f\}(\mathbf{x}_1)], \tag{3.74}$$

where we have expanded the right-hand side of Eq. (3.73) to the next-to-leading order at small y_1, $\{f\} = \{f_1, f_2, f_3\}$, and $\hat{L}_{1,E}$ is the integral operator defined by

$$\begin{aligned}\hat{L}_{1,E}\{f\}(\mathbf{x}_1) = 4\pi &\int G_E(|\mathbf{x}_1-\mathbf{x}'|)[f_1(\mathbf{x}_1)-f_1(\mathbf{x}')]\mathrm{d}^3x' \\ &- 4\pi\sum_{i=2,3}\int G_E\left[\sqrt{x_1^2+x'^2+2\sqrt{\mu_1/\mu_i}(1-\mu_i/m_1)\mathbf{x}_1\mathbf{x}'}\right] f_i(\mathbf{x}')\mathrm{d}^3x'.\end{aligned} \tag{3.75}$$

Comparing Eqs (3.74) and (3.9), we arrive at the integral equation

$$\hat{L}_{1,E}\{f\}(\mathbf{x}_1) + \left(\sqrt{-E} - 1/\sqrt{2\mu_1}a_1\right)f_1(\mathbf{x}_1) = 4\pi\Psi_0(\mathbf{x}_1, 0), \tag{3.76}$$

where a_1 is the scattering length corresponding to the interaction of atoms 2 and 3.

Applying the same procedure to the boundaries $\mathbf{y}_2 = 0$ and $\mathbf{y}_3 = 0$, we can derive two additional equations, the left-hand sides of which can be obtained from that of Eq. (3.76) by cyclically permuting subscripts 1,2, and 3. The right-hand sides of the new equations read, respectively, $4\pi\Psi_0[\mathbf{X}_{1\leftarrow 2}(\mathbf{x}_2, 0), \mathbf{Y}_{1\leftarrow 2}(\mathbf{x}_2, 0)]$ and $4\pi\Psi_0[\mathbf{X}_{1\leftarrow 3}(\mathbf{x}_3, 0), \mathbf{Y}_{1\leftarrow 3}(\mathbf{x}_3, 0)]$. We thus get three coupled integral equations for determining the boundary functions f_i and we have substantially reduced the configurational space of the problem. In fact, the operators $\hat{L}_{i,E}$ conserve angular momentum, and we can expand f_i in spherical harmonics to work only with a set of one-dimensional integral equations. Solution of these equations then gives us the wavefunction Ψ by virtue of Eq. (3.71). Vice versa, if we know Ψ, we can always obtain f_i by looking at the coefficients in front of the $1/y_i$-singularities of Ψ at small y_i. Namely,

$$f_i(\mathbf{x}_i) = 4\pi \lim_{y_i \to 0} y_i \Psi. \tag{3.77}$$

Thus, $\{f\}$ contains the same information about the system as the wavefunction Ψ. This is one of the key advantages of the zero-range approximation.

It is sometimes useful to use instead a momentum representation introducing the Fourier transform $f_k(\mathbf{p}) = \int f_k(\mathbf{x}) \exp(-i\mathbf{p}\mathbf{x}) \mathrm{d}^3x$. Then the operator $\hat{L}_{1,E}\{f\}$ reads

$$\begin{aligned}\hat{L}_{1,E}\{f\}(\mathbf{p}_1) &= \left(\sqrt{-E + p_1^2} - \sqrt{-E}\right) f_1(\mathbf{p}_1) \\ &- \frac{1}{2\pi^2} \sum_{i=2,3} \sqrt{\frac{\tilde{\mu}_1}{\mu_i}} \int \frac{f_i(\mathbf{p}')\,\mathrm{d}^3p'}{p'^2 + p_1^2 + 2\sqrt{\mu_1/\mu_i}(1 - \mu_i/m_1)\mathbf{p}_1\mathbf{p}' - (\mu_i/\tilde{\mu}_1)E}.\end{aligned} \tag{3.78}$$

Skorniakov and Ter-Martirosian derived Eq. (3.76) in momentum representation in the case of equal masses, in order to solve the problem of neutron-deuteron scattering (Skorniakov and Ter-Martirosian 1957). We will refer to Eq. (3.76) as the STM equation. Let us now illustrate the power of the method by considering some concrete examples.

3.4.2 The Efimov effect and determination of s_0

We return to the problem of Section 3.3.5 and show how to determine the parameter s_0 without relying on the Born–Oppenheimer approximation. The system we are interested in consists of two identical atoms (bosons or fermions) of mass $m_1 = m_2 = M$ interacting resonantly (scattering length a) with another atom of mass $m_3 = m$. Let us neglect the (non-resonant) interaction between the identical atoms. Then Ψ is not singular for $y_3 \to 0$ and, therefore, $f_3 \equiv 0$. By symmetry, the boundary functions f_1 and f_2 are equal to each other (bosons) or have different signs (fermions). Indeed, we see directly from Eq. (3.67) that the permutation $\mathbf{r}_1 \rightleftarrows \mathbf{r}_2$ is equivalent to $\mathbf{x}_1 \rightleftarrows \mathbf{x}_2$, and $\mathbf{y}_1 \rightleftarrows -\mathbf{y}_2$. Thus $\Psi(\mathbf{x}_1, \mathbf{y}_1) = \pm\Psi(\mathbf{x}_2, -\mathbf{y}_2)$, and by virtue of Eq. (3.77) we have $f(\mathbf{x}) := f_1(\mathbf{x}) = \pm f_2(\mathbf{x})$, where the upper sign stands for bosons and the lower for fermions. Clearly, Ψ_0 should have the same property. Finally, the three equations of type of Eq. (3.76) are reduced in this case to the single one,

$$\hat{L}_E f(\mathbf{x}) + (\sqrt{-E} - 1/\sqrt{2\mu a}) f(\mathbf{x}) = 4\pi\Psi_0(\mathbf{x}, 0), \tag{3.79}$$

where

$$\hat{L}_E f(\mathbf{x}) = 4\pi \int \left\{ G_E(|\mathbf{x} - \mathbf{x}'|)[f(\mathbf{x}) - f(\mathbf{x}')] \mp G_E\left(\sqrt{x^2 + x'^2 + 2\mathbf{x}\mathbf{x}' \sin\phi}\right) f(\mathbf{x}') \right\} \mathrm{d}^3x', \tag{3.80}$$

with $\mu = \mu_1 = \mu_2 = mM/(m+M)$, and $\sin\phi = M/(m+M)$.

At small $x \ll \{a, \sqrt{|E|}\}$, we can neglect all terms in Eq. (3.79) except the integral one and in the latter we can use the zero-energy asymptote of the Green's function (3.72). This means that the small-x asymptote of $f(\mathbf{x})$ should satisfy

$$\hat{L}_{E=0} f(\mathbf{x}) = 0. \tag{3.81}$$

We first consider spherically symmetric $f(\mathbf{x}) = f(x)$. In this case, averaging over the angles of $\mathbf{x}'$ (we integrate over the solid angle and divide by 4π) we obtain the zero momentum component of the operator $\hat{L}_{E=0}$:

$$\hat{L}_{E=0,l=0} f(x) = \frac{4}{\pi} \int_0^\infty \left[\frac{f(x) - f(x')}{(x^2 - x'^2)^2} \mp \frac{f(x')}{(x^2 + x'^2)^2 - 4x^2 x'^2 \sin^2\phi} \right] x'^2 \mathrm{d}x'. \tag{3.82}$$

The first integral in Eq. (3.82) (and also in Eqs (3.75) and (3.80)) is taken in the principal-value sense (Petrov *et al.* 2005a). We see that the operator in Eq. (3.82) is scaleless and has the property

$$\hat{L}_{E=0,l=0} x^\nu = \lambda_{l=0}(\nu) x^{\nu-1}. \tag{3.83}$$

We thus look for the solution of Eq. (3.81) in the form $f(x) = x^\nu$, where ν is a root of $\lambda_{l=0}(\nu)$. The region of convergence of the integral on the left-hand side of Eq. (3.83) is $-3 < \mathrm{Re}(\nu) < 1$ in the case of bosons and $-5 < \mathrm{Re}(\nu) < 3$ for fermions. The function $\lambda_{l=0}(\nu)$ is given by

$$\lambda_{l=0}(\nu) = -(\nu+1)\tan\frac{\pi\nu}{2} \mp \frac{2\sin[\phi(\nu+1)]}{\sin(2\phi)\cos(\pi\nu/2)}. \tag{3.84}$$

The integrals of type $\hat{L}_{E=0,l=0} x^\nu$ can be taken by using the complex analysis. We cut the complex x-plane along the positive real axis and choose the integration contour

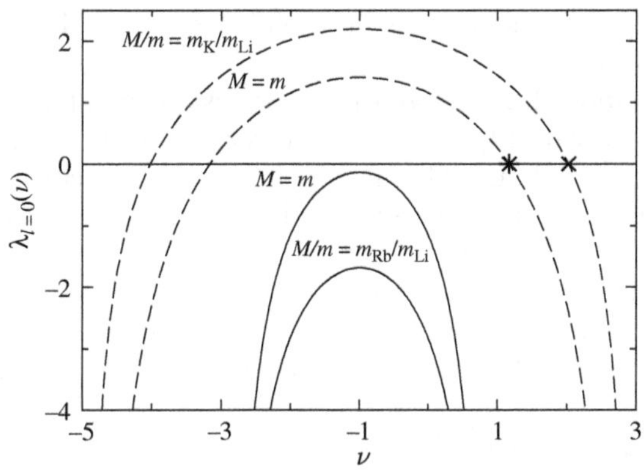

Figure 3.7 $\lambda_{l=0}(\nu)$ versus ν in the case of two identical bosons (solid lines) or two identical fermions (dashed lines) interacting resonantly with a third atom of mass m. The Efimov effect takes place only for bosons. The star and cross denote the value $\nu_1 \approx 1.16622$ for equal masses and $\nu_1 \approx 2.0193$ for the case of two ^{40}K and one ^{6}Li atoms.

embracing this cut. Then the contour can be blown to infinity, and the integral is determined by the pole residues of the integrand.

Figure 3.7 plots $\lambda_{l=0}(\nu)$ in the bosonic (solid lines) and fermionic (dashed lines) cases. In the fermionic case for any mass ratio, the equation $\lambda(\nu) = 0$ has two solutions $\nu_1 > \nu_2$ in the region of convergence mentioned previously. These roots are real, which means that we are in the non-Efimovian case discussed in Section 3.3.7. Accordingly, in the absence of a three-body resonance, $f(x) \propto x^{\nu_1}$. Note that we have just solved the problem that could not be solved in the Born–Oppenheimer approximation, even at very large mass ratios. Indeed, in the case of heavy identical fermions, the symmetry $l = 0$ corresponds to the light atom in the state ψ_-, which is unbound at these small distances.

The case of identical bosons in the Born–Oppenheimer language corresponds to the wavefunction ψ_+ for the light atom and zero total angular momentum (see Section 3.3.5). In this case the Born–Oppenheimer approach predicts the Efimov effect. Now, studying $\lambda_{l=0}(\nu)$ one can show that the effect takes place for any mass ratio as the roots of this function are complex conjugate, $\nu_{1,2} = -1 \pm i s_0$, where s_0 is real. Accordingly, the three-body parameter is necessary, to fix the ratio in between the coefficients in the linear superposition of x^{ν_1} and x^{ν_2}. The parameter s_0 plotted in Figure 3.4 (solid line) is found by solving the equation $\lambda_{l=0}(-1 + i s_0) = 0$. The increase of s_0 with the mass ratio is consistent with the Born–Oppenheimer picture.

As a historical remark, we note that the neutron–deuteron s-wave scattering problem considered by Skorniakov and Ter-Martirosian for total spin $I = 3/2$ corresponds to the fermionic case discussed here and that for total spin $I = 1/2$ corresponds to the bosonic case. Indeed, in the latter case, the neutron spins are antiparallel and their orbital wavefunction is symmetric, as for bosons. It was later pointed out by Danilov

(1961) that for $I = 1/2$ the solution of the STM equation (Eq. (3.79)) is not unique and an additional parameter is necessary.

We should also note that the bosonic case with equal masses is characterized by relatively small $s_0 \approx 0.414$, leading to a very large scaling factor, $\exp(\pi/s_0) \approx 1986.1$. This fact, although not disturbing for theorists, causes apparent practical difficulties. The reason for this 'weak' manifestation of the Efimov physics is that there are only two resonant interactions out of three. However, as we have already noticed in Section 3.3.5, the scaling factor rapidly decreases with the mass ratio (see the upper panel of Figure 3.4), which makes highly mass-imbalanced heteronuclear mixtures practically valuable for studies of the Efimov effect and discrete scaling invariance.

The case $l = 1$ is treated in the same manner as $l = 0$. We obtain the operator $\hat{L}_{E=0,l=1}$ by integrating out the angular dependence of $f(\mathbf{x}) = 3f(x)\hat{\mathbf{x}}\hat{\mathbf{n}}$ in $\hat{L}_{E=0}f(\mathbf{x})$ (we multiply it by $\hat{\mathbf{x}}\hat{\mathbf{n}}$, integrate over the solid angle, and divide by 4π):

$$\hat{L}_{E=0,l=1}f(x) = \frac{1}{\pi}\int_0^\infty \left\{ \frac{4x'^2[f(x)-f(x')]}{(x'^2-x^2)^2} - \frac{f(x')}{x^2}\left[\frac{2xx'}{(x+x')^2} - \log\frac{x+x'}{|x-x'|}\right] \mp \frac{f(x')}{2x^2\sin^2\phi} \right.$$
$$\left. \times \left[\frac{4xx'(x^2+x'^2)\sin\phi}{x^4+x'^4+2x^2x'^2\cos(2\phi)} + \log\frac{x^2+x'^2-2xx'\sin\phi}{x^2+x'^2+2xx'\sin\phi}\right]\right\} dx'. \tag{3.85}$$

The operator of Eq. (3.85) has the same scaling property of Eq. (3.83), with

$$\lambda_{l=1}(\nu) = \frac{\nu(\nu+2)}{\nu+1}\cot\frac{\pi\nu}{2} \mp \frac{\nu\sin(\phi)\cos[(\nu+1)\phi] - \sin(\nu\phi)}{(\nu+1)\sin^2(\phi)\cos(\phi)\sin(\pi\nu/2)}. \tag{3.86}$$

The region of convergence of the integrals in Eq. (3.85) is $-4 < \mathrm{Re}(\nu) < 2$ for bosonic and for fermionic symmetries.

In Figure 3.8, we plot the function $\lambda_{l=1}(\nu)$ for bosons and for fermions. In the Born–Oppenheimer language this symmetry ($l = 1$) corresponds to the antisymmetric heavy-atom wavefunction. In the bosonic case, besides leading to the centrifugal barrier, it also forces the light atom to be in state ψ_- (leading to an effective repulsion for the heavy atoms). This explains the absence of the Efimov effect for bosons in this case. For fermions we have the competition of the centrifugal barrier and the attraction due to the exchange of the light atom in state ψ_+. For mass ratios $M/m > 13.6$, the latter wins and the Efimov effect manifests itself in the complex roots of $\lambda_{l=1}(\nu)$, which have the same structure as in the case of zero angular momentum, $\nu_{1,2} = -1 \pm is_0$. The quantity s_0 for the $l = 1$ case is also plotted in Figure 3.4 along with the result of the Born–Oppenheimer approximation.

3.4.3 Analytical approach at zero total energy

In this section, we would like to present an approach that allows one to calculate all zero-energy three-body observables analytically. The approach was introduced for

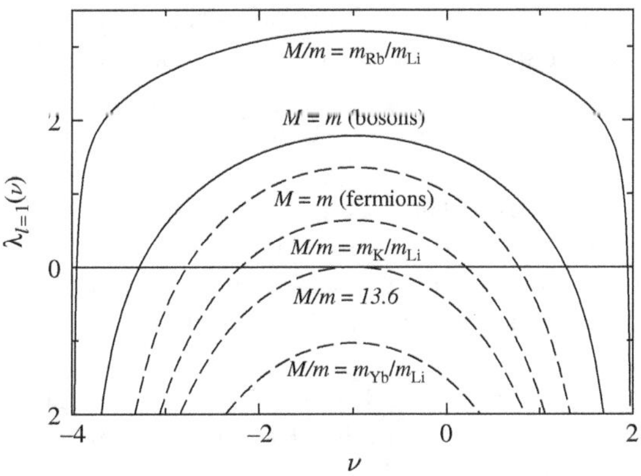

Figure 3.8 $\lambda_{l=1}(\nu)$ versus ν in the case of two identical bosons (solid lines) or two identical fermions (dashed lines) of mass M interacting resonantly with a third atom of mass m. For this symmetry ($l = 1$), the Efimov effect is absent in the bosonic case. For fermions, it takes place only for mass ratios larger than the critical one, $(M/m)_c \approx 13.6$.

the fermionic non-Efimovian case in (Petrov 2003) and has been recently generalized to Efimovian cases in heteronuclear mixtures (Helfrich *et al.* 2010). Here, we apply it to three identical bosons. Alternative analytical methods of treating this system can be found in (Macek *et al.* 2005; Macek and Ovchinnikov 2006; Gogolin *et al.* 2008; Mora *et al.* 2011a). It should be mentioned that our current understanding of the three-boson problem is strongly influenced by the papers of Esry *et al.* (1999), Nielsen and Macek (1999), Bedaque *et al.* (2000), and Braaten and Hammer (2001), who analysed the process of three-body recombination in this system numerically.

Close to a Feshbach resonance, a Bose gas suffers from three-body recombination—formation of weakly (for large positive a) or deeply bound molecules, the binding energy being transferred to the kinetic energy of the products—a molecule and the remaining atom. Let α_s, $\alpha_d(a > 0)$, and $\alpha_d(a < 0)$ stand for the rate constants for recombination to a weakly bound state and deeply bound states on the positive and negative sides of the resonance, respectively. In the low-temperature limit, as long as the de Broglie wavelengths are larger than $|a|$, these constants are practically temperature independent and can be written as products of $\hbar a^4/m$ and dimensionless log-periodic functions of $|a|/R_0$ (different for α_s, $\alpha_d(a > 0)$, and $\alpha_d(a < 0)$). To fix the phase of these log-periodic functions, the following practical notation is used in the literature: the value of $a > 0$ where α_s/a^4 reaches its minimum is denoted by a_{*0} (defined modulo $\exp(\pi/s_0)$). Another experimentally relevant reference point is the maximum of $\alpha_d(a < 0)/a^4$, which is, by definition, reached at $a = a_- < 0$. In fact, this is the point where a Efimov state crosses the three-atom threshold. We will show next that $|a_-|/a_{*0} = \exp(\pi/2s_0)$.

Let us first discuss the case $a > 0$ and temporarily adopt the units $\hbar = m = a = 1$. For the problem of three-body recombination, we now consider Eq. (3.76) and choose the incoming wave Ψ_0 as a symmetrized superposition of plain waves normalized to a volume V. In the region of space relevant for recombination it can be approximated by

$$\Psi_0 = \sqrt{6}V^{-3/2}, \tag{3.87}$$

where the factor $\sqrt{6}$ comes from the fact that we consider cold bosons but not in the same quantum state (not condensed). Let us note that Eq. (3.76) with $E = 0$ can also be used for the problem of atom–dimer scattering just below the break-up threshold ($E = -0$). In this case, the atoms cannot move freely at large distances and Ψ_0 should be set to zero.

Using the bosonic symmetry and Eq. (3.77), it is straightforward to show that $f(\mathbf{x}) := f_1(\mathbf{x}) = f_2(\mathbf{x}) = f_3(\mathbf{x})$ (we work in the first coordinate system, $\{\mathbf{x}_1, \mathbf{y}_1\}$, and omit the subscript). Equation (3.76) is thus a single three-dimensional integral equation. Moreover, we can set $f(\mathbf{x}) = f(x)$ consistent with the angular independence of Ψ_0 and with the fact that all processes with higher angular momenta are suppressed at low temperatures. The resulting STM equation reads:

$$(\hat{L}_{E=0,l=0} - 1)f(x) = 4\pi\Psi_0, \tag{3.88}$$

where the integral operator $\hat{L}_{E=0,l=1}$ is obtained in the same manner as Eq. (3.82). Omitting the subscripts, we get

$$\hat{L}f(x) = \frac{4}{\pi}\int_0^\infty \left[\frac{f(x) - f(x')}{(x^2 - x'^2)^2} - \frac{2f(x')}{(x^2 + x'^2)^2 - x^2x'^2}\right] x'^2 \mathrm{d}x'. \tag{3.89}$$

Let us now discuss the structure of possible solutions of Eq. (3.88). Obviously, $f(x)$ is a sum of a particular solution of the inhomogeneous equation (Eq. (3.88)) and a general solution of the homogeneous equation

$$(\hat{L} - 1)\chi(x) = 0. \tag{3.90}$$

Physically, Eq. (3.90) describes the atom–dimer channel just below the dimer break-up threshold ($\Psi_0 = 0$). Therefore, at distances $x \gg 1$ the function $\chi(x)$ is a linear combination of $\exp(\mathrm{i}x)/x$ and $\exp(-\mathrm{i}x)/x$. Indeed, consider the atom–dimer wavefunction $\Psi(x, y) = \phi_b(y)\exp(\pm\mathrm{i}x)/x$. Substituting it into Eq. (3.77), and using the correctly normalized dimer wavefunction

$$\phi_b(y) = \exp(-y)/\sqrt{2\pi}y \xrightarrow[y\to 0]{} (1/y - 1)/\sqrt{2\pi}, \tag{3.91}$$

we find that the corresponding contribution to $\chi(x)$ equals $(8\pi)^{1/2}\exp(\mathrm{i}x)/x$.

Our aim now is to solve Eq. (3.88), separate the large-x asymptote $f \propto \exp(\mathrm{i}x)/x$, and relate the coefficient in front of it to the three-body recombination rate constant. This problem can be solved analytically by using the property (Eq. (3.83)) of the operator $\hat{L}$. In the case of three identical bosons, the integral converges in the region $-3 < \mathrm{Re}(\nu) < 1$ and the function $\lambda_{l=0}(\nu)$ is given by (we again omit the subscript)

$$\lambda(\nu) = -(\nu+1)\tan\frac{\pi\nu}{2} - \frac{8}{\sqrt{3}}\frac{\sin[\pi(\nu+1)/6]}{\cos(\pi\nu/2)}. \tag{3.92}$$

The roots of Eq. (3.92) are complex conjugate, $\nu_{1,2} = -1 \pm \mathrm{i}s_0$, where s_0 satisfies

$$s_0\cosh(\pi s_0/2) - 8\sinh(\pi s_0/6)/\sqrt{3} = 0. \tag{3.93}$$

The solution is $s_0 \approx 1.00624$, which leads to the famous scaling factor $\exp(\pi/s_0) \approx 22.7$.

At short distances, the operator $\hat{L}$ in Eq. (3.90) dominates over the 1, and any solution of this equation should be a linear superposition of $\chi \propto x^{-1+\mathrm{i}s_0}$ and its complex conjugate. From now on, we will use the notation χ for the solution of Eq. (3.90) with the following asymptotes:

$$\chi(x) = \begin{cases} Ax^{\nu} = Ax^{-1+\mathrm{i}s_0}, & x \ll 1, \\ x^{-1}\mathrm{e}^{\mathrm{i}x+\mathrm{i}\sigma-h} + x^{-1}\mathrm{e}^{-\mathrm{i}x-\mathrm{i}\sigma+h}, & x \gg 1, \end{cases} \tag{3.94}$$

where A is a complex number, and σ and h are real numbers. The physical solution of Eq. (3.90), i.e., the one corresponding to a given three-body parameter, is expressed as

$$\chi_\theta(x) = \mathrm{e}^{\mathrm{i}\theta}\chi(x) + \mathrm{e}^{-\mathrm{i}\theta}\chi^*(x), \tag{3.95}$$

where we have introduced the three-body parameter θ (a complex number with imaginary part η_*).

The normalization in Eq. (3.94) is chosen such that

$$\langle p\chi(px)|p'\chi(p'x)\rangle = \int_0^\infty p\chi(px)p'\chi(p'x)x^2\mathrm{d}x = 2\pi\delta(p-p'). \tag{3.96}$$

The first equality in Eq. (3.96) is our definition of the scalar product (note the absence of the complex conjugation), and the second equality follows from the fact that $p\chi(px)$ and $p'\chi(p'x)$ are eigenfunctions of the symmetric operator $\hat{L}$, corresponding to the eigenvalues p and p'. They are orthogonal for $p \neq p'$ and their scalar product in the vicinity of $p = p'$ can be worked out in the same way as in (Landau and Lifshitz 1987, see §21). A simple change of the integration variable in Eq. (3.96) leads to the completeness condition

$$\int_0^\infty p^2\chi(px)\chi(px')\mathrm{d}p = 2\pi\delta(x-x')/x^2. \tag{3.97}$$

Equations (3.96) and (3.97) allow us to construct the integral operator $(\hat{L}-1)^{-1}$ needed to solve Eq. (3.88). To avoid problems with divergence of the corresponding integrals, let us introduce an auxiliary function $g_0(x)$, related to $f(x)$ by

$$f(x) = 4\pi\Psi_0[-1 - \lambda(0)/x + \lambda(0)\lambda(-1)g_0(x)]. \tag{3.98}$$

Substituting this expression into Eq. (3.88) and using Eq. (3.83), we find that $g_0(x)$ satisfies the equation $(\hat{L}-1)g_0(x) = x^{-2}$. Applying the operator $(\hat{L}-1)^{-1}$ to x^{-2}, we obtain the following particular solution:

$$g_0(x) = \frac{1}{2\pi x}\int_0^\infty \chi(z)\mathrm{d}z\left[\int_0^\infty \frac{\chi(y)y\mathrm{d}y}{y-x-\mathrm{i}0} - \frac{2\pi\mathrm{i}x\chi(x)}{1-\exp(-2\pi s_0)}\right], \tag{3.99}$$

where the first integral is defined as

$$\int_0^\infty \chi(z)\mathrm{d}z = \lim_{\epsilon\to+0}\int_0^\infty \chi(z)z^\epsilon\mathrm{d}z. \tag{3.100}$$

The rule of 'going around the pole' in the second integral and the numerical coefficient in front of the second term in the square brackets on the right-hand side of Eq. (3.99) regulate the entry of $\chi(x)$, which can be arbitrary, into the particular solution $g_0(x)$. Using this freedom, we choose these parameters in such a way that $g_0(x)$ does not contain oscillating terms proportional to $x^{-1+\mathrm{i}s_0}$ at small x. Direct calculation shows that in the limit $x\to 0$ the right-hand side of Eq. (3.99) equals $g_0(x)\approx[\int_0^\infty\chi(z)\mathrm{d}z]^2/2\pi x$ to the leading order in x. On the other hand, according to Eq. (3.83), the same quantity in the same limit can be written as $g_0(x) = (\hat{L}-1)^{-1}x^{-2}\approx\hat{L}^{-1}x^{-2} = 1/[\lambda(-1)x]$, which leads to the result

$$\int_0^\infty \chi(z)\mathrm{d}z = \sqrt{2\pi/\lambda(-1)}. \tag{3.101}$$

Another consequence of our choice of the particular solution of Eq. (3.99) is that removing the oscillating terms from $g_0(x)$ makes it real, since any imaginary part of g_0 would necessarily be a solution of the homogeneous Eq. (3.90). Therefore, g_0 would have oscillations at short x, the absence of which we have ensured. Clearly, the function f obtained by virtue of Eq. (3.98) is also real. Moreover, the property of Eq. (3.83) ensures that $f = o(1)$ at small x, i.e., its Taylor expansion starts with x^1, at least. Therefore, this solution of Eq. (3.88) is not sensitive to the short-range physics and does not depend on the three-body parameter.

Integrating Eq. (3.99) in the limit $x\gg 1$, we get

$$g_0(x)\xrightarrow[x\to\infty]{}\frac{-\mathrm{i}}{\sinh(\pi s_0)}\sqrt{\frac{2\pi}{\lambda(-1)}}\frac{\cos[x+\sigma+\mathrm{i}(h+\pi s_0)]}{x}. \tag{3.102}$$

It can be real only if $h=-\pi s_0$ (note that $\lambda(-1)<0$), cf. (Macek *et al.* 2005).

Finally, the result that we are interested in is the linear combination

$$f_\theta(x) = f(x) + \gamma\chi_\theta(x), \tag{3.103}$$

where the complex number γ is chosen such that $f_\theta(x)$ contains only an outgoing wave at large x (this corresponds to an atom and a dimer flying apart after the three-body recombination event). This condition gives

$$\gamma = \mathrm{i}\frac{\pi\Psi_0\lambda(0)\sqrt{2\pi\lambda(-1)}}{\sinh(\pi s_0)\cosh(\pi s_0 - \mathrm{i}\theta)}. \tag{3.104}$$

Keeping only the relevant oscillating term at large x, we obtain

$$f_\theta(x) \xrightarrow[x\to\infty]{} \mathrm{i}4\gamma\sin\theta\sinh(\pi s_0)\exp(\mathrm{i}x + \mathrm{i}\sigma)/x. \tag{3.105}$$

So, we have found the coefficient in front of the outgoing atom–dimer wave, which is enough to calculate the atom–dimer outgoing flux. Indeed, the large-x asymptote $f = \xi\exp(\mathrm{i}x)/x$, where ξ is any complex amplitude, is accompanied by the flux $|\xi|^2\Phi_\infty$, where

$$\Phi_\infty = 3 \times (8\pi)^{-1} \times (4\pi) \times 2 = 3. \tag{3.106}$$

Here, we have explicitly written out the following factors: the factor of three reflects the three symmetric possibilities of forming the dimer (corresponds to the interchange $\mathbf{r}_1 \rightleftarrows \mathbf{r}_2 \rightleftarrows \mathbf{r}_3$), the factor of $(8\pi)^{-1}$ arises from the relation in between Ψ and f (see the discussion after Eq. (3.90)), the factor of 4π is the solid angle in the outgoing atom–dimer channel, and the last factor of two is the atom–dimer relative velocity in the x, y-coordinates. The three-body recombination rate constant α_s is obtained by taking the squared modulus of the prefactor in front of $\exp(\mathrm{i}x + \mathrm{i}\sigma)/x$ in Eq. (3.105) and by multiplying it by Φ_∞, by the factor of $1/6$, reflecting the fact that the number of triples in the gas is $n^3/3!$, and by the factor $\hbar a^4/m$, to restore the original physical units. We should also mention that the nine-dimensional volume V^3 is taken to be a unit volume in the original system of coordinates $\{\mathbf{r}_1, \mathbf{r}_2, \mathbf{r}_3\}$. In the new coordinates $\{\mathbf{x}, \mathbf{y}, \mathbf{R}_{\rm cm}\}$, where $\mathbf{R}_{\rm cm}$ is the centre-of-mass coordinate, this volume equals $V^3 = 8/3\sqrt{3}$. The final result for the three-body recombination rate constant reads

$$\alpha_s = 128\pi^2(4\pi - 3\sqrt{3})\frac{\sin^2[s_0\ln(a/a_{*0})] + \sinh^2\eta_*}{\sinh^2(\pi s_0 + \eta_*) + \cos^2[s_0\ln(a/a_{0*})]}\frac{\hbar a^4}{m}, \tag{3.107}$$

where we have expressed the three-body parameter θ through the original physical units:

$$\theta = s_0\ln(a/a_{*0}) + \mathrm{i}\eta_*. \tag{3.108}$$

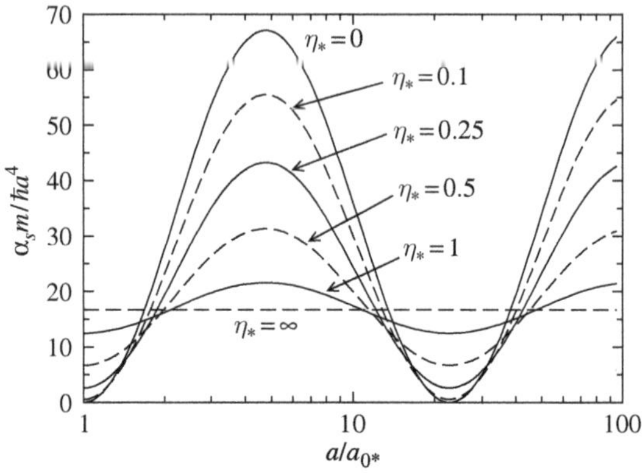

Figure 3.9 The rate constant α_s for the recombination into a weakly bound state for three identical bosons. The function is log-periodic with the scaling factor $\exp(\pi/s_0) \approx 22.7$.

In Figure 3.9 we plot the quantity $m\alpha_s/\hbar a^4$ as a function of a for different values of the elasticity parameter η_*.

The rate of recombination into a weakly bound level is only a fraction of the total loss rate. The rest is due to the formation of deeply bound molecular states. We have already discussed this mechanism in Section 3.3.6. In contrast with the recombination into shallow states, we now have to look at the balance of the incoming and outgoing fluxes of atoms corresponding to the short distance asymptote of $f_\theta(x)$ given by Eq. (3.103)

$$f_\theta(x) \xrightarrow[x\to 0]{} \gamma\chi_\theta(x) = \gamma(Ae^{i\theta}x^{-1+is_0} + A^*e^{-i\theta}x^{-1-is_0}), \tag{3.109}$$

In analogy with Φ_∞, let Φ_0 denote the number of atom triples disappearing at the origin ($\mathbf{x} = 0$, $\mathbf{y} = 0$), provided the function f takes the form of the incoming wave x^{-1-is_0} with unit weight. With this definition, the recombination rate constant follows from Eq. (3.109):

$$\alpha_d = (1/3!)(\hbar a^4/m)|\gamma|^2\Phi_0|A|^2 2\sinh(2\eta_*), \tag{3.110}$$

where the prefactor 1/3! is related to the number of triples in the gas. The product $\Phi_0|A|^2$ can easily be found from the definition of Eq. (3.94) by equating the fluxes at $x \to 0$ and at $x \to \infty$ and using Eq. (3.106):

$$\Phi_0|A|^2 = 2\Phi_\infty \sinh(2\pi s_0) = 6\sinh(2\pi s_0). \tag{3.111}$$

Substituting Eqs (3.111) and (3.104) into Eq. (3.110), we obtain

$$\alpha_d(a>0) = 128\pi^2\left(4\pi - 3\sqrt{3}\right)\frac{\coth(\pi s_0)\cosh(\eta_*)\sinh(\eta_*)}{\sinh^2(\pi s_0 + \eta_*) + \cos^2[s_0\ln(a/a_{0*})]}\frac{\hbar a^4}{m}. \tag{3.112}$$

Since in the case of identical bosons the product πs_0 is rather large, the function $\alpha_d(a>0)/a^4$ is almost flat (we can neglect the a-dependent term in the denominator) in contrast with α_s/a^4. The constant $\alpha_d(a>0)$ monotonically increases with η_* and in the extreme limit $\eta_* \to \infty$ the ratio between the two rate constants equals $\alpha_d(a>0)/\alpha_s = \coth(\pi s_0) \approx 1.0036$.

Let us now discuss the negative side of the resonance and derive $\alpha_d(a<0)$. In this case we use the units $\hbar = |a| = m = 1$, and some equations described previously should be modified accordingly. In particular, Eq. (3.88) reads

$$(\hat{L}+1)\tilde{f}(x) = 4\pi\Psi_0 \tag{3.113}$$

and we now introduce an auxiliary function $\tilde{g}_0$ related to $\tilde{f}$ by

$$\tilde{f}(x) = 4\pi\Psi_0[1-\lambda(0)/x+\lambda(0)\lambda(-1)\tilde{g}_0(x)], \tag{3.114}$$

where $\tilde{g}_0$ satisfies $(\hat{L}+1)\tilde{g}_0(x) = x^{-2}$. We write the solution in the form

$$\tilde{g}_0(x) = \frac{1}{2\pi x}\int_0^\infty \chi(z)\mathrm{d}z \int_0^\infty \frac{\chi(y)y\mathrm{d}y}{y+x} \tag{3.115}$$

and integrating it in the small-x limit we find the asymptote

$$\tilde{f}(x) \xrightarrow[x\to 0]{} \mathrm{i}\frac{2\pi\Psi_0\lambda(0)\sqrt{2\pi\lambda(-1)}}{\sinh(\pi s_0)}Ax^{-1+\mathrm{i}s_0}. \tag{3.116}$$

The function $\tilde{f}$ is a solution of Eq. (3.113), but its oscillations at small x do not have (in general) the correct phase imposed by Eq. (3.95). This difficulty is resolved by observing that $\tilde{f}^*$ also satisfies Eq. (3.113). The correctly behaving solution reads

$$\tilde{f}_\theta(x) = \frac{\exp(\mathrm{i}\theta)\tilde{f}(x)+\exp(-\mathrm{i}\theta)\tilde{f}^*(x)}{\exp(\mathrm{i}\theta)+\exp(-\mathrm{i}\theta)}, \tag{3.117}$$

and by subtracting the outgoing flux from the incoming one at small x we obtain the result

$$\alpha_d(a<0) = 128\pi^2(4\pi-3\sqrt{3})\frac{\coth(\pi s_0)\cosh(\eta_*)\sinh(\eta_*)}{\cos^2[s_0\ln(|a|/a_{0*})]+\sinh^2(\eta_*)}\frac{\hbar a^4}{m}. \tag{3.118}$$

For small η_*, Eq. (3.118) is characterized by resonances at $a = -a_{0*}\exp(\pi/2s_0)$ modulo $\exp(\pi/s_0)$ (see Figure 3.10). These points mark the passage of Efimov trimers across the three-atom threshold with the usual consequences: the contribution of the trimer state (of size $\sim |a|$) in the three-body wavefunction becomes very large (close to $\theta = \pi/2$, the right-hand side of Eq. (3.117) diverges) and atoms spend a lot of time close to each other, which leads to enhanced recombination losses. Note that on average $\alpha_d(a<0)$ is significantly larger (by about three orders of magnitude) than

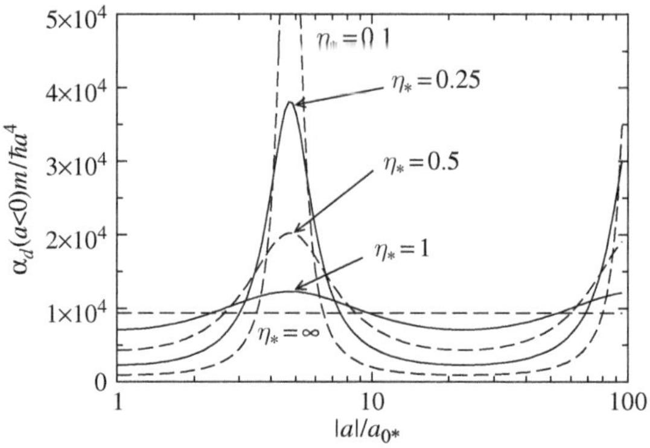

Figure 3.10 The rate constant α_d for the recombination into deeply bound states for three identical bosons on the negative side of the Feshbach resonance. The peaks correspond to Efimov states crossing the three-atom threshold.

α_s and $\alpha_d(a > 0)$ for the same $|a|$. In the limit $\eta_* \to \infty$, we have $\alpha_d(a < 0)/\alpha_s = \exp(2\pi s_0)\coth(\pi s_0) \approx 558.9$.

Recalling that a_- is defined as the value of a where $\alpha_d(a < 0)/a^4$ reaches its maximum, Eq. (3.118) implies

$$|a_-|/a_{0*} = \exp(\pi/2s_0). \tag{3.119}$$

The zero-range theory thus predicts that the maxima of α_s and $\alpha_d(a > 0)$ and the maxima of $\alpha_d(a < 0)$ are placed at $a = \pm|a_-|$, i.e., they are symmetric with respect to the centre of the Feshbach resonance.

This analytical approach can be extended to homonuclear and heteronuclear mixtures of fermions and bosons by taking into account the mass imbalance and non-zero angular momentum. It can be used as a zero-energy reference point for numerical approaches, which are supposed to give results also at finite energies (three-body recombination at finite temperatures, atom–dimer scattering, etc.) It is also of immense help in the cases of small s_0 where the exponentially large scaling parameter impedes numerical calculations.

3.4.4 STM equation near a narrow resonance

Let us now discuss how one can include the finite width of the resonance into the STM approach. In fact, the derivation of Section 3.1 is also valid in the narrow resonance case, except that the STM equation itself, Eq. (3.76), should be modified. Indeed, the asymptote of Eq. (3.74) should be compared with the narrow resonance Bethe–Peierls condition, Eq. (3.17), which in current notation reads

$$\Psi(\mathbf{x}_1, \mathbf{y}_1) \propto 1/y_1 - 1/\sqrt{2\mu_1}a_1 - \sqrt{2\mu_1}R_1^* E_1, \; y_1 \to 0, \tag{3.120}$$

where E_1 is the collision energy of atoms 2 and 3, i.e., the kinetic energy of motion 'along' the $\mathbf{y}_1$-direction of the six-dimensional $\{\mathbf{x}_1, \mathbf{y}_1\}$-space. Using Eq. (3.70) and the definition of Eq. (3.77), we write

$$E_1 = \lim_{\mathbf{y}_1 \to 0} \frac{-\nabla^2_{\mathbf{y}_1} \Psi}{\Psi} = E + \lim_{\mathbf{y}_1 \to 0} \frac{\nabla^2_{\mathbf{x}_1} \Psi}{\Psi} = E + \frac{\nabla^2_{\mathbf{x}_1} f_1}{f_1}. \tag{3.121}$$

Substituting Eq. (3.121) into Eq. (3.120) and comparing the latter with (3.74), we obtain the narrow resonance STM equation

$$\hat{L}_{1,E}\{f\}(\mathbf{x}_1) + \left[\sqrt{-E} - 1/\sqrt{2\mu_1}a_1 - \sqrt{2\mu_1}R_1^*(E + \nabla^2_{\mathbf{x}_1})\right] f_1(\mathbf{x}_1) = 4\pi\Psi_0(\mathbf{x}_1, 0). \tag{3.122}$$

The equations for f_2 and f_3 should also be modified by including similar terms proportional to R_2^* and R_3^*, respectively. Equation (3.122) can be, of course, written in momentum space. The operator $-\nabla^2_{\mathbf{x}_1}$ is then substituted by p_1^2 and the corresponding representation of $\hat{L}_{1,E}$ is given by Eq. (3.78).

At sufficiently short distances, the operator $\nabla^2_{\mathbf{x}_1}$ in Eq. (3.122) dominates and the short-distance asymptote of f_1 satisfies $-\nabla^2_{\mathbf{x}_1} f_1(\mathbf{x}_1) = 0$. This is consistent with Section 3.3.8, where we argued that at these distances the three-body wavefunction describes the free motion of an atom and a closed-channel molecule. In fact, according to Eq. (3.27), the probability of atoms 2 and 3 being in the closed channel equals $4\pi R_1^* |\lim_{y_1 \to 0} y_1 \Psi|^2 = 4\pi R_1^* |f_1(\mathbf{x}_1)/4\pi|^2$, and, therefore, $\sqrt{R_1^*/4\pi} f_1(\mathbf{x}_1)$ can be considered as the wavefunction of the relative motion of an atom and a bare molecule. In the narrow resonance limit, they interact only by virtually breaking up the molecule and exchanging one of its constituents with the free atom. This exchange interaction is given by the integral operator $\hat{L}_{1,E}$ in Eq. (3.122) and can be treated perturbatively in the limit $R^* \to \infty$. Obviously, we can also introduce the direct interaction between the closed-channel molecule and atom by imposing another zero-range Bethe–Peierls boundary condition, this time on the function f_1.

3.4.5 Atom–dimer scattering near a narrow resonance

To illustrate how the approach of Section 3.4.4, can be implemented in practice, let us consider a concrete example. Namely, we calculate the atom–dimer scattering length for a system of two statistically identical atoms of mass M and another atom of mass m. We use the notation of Section 3.4.2 with the upper sign for identical bosons and the lower one for fermions. To account for the finite resonance width, we now add the term $-\sqrt{2\mu}R^*(E + \nabla^2_{\mathbf{x}})f(\mathbf{x})$ to the left-hand side of Eq. (3.79) and look for its solution at zero atom–molecule collision energy, i.e., we set $E = \epsilon_0 < 0$, where ϵ_0 is the energy of the molecular state (see Eq. (3.31)). Since E is negative, the free solution Ψ_0 vanishes, and Eq. (3.79) takes the form

$$(-\sqrt{2\mu}R^*\nabla^2_{\mathbf{x}} + \hat{L}_{E=\epsilon_0})f(\mathbf{x}) = 0. \tag{3.123}$$

The algorithm of calculating the atom–dimer scattering length a_{ad} is straightforward; expand Eq. (3.123) in spherical harmonics, solve it for $l = 0$ (s-wave symmetry), and deduce a_{ad} from the large-x asymptote of $f(x)$. Indeed, at distances $x \gg 1/\kappa$, where $\kappa = \sqrt{2\mu|\epsilon_0|}$, Eq. (3.123) describes the free atom–molecule motion, i.e., the s-wave symmetric solution is a linear superposition of x^{-1} and x^0. The relation between the corresponding coefficients is fixed by the atom–dimer scattering length:

$$f(x) \xrightarrow[x\to\infty]{} N\left(1 - \sqrt{2\tilde{\mu}}a_{ad}/x\right), \tag{3.124}$$

where $\tilde{\mu} = m(M+m)/(2M+m)$ is the atom–molecule reduced mass and N is the normalization prefactor; this is actually not needed for determining a_{ad}, but we will return to it when discussing the inelastic atom–dimer relaxation.

Alternatively, one can follow the same procedure in momentum space. Introducing the Fourier transform $f(\mathbf{p}) = \int f(\mathbf{x})\exp(-i\mathbf{p}\mathbf{x})d^3x$ we look for the solution in the form

$$f(p) = N\left[(2\pi)^3\delta(\mathbf{p}) + 4\pi g(p/\sqrt{-\epsilon_0})/(\sqrt{-\epsilon_0}p^2)\right]. \tag{3.125}$$

We substitute Eq. (3.125) into the momentum space STM equation, integrate over the angles of $\mathbf{p}$ and arrive at the following equation for the function $g(k)$,

$$\begin{aligned}\sqrt{-2\mu\epsilon_0}R^* g(k) = &\pm \frac{1}{(k^2+\cos^2\phi)\cos\phi} - \frac{\sqrt{1+k^2}-1}{k^2}g(k)\\ &\pm \frac{1}{\pi\sin(2\phi)}\int_0^\infty \frac{g(k')}{kk'}\log\frac{k^2+k'^2+2kk'\sin\phi+\cos^2\phi}{k^2+k'^2-2kk'\sin\phi+\cos^2\phi}dk',\end{aligned} \tag{3.126}$$

the solution of which gives the atom–dimer scattering length: $a_{ad} = -g(0)/\sqrt{-2\tilde{\mu}\epsilon_0}$.

In Figure 3.11 we plot a_{ad} in units of a calculated from Eq. (3.126) for collisions of K atoms with KLi molecules formed near a narrow interspecies resonance. The solid line shows the case of identical bosonic ^{39}K atoms and the dashed one shows identical fermionic ^{40}K. We see that in the limit $R^*/a \ll 1$ the bosonic case is Efimovian and is characterized by the log-periodic dependence of a_{ad}/a on a of the form of Eq. (3.57), the parameter R^* plays the role of the three-body parameter, as discussed in Section 3.3.8. In contrast, in the fermionic case, the atom–dimer scattering length has a well-defined limit for $R^* = 0$ and the left-hand side of Eq. (3.126) can be considered as a weak perturbation for small R^*/a.

In the opposite limit, $R^*/a \gg 1$, which we call the regime of intermediate detuning, we treat Eq. (3.126) perturbatively in the small parameter $\eta = 1/\sqrt{-2\mu\epsilon_0}R^* \ll 1$. Namely, we look for the solution in the form $g(k) = g^{(0)}(k) + g^{(1)}(k) + g^{(2)}(k) + \ldots$, where the functions $g^{(i)}$ are of order η^i and can be found recursively: $g^{(0)} \equiv 0$, $g^{(1)}(k) = \pm\eta\cos^{-1}(\phi)/(k^2+\cos^2\phi)$. Then $g^{(i)}$ for $i > 1$ is obtained by substituting $g^{(i-1)}$ into

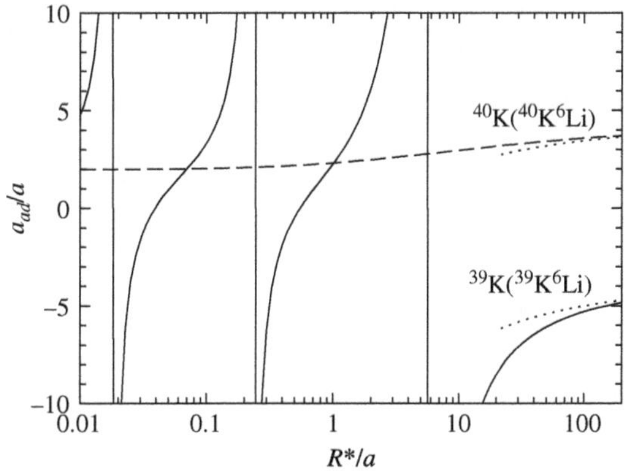

Figure 3.11 The atom–dimer s-wave scattering length for ^{39}K(^{39}K^{6}Li) (solid line) and ^{40}K(^{40}K^{6}Li) (dashed line) collisions near a narrow interspecies resonance. The dotted lines show the large R^*/a asymptotes of Eq. (3.129).

the g-dependent terms in the right hand side of Eq. (3.126) and by multiplying the result by η. In this way, we obtain the first two leading terms:

$$g(0) \approx \pm\frac{\eta}{\cos^3\phi}\left[1+\frac{-1\pm\cos^{-2}\phi}{2}\eta\right]. \tag{3.127}$$

Then, expressing the small parameter in terms of $\sqrt{a/R^*}$ and using

$$\eta = \frac{2}{\sqrt{1+4R^*/a}-1} \approx \sqrt{\frac{a}{R^*}}\left(1+\frac{1}{2}\sqrt{\frac{a}{R^*}}\right), \tag{3.128}$$

we finally obtain the atom–dimer scattering length up to the next-to-leading order in $\sqrt{a/R^*}$:

$$\frac{a_{ad}}{a} \approx \mp\frac{1}{\cos^2\phi}\left(1+\frac{1\pm\cos^{-2}\phi}{2}\sqrt{\frac{a}{R^*}}\right). \tag{3.129}$$

The dotted lines in Figure 3.11 show the asymptotes of Eq. (3.129).

Let us now discuss how the atom–dimer relaxation can be calculated for molecules formed near a narrow resonance. The process is local; it happens at distances of the order of the size of the closed-channel molecule, which is much smaller than R^*, according to our definition of narrow resonances. We have already mentioned that the probability of finding three- atoms in a small volume of size $\ll R^*$ is dominated by the probability of finding there an atom and a closed-channel molecule. Therefore, the probability density of finding three atoms in the recombination region equals $(R^*/4\pi)|f(x=0)|^2$, which is proportional to the relaxation rate if we treat

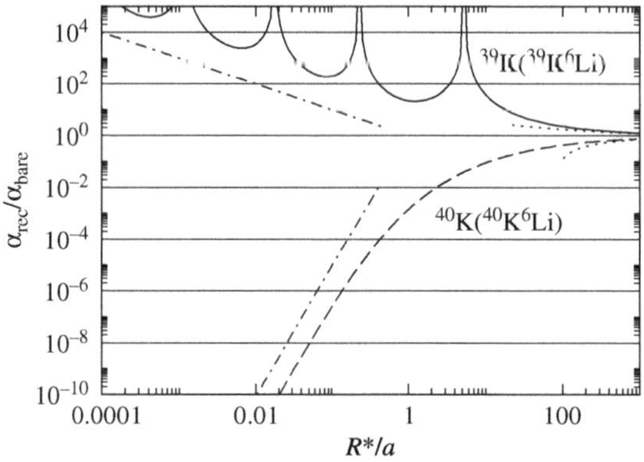

Figure 3.12 The atom–dimer s-wave relaxation rate constant versus the detuning. The physical systems and notations are the same as in Figure 3.11. The dash-dotted lines reflect the power scaling of $\alpha_{\rm rel}$ at large a (see text).

the process perturbatively. The proportionality prefactor requires that we solve the three-body problem at short distances, which is a challenging task. However, since the shape of the closed-channel wavefunction is not dramatically sensitive to the magnetic field, we can assume that this microscopic prefactor stays approximately constant close to a given Feshbach resonance. Therefore, the quantity that can be calculated in the zero-range approximation is the ratio $\alpha_{\rm rel}(a)/\alpha_{\rm bare}$ of the relaxation rate constant at a given a to the relaxation rate constant for collisions of atoms and closed-channel molecules. This ratio should tend to 1 in the limit $R^* \gg a$.

In Figure 3.12, we plot the ratio $\alpha_{\rm rel}(a)/\alpha_{\rm bare}$ for the same physical systems and using the same notation as in Figure 3.11, i.e., the solid line stands for bosonic ^{39}K and the dashed one for fermionic ^{40}K. The results are obtained by substituting the already calculated function $g(k)$ into Eq. (3.125) and evaluating $f(x = 0)$ by integrating $f(k)$ over momenta. One should keep track of the a-dependence of the normalization prefactor $N \propto \alpha_{\rm norm}$ given by Eq. (3.35). This dependence follows from the correct normalization of the dimer wavefunction in the narrow resonance case (see Section 3.2.3). The final result reads

$$\frac{\alpha_{\rm rel}(a)}{\alpha_{\rm bare}} = \frac{\sqrt{1+4R^*/a}-1}{\sqrt{1+4R^*/a}} \left| 1 + \frac{2}{\pi} \int_0^\infty g(k) {\rm d}k \right|^2 . \tag{3.130}$$

The dotted lines in Figure 3.12 show the large R^*/a-asymptotes and are obtained by approximating $g(k) \approx g^{(1)}(k)$ in Eq. (3.130).

In the regime of small detuning, $R^*/a \ll 1$, the dependence of $\alpha_{\rm rel}/\alpha_{\rm bare}$ on R^*/a can be estimated qualitatively. Consider first the fermionic case. The function $f(x)$ is of order $N \sim 1/\sqrt{a}$ at distances $\sim a$, where it should be matched with the power law $f(x) \sim a^{-1/2}(x/a)^{\nu_1}$, which is approximately valid in the window $R^* \lesssim x \lesssim a$. The power ν_1 is the root of $\lambda_{l=0}(\nu)$ (see Eq. (3.84) and Figure 3.7). For the system

^{40}K^{40}K^{6}Li, it equals $\nu_1 \approx 2.0193$. Then at distances $x \lesssim R^*$ the function f is approximately constant since the interaction in between the atom and the closed-channel molecule is not resonant. We finally obtain

$$\alpha_{\rm rel}/\alpha_{\rm bare} \propto R^*|f(x=0)|^2 \propto (R^*/a)^{-2\nu_1-1}, \; R^* \ll a. \tag{3.131}$$

This qualitative scaling is shown in Figure 3.12 as the lower dash-dotted line.

We see that the s-wave atom-dimer relaxation is highly suppressed in the limit $R^* \ll a$. Of good news is that even for $a = R^*$, where the qualitative scaling of Eq. (3.131) shows no suppression, the quantitative result predicts the suppression of the relaxation by three orders of magnitude compared with the 'bare' atom–molecule case. Apparently, for this relatively large mass ratio, the exchange of the light atom leads to an appreciable repulsion between the heavy ones (remember that according to the Born–Oppenheimer approach, the light atom is in the 'repulsive' ψ_- state). These effects are not taken into account in the derivation of Eq. (3.131).

As far as the bosonic case is concerned, we can repeat the same speculations, disregarding the atom–dimer resonances and the oscillatory part of $f(x) \sim a^{-1/2}x^{-1}\cos[s_0 \ln(x/r_0)]$ at distances $R^* \lesssim x \lesssim a$. We then obtain for bosons $\alpha_{\rm rel}/\alpha_{\rm bare} \sim (R^*/a)^{-1}$, i.e., Eq. (3.131) with $\nu_1 = -1$ (the upper dash-dotted line in Figure 3.12). Figure 3.12 clearly shows that in this case there is no suppression of the relaxation.

3.5 Final remarks

3.5.1 Outlook

Let us now briefly consider other rapidly developing directions in theoretical few-atom physics. Molecular regimes in fermionic and bosonic mixtures require a better understanding of collisional properties of molecules, which is a four-atom problem. The transition from three-body to four-body is conceptually straightforward, it requires just a bit more space for formulae, although much more computing power. The four-body STM equation for fermions in coordinate and momentum space has been derived along the lines of Section 3.4 by Petrov *et al.* (2004; 2005a; 2005b). The three-body STM approach has also been generalized to the four-body case by using diagrammatic techniques (Brodsky *et al.* 2005; Levinsen and Gurarie 2006). The Born–Oppenheimer approximation has been applied to the calculation of collisional properties of highly mass-imbalanced heteronuclear molecules (Marcelis *et al.* 2008). Various aspects of the four-boson problem with short-range interactions (universality, tetramer states, necessity of an additional four-body parameter) have been actively studied in momentum space using the Faddeev–Yakubovsky equations (Platter *et al.* 2004a; Yamashita *et al.* 2006, Hammer and Platter 2007; Deltuva 2010), the adiabatic hyperspherical approach (von Stecher *et al.* 2009; see also review of Rittenhouse *et al.* 2011), and by other methods (Hanna and Blume 2006; von Stecher 2010; Yamashita *et al.* 2010). The problem of three heavy fermions interacting with a light one has recently attracted attention (Castin *et al.* 2010; Blume and Daily 2010; Gandolfi and Carlson 2010;

Mora *et al.* 2011b). In particular, Castin *et al.* (2010) have argued that in this system the four-body Efimov effect can occur for mass ratios $M/m < 13.6$, i.e., in the absence of the three-body one.

Another direction of research, obviously relevant for ultracold gases, is the few-body problem in a trap. The spectrum of three trapped bosons has been calculated by Jonsell *et al.* (2002). The relation of the trapped problem at unitarity ($a = \infty$) and the free-space problem at zero energy has been established by Tan (2004) and by Werner and Castin (2006). Various results on the spectrum and correlation functions of few-fermion systems in a harmonic potential have been obtained by Blume and co-workers (Blume *et al.* 2007; von Stecher *et al.* 2007, 2008). For an overview of the few-body problem in a trap, see (Blume 2012).

Notable progress has been made in solving the few-body problem in low dimensions. The hyperspherical approach has been applied to the two-dimensional three-boson problem by Nielsen *et al.* (1999) and by Kartavtsev and Malykh (2006). Platter *et al.* (2004b) have calculated the bound state energies of four two-dimensional bosons using the Faddeev–Yakubovsky equations. Brodsky *et al.* (2005) have considered the two-dimensional few-body problem in the diagrammatic framework. Heteronuclear two-dimensional trimers have been discussed by Pricoupenko and Pedri (2010), using the two-dimensional STM equation in momentum space. Liu *et al.* (2010b) have calculated the second and third virial coefficients for a two-dimensional strongly interacting Fermi gas.

As far as the one-dimensional case is concerned, the problem of N particles is exactly solvable in some cases (Lieb, 1963; Lieb and Liniger 1963; McGuire 1965; Gaudin 1967; Yang 1967) and the corresponding wavefunctions can be written out explicitly. The exact results can be compared with the results of various few-body methods (Dodd 1970; Thacker 1975; Amaya-Tapia *et al.* 2004). Speaking of non-integrable cases, the atom–dimer and dimer–dimer scattering problem in quasi-one-dimensional geometry has been studied by Mora and co-workers (Mora *et al.* 2004; 2005a; 2005b). Muryshev *et al.* 2002, Sinha *et al.* 2006, and Mazets *et al.* (2008) have perturbatively estimated the three-body diffraction effect originating from the finite transversal size in a quasi-1D system of bosons. Deviations from integrability originating from the finite width of a Feshbach resonance have been investigated by Yurovsky *et al.* (2006). The spectral statistics and the response of 1D few-body systems with and without mass imbalance (non-integrable and integrable, respectively) have been studied by Colomè-Tatchè and Petrov (2011). The one-dimensional three-body problem on a lattice has recently been discussed by Orso *et al.* (2011) and Valiente *et al.* (2010), see also (Keilmann *et al.* 2009).

Different components of a heteronuclear mixture can feel different external forces. For example, we can imagine that one of the components is confined to a quasi-one-dimensional or quasi-two-dimensional geometry whereas the other component remains free. Besides, by adding an optical lattice, one can separately control the effective masses of the components. This leads to interesting peculiarities of many-, few-, and even two-body problems in such mixed-dimensional systems (Petrov *et al.* 2007; Nishida and Tan 2008; Levinsen *et al.* 2009; Lamporesi *et al.* 2010).

We should mention the analysis of the few-fermion problem near a p-wave resonance (Levinsen *et al.* 2007; Jona-Lasinio *et al.* 2008), which is interesting in view of the realization of non-trivial superfluid phases in strongly interacting polarized fermionic gases. Finally, the progress in preparing cold dipolar gases (see Chapter 7) motivates studies of the few-body problem with dipolar interactions (Wang *et al.* 2011b; 2011c).

Acknowledgements

I would like to thank the organizers of this summer school, Gora Shlyapnikov and Christophe Salomon, for giving me the opportunity to teach at Les Houches. I also thank them as my collaborators on the topic of these lectures. I am very much indebted to people with whom I collaborated or deeply discussed few-body physics, Maria Colomè, Kerstin Helfrich, Jesper Levinsen, Hans-Werner Hammer, Tobias Tiecke, Jook Walraven, Bout Marcelis, Servaas Kokkelmans, Gregory Astrakharchik, David Papoular, Victor Gurarie, Maxim Efremov, Mikhail Baranov, Felix Werner, Yvan Castin, Ludovic Pricoupenko, Maxim Olshanii, Vladimir Yurovsky, Doerte Blume, Dmitri Fedorov, Mikhail Zvonarev, Pietro Massignan, Roman Krems, and many others. The work on these lectures was financially supported by the EuroQUAM-FerMix program, by the ANR grant BLAN0165, and by the IFRAF Institute.

References

Amaya-Tapia, A, Gasaneo, G, Ovchinnikov, S, Macek, JH, and Larsen, SY (2004). Integral representation of one-dimensional three particle scattering for δ function interactions. *J. Math. Phys.* **45**, 3533.

Barontini, G, Weber, C, Rabatti, F, *et al.* (2009). Observation of heteronuclear atomic Efimov resonances. *Phys. Rev. Lett.* **103**, 043201.

Bedaque, PF, Hammer, H-W, and van Kolck, U (1999). Renormalization of the three-body system with short-range interactions. *Phys. Rev. Lett.* **82**, 463.

Bedaque, PF, Braaten, E, and Hammer, H-W (2000). Three-body recombination in Bose gases with large scattering length. *Phys. Rev. Lett.* **85**, 908.

Berry, MV (1981). Quantizing a classically ergodic system: Sinai's billiard and the KKR method. *Ann. Phys. (N.Y.)* **131** 163.

Bohigas, O, Giannoni, MJ, and Schmit, C (1984). Characterization of chaotic quantum spectra and universality of level fluctuation laws. *Phys. Rev. Lett.* **52**, 1.

Blume, D (2012). Few-body physics with ultracold atomic and molecular systems in traps. *Rep. Prog. Phys.* **75**, 046401.

Blume, D and Daily, KM (2010). Breakdown of universality for unequal-mass Fermi gases with infinite scattering length. *Phys. Rev. Lett.* **105**, 170403.

Blume, D, von Stecher, J, and Greene, CH (2007). Universal properties of a trapped two-component Fermi gas at unitarity. *Phys. Rev. Lett.* **99**, 233201.

Braaten, E, and Hammer, H-W (2001). Three-body recombination into deep bound states in a Bose gas with large scattering length. *Phys. Rev. Lett.* **87**, 160407.

Braaten, E, and Hammer, H-W (2006). Universality in few-body systems with large scattering length. *Phys. Rept.* **428**, 259.

Braaten, E, and Hammer, H-W (2007). Efimov physics in cold atoms. *Ann. Phys.* **322**, 120.

Brodsky, IV, Klaptsov, AV, Kagan, MY, Combescot, R, and Leyronas, X (2005). Bound states of three and four resonantly interacting particles. *JETP Lett.* **82**, 273.

Castin, Y, Mora, C, and Pricoupenko, L (2010) Four-body Efimov effect for three fermions and a lighter particle. *Phys. Rev. Lett.* **105**, 223201.

Chin, C, Grimm, R, Julienne, P, and Tiesinga, E (2010). Feshbach resonances in ultracold gases. *Rev. Mod. Phys.* **82**, 1225.

Colomè-Tatchè, M and Petrov, DS (2011). Parametric excitation of a 1D gas in integrable and nonintegrable cases. *Phys. Rev. Lett.* **106**, 125302.

Danilov, GS (1961). On the three-body problem with short-range forces. *Sov. Phys. JETP* **13**, 349.

Deltuva, A (2010). Efimov physics in bosonic atom-trimer scattering. *Phys. Rev. A* **82**, 040701(R).

Dodd, LR (1970). Exact solution of the Faddeev equations for a one-dimensional system. *J. Math. Phys.* **11**, 207.

Efimov, V (1970). Energy levels arising from resonant two-body forces in a three-body system. *Phys. Lett.* **33B**, 563.

Esry, BD, Greene, CH, and Burke, JP (1999). Recombination of three atoms in the ultracold limit. *Phys. Rev. Lett.* **83**, 1751.

Faddeev, LD (1961). Scattering theory for a three particle system. *Sov. Phys. JETP* **12**, 1014.

Faddeev, LD and Merkuriev, SP (1993). *Quantum Scattering Theory for Several Particle Systems.* Kluwer Academic, Dordrecht.

Ferlaino, F, Zenesini, A, Berninger, M, Huang, B, Nägerl, H-C, and Grimm, R (2011). Efimov resonances in ultracold quantum gases. *Few-Body Syst.* **51**, 113.

Fonseca, AC, Redish, EF, and Shanley, PE (1979). Efimov effect in an analytically solvable model. *Nucl. Phys. A* **320**, 273.

Gandolfi, S and Carlson, J (2010). Heavy-light few fermion clusters at unitarity. arXiv:1006.5186.

Gaudin, M (1967). Un systeme a une dimension de fermions en interaction. *Phys. Lett. A* **24**, 55.

Giorgini, S, Pitaevskii, LP, and Stringari, S (2008). Theory of ultracold atomic Fermi gases. *Rev. Mod. Phys.* **80**, 1215.

Gogolin, AO, Mora, C, and Egger, R (2008). Analytic solution of the bosonic three-body problem. *Phys. Rev. Lett.* **100**, 140404.

Gross, N, Shotan, Z, Kokkelmans, S, and Khaykovich, L (2009). Observation of universality in ultracold ^{7}Li three-body recombination. *Phys. Rev. Lett.* **103**, 163202.

Gross, N, Shotan, Z, Kokkelmans, S, and Khaykovich, L (2010). Nuclear-spin-independent short-range three-body physics in ultracold atoms. *Phys. Rev. Lett.* **105**, 103203.

Hammer, H-W and Platter, L (2007). Universal properties of the four-body system with large scattering length. *Eur. Phys. J. A* **32**, 113.

Hanna, GJ and Blume, D (2006). Energetics and structural properties of three-dimensional bosonic clusters near threshold. *Phys. Rev. A* **74**, 063604.

Helfrich, K, Hammer, H-W, and Petrov, DS (2010). Three-body problem in heteronuclear mixtures with resonant interspecies interaction. *Phys. Rev. A* **81**, 042715.

Huang, K (1963). *Statistical Mechanics*. Wiley, New York.

Huckans, JH, Williams, JR, Hazlett, EL, Stites, RW, and O'Hara, KM (2009). Three-body recombination in a three-state Fermi gas with widely tunable interactions. *Phys. Rev. Lett.* **102**, 165302.

Jona-Lasinio, M, Pricoupenko, L, and Castin, Y (2008). Three fully polarized fermions close to a p-wave Feshbach resonance. *Phys. Rev. A* **77**, 043611.

Jonsell, S, Heiselberg, H, and Pethick, CJ (2002). Universal behavior of the energy of trapped few-boson systems with large scattering length. *Phys. Rev. Lett.* **98**, 250401.

Kartavtsev, OI and Malykh, AV (2006). Universal low-energy properties of three two-dimensional bosons. *Phys. Rev. A* **74**, 042506.

Kartavtsev, OI and Malykh, AV (2007). Low-energy three-body dynamics in binary quantum gases. *J. Phys. B* **40**, 1429.

Keilmann, T, Cirac, I, and Roscilde, T (2009). Dynamical creation of a supersolid in asymmetric mixtures of bosons. *Phys. Rev. Lett.* **102**, 255304.

Kohn, W and Rostoker, N (1954). Solution of the Schrödinger equation in periodic lattices with an application to metallic lithium. *Phys. Rev.* **94**, 1111.

Korringa, J (1947). On the calculation of the energy of a Bloch wave in a metal. *Physica* **13**, 392.

Kraemer, T, Mark, M, Waldburger, P, *et al.* (2006). Evidence for Efimov quantum states in an ultracold gas of caesium atoms. *Nature* **440**, 315.

Lamporesi, G, Catani, J, Barontini, G, Nishida, Y, Inguscio, M, and Minardi, F (2010). Scattering in mixed dimensions with ultracold gases. *Phys. Rev. Lett.* **104**, 153202.

Landau, LD, and Lifshitz, EM (1987). *Quantum Mechanics* (3rd edn). Pergamon, Oxford.

Levinsen, J, and Gurarie, V (2006). Properties of strongly paired fermionic condensates. *Phys. Rev. A* **73**, 053607.

Levinsen, J and Petrov, DS (2011). Atom–dimer and dimer–dimer scattering in fermionic mixtures near a narrow Feshbach resonance. *Eur. Phys. J. D* **65**, 67.

Levinsen J, Cooper, N, and Gurarie, V (2007). Strongly resonant p-wave superfluids. *Phys. Rev. Lett.* **99**, 210402.

Levinsen, J, Tiecke, TG, Walraven, JTM, and Petrov, DS (2009). Atom–dimer scattering and long-lived trimers in fermionic mixtures. *Phys. Rev. Lett.* **103**, 153202.

Lieb, EH (1963). Exact analysis of an interacting Bose gas. II. The excitation spectrum. *Phys. Rev.* **130**, 1616.

Lieb, EH and Liniger, W (1963). Exact analysis of an interacting Bose gas. I. The general solution and the ground state. *Phys. Rev.* **130**, 1605.

Lin, CD (1995). Hyperspherical coordinate approach to atomic and other coulombic 3-body systems. *Phys. Rep.* **257**, 2.

Liu, X-J, Hu, H, and Drummond, PD (2010a). Virial expansion for a strongly correlated Fermi gas. *Phys. Rev. Lett.* **102**, 160401.

Liu, X-J, Hu, H, and Drummond, PD (2010b). Exact few-body results for strongly correlated quantum gases in two dimensions. *Phys. Rev. B* **82**, 054524.

Lompe, T, Ottenstein, TB, Serwane, F, *et al.* (2010). Atom–dimer scattering in a three-component Fermi gas. *Phys. Rev. Lett.* **105**, 103201.

Macek, JH and Ovchinnikov, S (2006). Exact solution for three particles interacting via zero-range potentials. *Phys. Rev. A* **73**, 032704.

Macek, JH, Ovchinnikov, S, and Gasaneo, G (2005). Solution for boson–diboson elastic scattering at zero energy in the shape-independent model. *Phys. Rev. A* **72**, 032709.

Marcelis, B, Kokkelmans, SJJMF, Shlyapnikov, GV, and Petrov, DS (2008). Collisional properties of weakly bound heteronuclear dimers. *Phys. Rev. A* **77**, 032707.

Mazets, IE, Schumm, T, and Schmiedmayer, J (2008). Breakdown of integrability in a quasi-one-dimensional ultracold bosonic gas. *Phys. Rev. Lett.* **100**, 210403.

McGuire, JB (1965). Interacting fermions in one dimension. I. Repulsive potential. *J. Math. Phys.* **6**, 432.

Mora, C, Egger, R, Gogolin, AO, and Komnik, A (2004). Atom–dimer scattering for confined ultracold fermion gases. *Phys. Rev. Lett.* **93**, 170403.

Mora, C, Egger, R, and Gogolin, AO (2005a). Three-body problem for ultracold atoms in quasi-one-dimensional traps. *Phys. Rev. A* **71**, 052705.

Mora, C, Komnik, A, Egger, R, and Gogolin, AO (2005b). Four-body problem and BEC–BCS crossover in a quasi-one-dimensional cold fermion gas. *Phys. Rev. Lett.* **95**, 080403.

Mora, C, Gogolin, AO, and Egger, R (2011a). Exact solution of the three-boson problem at vanishing energy. *C. R. Physique* **12**, 27.

Mora, C, Castin, Y, and Pricoupenko, L (2011b). Integral equations for the four-body problem. *C. R. Physique* **12**, 71.

Muryshev, A, Shlyapnikov, GV, Ertmer, W, Sengstock, K, and Lewenstein, M (2002). Dynamics of dark solitons in elongated Bose-Einstein condensates. *Phys. Rev. Lett.* **89**, 110401.

Nakajima, S, Horikoshi, M, Mukaiyama, T, Naidon, P, and Ueda, M (2010). Nonuniversal Efimov atom–dimer resonances in a three-component mixture of ^{6}Li. *Phys. Rev. Lett.* **105**, 023201.

Nakajima, S, Horikoshi, M, Mukaiyama, T, Naidon, P, and Ueda, M (2011). Measurement of an Efimov trimer binding energy in a three-component mixture of ^{6}Li. *Phys. Rev. Lett.* **106**, 143201.

Nielsen, E and Macek, JH (1999). Low-energy recombination of identical bosons by three-body collisions. *Phys. Rev. Lett.* **83**, 1566.

Nielsen, E, Fedorov, DV, and Jensen, AS (1999). Structure and occurrence of three-body halos in two dimensions. *Few-Body Syst.* **27**, 15.

Nielsen, E, Fedorov, DV, Jensen, AS, and Garrido, E (2001). The three-body problem with short-range interactions. *Phys. Rep.* **347**, 374.

Nishida, Y and Tan, S (2008). Universal Fermi gases in mixed dimensions. *Phys. Rev. Lett.* **101**, 170401.

Orso, G, Burovski, E, Jolicoeur, T (2011). Fermionic trimers in spin-dependent optical lattices. *C. R. Physique* **12**, 39.

Ottenstein, TB, Lompe, T, Kohnen, M, Wenz, AN, and Jochim, S (2008). Collisional stability of a three component degenerate Fermi gas. *Phys. Rev. Lett.* **101**, 203202.

Petrov, DS (2003). Three-body problem in Fermi gases with short-range interparticle interaction. *Phys. Rev. A* **67**, 010703(R).

Petrov, DS (2004). Three-boson problem near a narrow Feshbach resonance. *Phys. Rev. Lett.* **93**, 143201.

Petrov, DS, Salomon, C, and Shlyapnikov, GV (2004). Weakly bound dimers of fermionic atoms. *Phys. Rev. Lett.* **93**, 090404.

Petrov, DS, Salomon, C, and Shlyapnikov, GV (2005a). Scattering properties of weakly bound dimers of fermionic atoms. *Phys. Rev. A* **71**, 012708.

Petrov, DS, Salomon, C, and Shlyapnikov, GV (2005b). Diatomic molecules in ultracold Fermi gases — novel composite bosons. *J. Phys. B: At. Mol. Opt. Phys.* **38**, S645.

Petrov, DS, Astrakharchik, GE, Papoular, DJ, Salomon, C, and Shlyapnikov, GV (2007) Crystalline phase of strongly interacting Fermi mixtures. *Phys. Rev. Lett.* **99**, 130407.

Platter, L, Hammer, H-W, and Meißner, U-G (2004a). The four-boson system with short-range interactions. *Phys. Rev. A* **70**, 052101.

Platter, L, Hammer, H-W, and Meißner, U-G (2004b). Universal properties of the four-boson system in two dimensions. *Few Body Syst.* **35**, 169.

Pollack, SE, Dries, D, and Hulet, RG (2009). Universality in three- and four-body bound states of ultracold atoms. *Science* **326**, 1686.

Pricoupenko, L and Pedri, P (2010). Universal (1+2)-body bound states in planar atomic waveguides. *Phys. Rev. A* **82**, 033625.

Rittenhouse, ST, von Stecher, J, D'Incao, JP, Mehta, NP, and Greene, CH (2011). The hyperspherical four-fermion problem. *J. Phys. B* **44**, 172001.

Sinha, S, Cherny, AYu, Kovrizhin, D, and Brand, J (2006). Friction and diffusion of matter-wave bright solitons. *Phys. Rev. Lett.* **96**, 030406.

Skorniakov, GV and Ter-Martirosian, KA (1957). Three body problem for short range forces. I. Scattering of low energy neutrons by deuterons. *Sov. Phys. JETP* **4**, 648.

Tan, S (2004). Short range scaling laws of quantum gases with contact interactions. arXiv:cond-mat/0412764.

Thacker, HB (1975). Bethe's hypothesis and Feynman diagrams: exact calculation of a three-body scattering amplitude by perturbation theory. *Phys. Rev. D* **11**, 838.

Thomas, LH (1935). The interaction between a neutron and a proton and the structure of H^3. *Phys. Rev.* **47**, 903.

Tiecke, TG, Goosen, MR, Ludewig, A, *et al.* (2010). Broad Feshbach resonance in the ^{6}Li–^{40}K mixture. *Phys. Rev. Lett.* **104**, 053202.

Valiente, M, Petrosyan, D, and Saenz, A (2010). Three-body bound states in a lattice. *Phys. Rev. A* **81**, 011601(R).

Volz, T, Dürr, S, Ernst, S, Marte, A, and Rempe, G (2003). Characterization of elastic scattering near a Feshbach resonance in rubidium 87. *Phys. Rev. A* **68**, 010702(R).

von Stecher, J (2010). Weakly bound cluster states of Efimov character. *J. Phys. B* **43**, 101002.

von Stecher, J, Greene, CH, and Blume, D (2007). BEC–BCS crossover of a trapped two-component Fermi gas with unequal masses. *Phys. Rev. A* **76**, 053613.

von Stecher, J, Greene, CH, and Blume, D (2008). Energetics and structural properties of trapped two-component Fermi gases. *Phys. Rev. A* **77**, 043619.

von Stecher, J, D'Incao, JP, and Greene, CH (2009). Four-body legacy of the Efimov effect. *Nature Phys.* **5**, 417.

Wang, Y, D'Incao, JP, and Esry, BD (2011a). Ultracold three-body collisions near narrow Feshbach resonances. *Phys. Rev. A* **83**, 042710.

Wang, Y, D'Incao, JP, and Greene, CH (2011b). The Efimov effect for three interacting bosonic dipoles. *Phys. Rev. Lett.* **106**, 233201.

Wang, Y, D'Incao, JP, and Greene, CH (2011c). Universal three-body physics for fermionic dipoles. *Phys. Rev. Lett.* **107**, 233201.

Werner, F and Castin, Y (2006). The unitary gas in an isotropic harmonic trap: symmetry properties and applications. *Phys. Rev. A* **74**, 053604.

Wille, E, Spiegelhalder, FM, Kerner, G, *et al.* (2008). Exploring an ultracold Fermi–Fermi mixture: interspecies Feshbach resonances and scattering properties of ^{6}Li and ^{40}K. *Phys. Rev. Lett.* **100**, 053201.

Yakubovskii, OA (1967). On the integral equations in the theory of N particle scattering. *Sov. J. Nucl. Phys.* **5**, 937.

Yamashita, MT, Tomio, L, Delfino, A, and Frederico, T (2006). Four-boson scale near a Feshbach resonance. *Europhys. Lett.* **75**, 555.

Yamashita, MT, Fedorov, DV, and Jensen, AS (2010). Universality of Brunnian (N-body Borromean) four- and five-body systems. *Phys. Rev. A* **81**, 063607.

Yang, CN (1967). Some exact results for the many-body problem in one dimension with repulsive delta-function interaction. *Phys. Rev. Lett.* **19**, 1312.

Yurovsky, VA, Ben-Reuven, A, and Olshanii, M (2006). One-dimensional Bose chemistry: effects of non-integrability. *Phys. Rev. Lett.* **96**, 163201.

Zaccanti, M, Deissler, B, D'Errico, C, *et al.* (2009). Observation of an Efimov spectrum in an atomic system. *Nature Phys.* **5**, 586.

4
Entanglement in many-body quantum systems

J. Ignacio CIRAC

Max-Planck-Institut für Quantenoptik, Hans-Kopferman-Strasse 1,
D-85788 Garching, Germany

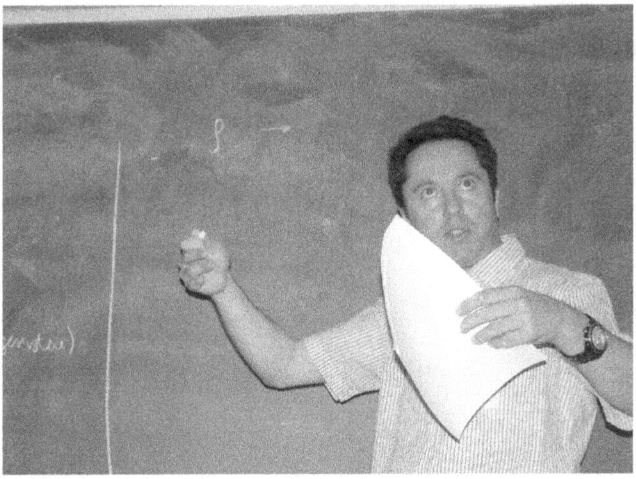

Chapter Contents

4.1 Introduction

Entanglement is a pure quantum property that may appear when we have a composite object. It features the existence of a very special kind of correlation, which cannot occur for product states or mixtures thereof. These correlations, sometimes referred to as quantum correlations, give rise to a great variety of phenomena and form the basis of many applications in quantum information science [1]. This is why, during the last fifteen years, a great deal of theoretical effort has been devoted to define and characterize this intriguing property. More recently, some of the ideas developed in this context have been used to get a novel perspective into the many-body quantum systems that appear in nature. In particular, the role of entanglement has been analysed in many papers, and its implications have been used to introduce novel ways of describing such systems. In this chapter, we review some of the basic concepts coming from the theory of entanglement, their applications to many-body systems, and some of the new theoretical methods that have come up in this context.

Entanglement appears whenever we have two or more quantum objects. It highlights the appearance of certain kinds of correlation that cannot appear in classical theories. By object, we may mean a particle, a bosonic or fermionic mode, etc. Strictly speaking, the Hilbert space corresponding to the whole system must decompose as a tensor product of several Hilbert spaces, one for each and every object. Entanglement will depend on how we understand this decomposition and thus we must specify to what (particles, modes, etc.) we refer when we discuss entanglement properties of systems. Entanglement can also be defined for both pure or mixed states. In the first case, the definition of entanglement poses no difficulty and can be very naturally understood in terms of standard correlations. In the second, however, the definition is subtle, as correlations do not directly imply entanglement (in fact, correlations are everywhere in the classical world!). For two objects, entanglement can easily be defined, qualified, and quantified. For many objects, however, this is not the case. New possibilities (and correlations) may appear, which cannot be reduced to the case of two objects. In fact, as of today, the theory of entanglement has not been fully developed: there exist many open questions and some definitions need to be sharpened. In the first two sections of this chapter, we attempt to give a rather superficial introduction to this topic, to prepare the reader for the following sections. We will first consider bipartite systems (i.e., two objects), and later on multipartite ones. I will introduce entanglement separately for pure and mixed states. I will highlight two quantities: the entropy of entanglement and the quantum mutual information. The first one measures the bipartite entanglement for pure states, whereas the second one measures correlations for both pure and mixed states. A thorough review of the theory of entanglement can be found in [2].

In the next section, we apply the ideas previously introduced to many-body quantum systems as they appear in typical physical scenarios. In particular, we will consider spin lattices with short-range interactions and in thermal equilibrium. I will first show that certain entanglement measures display special features whenever a phase transition occurs, and explain that this is simply because those measures display correlations, which are known to change abruptly under those circumstances. A thorough review of the behaviour of entanglement under phase transitions can be found in [3]. Then we will consider area laws in those systems. For that, we will use the quantities previously defined, namely the entropy of entanglement and the quantum mutual information. When we select a region of the lattice and consider the existing entanglement and correlations between that region and its complement we find that they scale with the number of spins lying at the boundary of such a region. I will explain where this peculiar property comes from, and derive it for Gibbs states. There are many reviews of the area law and its application, for example, [4; 5].

In the last section, we will use the intuition developed through the area law to introduce an efficient way of describing many-body quantum systems. It is based on a construction in which each spin in the lattice is replaced by several auxiliary spins, which are maximally entangled to their nearest neighbours. A map is then applied to each lattice site, which transforms the auxiliary spins into the original ones, giving rise to the quantum many-body state. The resulting states are known as projected entangled-pair states (PEPS), and provide such an effective description. I will highlight some of their properties, and present some explicit examples in 1D, where PEPS reduce to the so-called matrix product states (MPS), which play a central role in renormalization procedures and algorithms. Reviews on PEPS, MPS, and other tensor network states can be found in [6; 7].

In this chapter, we will not make any reference to the physical system we are dealing with. This is standard in quantum information, where the theory is valid for all physical quantum systems with some reachable quantum levels. Thus, many of the ideas reviewed here may well apply to atoms in optical lattices, spins in magnetic materials, or electrons in solids.

4.2 Entanglement in many-body systems: pure states

We start by considering the simplest case, namely the entanglement present in a many-body quantum state, Ψ, whenever we have a pure state. This is the case, for instance, at zero temperature (if there is no degeneracy), or in most applications in quantum information science. Unfortunately, it is an idealized case, since zero temperature cannot be reached in practice and, in most of the systems, interaction with the environment leads to decoherence, which reflects itself in the state becoming mixed. Nevertheless, pure states are much easier to deal with than mixed states, and already display many of the features characteristic of entanglement. In this section, we will concentrate on this case and consider first the situation in which we only have two systems. Later on, we will analyse the more general case of multipartite systems.

4.2.1 Bipartite systems

We consider here two systems, A and B. We denote by H_A and H_B the corresponding Hilbert spaces, and by $\{|n\rangle_X\}$ an orthonormal basis in H_X, where $n = 1, 2, \ldots, d_X$, with $d_X = \dim(H_X)$. Most of the time we will concentrate on the simplest systems, qubits, where $d_A = d_B = 2$. In that case, to keep the standard notation, we will take as a basis $\{|0\rangle_X, |1\rangle_X\}$. Unless stated otherwise, we will always work with qubits.

The Hilbert space corresponding to the whole system, H, is the tensor product of H_A and H_B, which we write $H = H_A \otimes H_B$. An orthonormal basis in that space is $\{|n\rangle_A \otimes |m\rangle_B\}$. To simplify the notation, we will typically omit the symbol '$\otimes$', and the subindices A, B whenever there is no possible confusion. For instance, any state for two qubits can be written as

$$|\Psi\rangle = \sum_{n,m=1}^{1} c_{n,m}|n, m\rangle, \qquad \sum_{n,m=0}^{1} |c_{n,m}|^2 = 1. \tag{4.1}$$

We will also omit the limits in the sum whenever it is obvious.

We will consider observables for each of the systems, which will be represented as operators acting on the corresponding spaces. For instance, $\sigma_1^A \otimes \sigma_2^B$ denotes an operator σ_1 acting on A and σ_2 on B. As before, we will omit the symbol for the tensor product; additionally, σ_1^A will stand for $\sigma_1^A \otimes 1^B$, where 1 is the identity operator. Pauli operators acting on qubits will often appear. They are defined as follows:

$$\begin{aligned}
\sigma_x &= |0\rangle\langle 1| + |1\rangle\langle 0|, \\
\sigma_y &= \mathrm{i}(|0\rangle\langle 1| - |1\rangle\langle 0|), \\
\sigma_z &= |1\rangle\langle 1| - |0\rangle\langle 0| = -\mathrm{i}\sigma_x\sigma_y,
\end{aligned}$$

We say that $\Psi \in H_A \otimes H_B$ is a *product state* if there exist two vectors $\varphi_1 \in H_A$ and $\varphi_2 \in H_B$ such that $|\Psi\rangle = |\varphi_1\rangle_A \otimes |\varphi_2\rangle_B$. Otherwise we say that Ψ is an *entangled state*. Examples of product states are those forming the orthonormal basis $|n, m\rangle$. Examples of entangled states are the so-called Bell states

$$|\Phi^{\pm}\rangle = \frac{1}{\sqrt{2}}(|0,0\rangle \pm |1,1\rangle), \tag{4.2}$$

$$|\Psi^{\pm}\rangle = \frac{1}{\sqrt{2}}(|0,1\rangle \pm |1,0\rangle). \tag{4.3}$$

Entanglement and correlations The difference between entangled and product states is that the first ones give rise to correlations. If we have a product state $|\Psi\rangle = |\varphi_1, \varphi_2\rangle$, then the expectation value $\langle\Psi|\sigma_1^A \otimes \sigma_2^B|\Psi\rangle$ factorizes into $\langle\psi_1|\sigma_1|\psi_1\rangle\langle\psi_2|\sigma_1|\psi_2\rangle$, and so the results of measurements in both systems will be uncorrelated. For any entangled state, on the contrary, there always exist observables in A and B for which the expectation value does not factorize and thus for which the results of measurements will be correlated. For example, considering the Pauli operator along the direction defined by a unit vector (in the xz-plane), $\vec{n}$, as $\sigma_{\vec{n}} = n_x\sigma_x + n_z\sigma_z$, we have

$$\langle\Phi^+|\sigma_{\vec{n}}^A \otimes \sigma_{\vec{m}}^B|\Phi^+\rangle = \vec{n}\cdot\vec{m}, \tag{4.4}$$

whereas $\langle\Phi^+|\sigma_{\vec{n}}^A|\Phi^+\rangle = 0$. Thus, whenever we measure the same Pauli operators in A and B (i.e., when $\vec{n}$ and $\vec{m}$ are parallel), the results are random but completely correlated (i.e., the same outcome in A and B).

To highlight the power of the correlations arising in entangled states, one can consider the following game [8]. Two people, Alice and Bob, after meeting to discuss their strategy, are isolated in two rooms and each given a different number, $x, y = 0, 1$. Their goal is to output another number, $a, b = 0, 1$, respectively, such that $a \oplus b = xy$ (where $\oplus$ denotes addition modulo 2). That is, if $(x, y) = (0, 0), (0, 1)$, or $(1, 0)$, they must output $a = b$, whereas for $(x, y) = (1, 1)$ they must give a different output. It is very easy to show that since Alice does not know the number y that Bob will receive (and vice versa), she does not know which a she has to output. For instance, if Alice is given $x = 1$, Bob could have $y = 0$ or $y = 1$, in which case she would have to give a different value of a. But since she does not know y, there is no way they can always guess. The best strategy can be easily shown to give the right answer with probability $P = 0.75$. Now, if Alice and Bob share an entangled state Φ^+ while discussing the strategy, then they can guess with a probability $P = 0.853\ldots > 0.75$. The strategy is: (i) for Alice to measure on her qubit the observable $\sigma_{\vec{n}_x}$, whenever she is given x, where $\vec{n}_0 = (0, 1)$, $\vec{n}_1 = (1, 0)$; (ii) for Bob, the observable $\sigma_{\vec{m}_y}$ whenever he is given y, where $\vec{m}_0 = (1/\sqrt{2}, 1/\sqrt{2})$ and $\vec{m}_1 = (1/\sqrt{2}, -1/\sqrt{2})$. It is very simple, using Eq. (4.4), to determine that the probability of guessing is indeed $(1 + \sqrt{2})/(2\sqrt{2})$. Thus, in a world where entangled states exist, one can do things that are otherwise impossible, such as play this game with a higher probability. Actually, this is the idea behind quantum information science, where by using quantum states and correlations one can perform tasks (in the context of cryptography or computation) that are classically impossible. Furthermore, if one performs this game experimentally (for instance, using photons) one obtains a result that is incompatible with local realistic theories (i.e., theories where the outcome of measurements do not depend on what is being measured somewhere else, and where the properties we measure are already well defined prior to the act of measuring). This is the essence of Bell's theorem [9], which states that local realistic theories are incompatible with quantum mechanics. Note that the Bell states are the ones that give a highest probability of winning the game, and thus they give rise to the maximal quantum correlations. They are usually referred to as maximally entangled states.

Schmidt decomposition To analyse the entanglement in bipartite systems it is useful to introduce the Schmidt decomposition (SD). Given the state Ψ, Eq. (4.1), it is always possible to find an orthonormal basis $\{|u_n\rangle\}$ in H_A and $\{|v_m\rangle\}$ in H_B, such that we can write

$$|\Psi\rangle = \sum_{k \leq d_A, d_B} d_k |u_k, v_k\rangle. \tag{4.5}$$

This is because the matrix C can always be written as $C = UDV$, where U and V are isometries and D is diagonal with positive elements. This way of writing C is called singular decomposition, and it is valid for any matrix [10]. When $d_A = d_B$, the square matrices U, V are unitary ($UU^\dagger = U^\dagger U = 1$). Otherwise, they are rectangular but still fulfil the required properties, such that we can always find the Schmidt decomposition. Note that U and V can be found by diagonalizing $CC^\dagger$ and $C^\dagger C$, respectively.

The matrix D can be found by taking the square root of the resulting diagonal matrix, which coincides in both cases. The diagonal elements of D are the d_k that appear in Eq. (4.5) and are called Schmidt coefficients.

Once equipped with the SD, we can easily figure out whether a state is entangled or not. If one of the Schmidt coefficients is one and the rest are zero, we have a product state. Otherwise, our state is entangled. The first statement is obvious, while the second follows immediately when one tries to write the state as a product state.

The SD is also very useful for determining the reduced density operators for subsystems A and B alone. Using Eq. (4.5) we find

$$\rho_A = \mathrm{tr}_B(|\Psi\rangle\langle\Psi|) = \sum_k d_k^2 |u_k\rangle\langle u_k|$$
$$\rho_B = \mathrm{tr}_A(|\Psi\rangle\langle\Psi|) = \sum_k d_k^2 |v_k\rangle\langle v_k|,$$

where by tr_X we mean the trace with respect to system X. The reduced density operators so obtained are automatically diagonalized, and we observe that their eigenvalues are nothing but the square of the Schmidt coefficients.

Let us consider the following example:

$$|\Psi\rangle = \cos(\theta)|0,0\rangle + \sin(\theta)|1,1\rangle, \tag{4.6}$$

for $\theta \in [0, \pi/4]$. This is already written in the SD form. The eigenvalues of the reduced density operator are $\cos^2(\theta)$ and $\sin^2(\theta)$. For $\theta = 0$ we have a product state, whereas for $\theta = \pi/4$ we have a Bell state, which is the one that gives the largest quantum correlations, as explained in the context of the game. In parallel, the reduced density operators become more and more mixed as one increases θ from 0 to $\pi/4$ (note that the purity of a mixed state, ρ, is related to the distribution of its eigenvalues when considered as probabilities).

Entropy of entanglement The previous example indicates that entanglement is related to the mixedness of the reduced density operators. In fact, this relation suggests that we can introduce a measure of entanglement by using any measure of mixedness of a state. A very natural measure of the latter is the von Neumann entropy $S(\rho) = -\mathrm{tr}[\rho \log_2(\rho)]$ (in the context of quantum information, one defines the logarithm in base 2). Thus, we define the *entropy of entanglement* of a state [11], Ψ,

$$E(\Psi) = -\mathrm{tr}[\rho_A \log_2(\rho_A)] = -\mathrm{tr}[\rho_B \log_2(\rho_B)] = -\sum_k d_k^2 \log_2(d_k^2). \tag{4.7}$$

As we see, this quantity can easily be determined through the SD. Note that in this expression $0\log(0) = 0$, by definition. For a product state, $E = 0$, whereas the maximum entanglement is $E = \log[\min(d_A, d_B)]$, which is reached for the state for which all the d_k are equal. These are thus called maximally entangled states (even for $d > 2$).

To give a physical interpretation of the entropy of entanglement, we will now define two key concepts in the context of quantum information: entanglement concentration and distillation [12]. But, before doing that, we digress, to introduce yet another basic feature in quantum information, which is needed to explain the aforementioned concepts, namely *generalized measurements*. The postulates of quantum mechanics specify what occurs when we perform a so-called filtering measurement of an observable. One just has to consider the associate operator, say O, and its spectral decomposition. Let us denote by o_i its eigenvalues, and by Π_i the projector onto the corresponding eigenspaces. If o_i is not degenerate, $\Pi_i = |\phi_i\rangle\langle\phi_i|$, where $O|\phi_i\rangle = o_i|\phi_i\rangle$. Note that $\sum_i \Pi = 1$. Given a, in general, mixed state represented by a density operator ρ, the probability of obtaining the outcome o_i is $P_i = \mathrm{tr}(\Pi_i\rho)$, and the state after the measurement is $\rho_i = \Pi_i\rho\Pi_i/P_i$. Thus, a filtering measurement is characterized by a set of operators Π_i that are positive (since they are projectors, their eigenvalues are zero or one) and add up to the identity operator. In fact, there is a more general scenario representing a measurement. We may consider bringing a measuring apparatus, letting it interact with our system, and then reading off the apparatus. According to the laws of quantum mechanics, we will describe the measurement as follows. First, the initial state of the system and the apparatus will be $\rho \otimes |A\rangle\langle A|$, where $|A\rangle$ denotes the (pure) sate of the latter. The interaction can be described in terms of a unitary operator, U, so that the state becomes $U(\rho \otimes |A\rangle\langle A|)U^\dagger$. Denoting the projector operators defining the measurement on the apparatus by $p_\mu = |A_\mu\rangle\langle A_\mu|$, we have that the probability, P_μ, of obtaining the outcome labelled by μ, and the state of our system after the measurement, ρ_μ, can be written as

$$p_\mu = \mathrm{tr}\left(A_\mu \rho A_\mu^\dagger\right), \quad \rho_\mu = \frac{A_\mu \rho A_\mu^\dagger}{p_\mu}, \tag{4.8}$$

where $A_\mu = \langle A_\mu|U|0\rangle$ is an operator acting on our system. Note that

$$\sum_\mu A_\mu^\dagger A_\mu = \langle A|U\left[\sum_\mu A_\mu\rangle\langle A_\mu\right]U^\dagger|A\rangle = 1, \tag{4.9}$$

where we have used the fact that $|A_\mu\rangle$ is an orthonormal basis. Thus, a generalized measurement is described in terms of a set of operators, A_μ, fulfilling Eq. (4.9). In fact, it can easily be shown that for any set of operators fulfilling that condition, there exists a measurement scheme (i.e., states $|A\rangle$ and $|A_\mu\rangle$) that renders Eq. (4.8) [1]. Here, for the sake of simplicity, we have restricted ourselves to pure states of the measurement apparatus and non–degenerate measurements. The extension to other cases is straightforward.

Now that we have defined generalized measurements, we can introduce the concept of entanglement distillation (Figure 4.1(a)). Let us first consider the state of Eq. (4.6), and let us assume that our goal is to create a maximally entangled state (i.e., a state in the same form but with $\theta = \pi/4$) by acting locally on each of the particles. To do that, we can try to apply a generalized measurement to the first particle. We choose $A_0 = \tan(\theta)|0\rangle\langle 0| + |1\rangle\langle 1|$ and $A_1 = (1 - A_0^\dagger A_0)^{1/2}$. These two operators fulfil Eq. (4.9) and this ensures that there is a physical measurement scheme associated with them

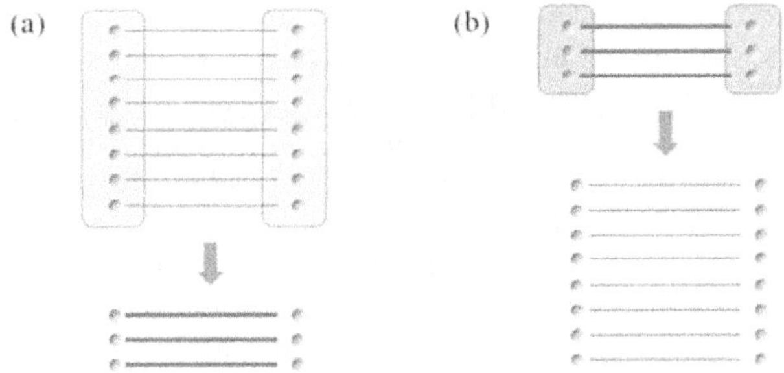

Figure 4.1 (a) Entanglement distillation. Out of many copies of a weakly entangled states, by LOCC we obtain fewer copies of a maximally entangled state (b) Entanglement dilution. The reverse process.

that can easily be determined. If we measure and obtain the outcome associated with A_0, we will achieve our goal. Otherwise, we will produce a product state instead. The probability of succeeding is $p_0 = 2\sin^2(\theta)$. Note that, if A and B are spatially separated and are held by Alice and Bob, respectively, in order to know whether the measurement has been successful, the outcome (i.e., a classical bit of information) has to be transmitted from Alice to Bob. One says that by local operations and classical communication (LOCC) one can distill a maximally entangled state out of the state Ψ with probability p_0. One may wonder if there is another generalized measurement applied to A and B (individually) giving a higher probability of success. In fact, this is not the case, since the generalized measurement we just chose is the optimal one. Now one can consider the case in which Alice and Bob possess two identical copies of the state Ψ, and they try to obtain maximally entangled states by LOCC (in which joint measurements on both qubits of Alice, or both qubits of Bob, are authorized). In general, they may get as outcome a maximally entangled state in a Hilbert space of dimension $d = 2, 3, 4$. For instance, if they are completely successful, they will get two copies of a maximally entangled state, which is equivalent to a single copy of such a state in a space of dimension $d = 4$. Or if they get a single copy, they will have $d = 2$. One can, indeed, show that in the case of two copies the average entanglement is strictly larger than twice that for a single copy if one chooses the optimal strategy. Now we can consider what happens when we take n copies and allow for the optimal LOCC to optimize

$$\bar{E} = \frac{1}{n}\sum_{d=1}^{2^n} p_d \log_2(d), \tag{4.10}$$

where p_d is the probability that they end up with an entangled state in a space of dimension d (for $d = 1$ they end up in a product state). The logarithm is the right quantity such that n copies of a maximally entangled state (which corresponds to a dimension 2^n) exactly gives a factor n. It turns out [13] that in the limit $n \to \infty$ the result precisely coincides with the entropy of entanglement $E(\Psi)$ of Eq. (4.7). This occurs not only for qubits, but for any d-level systems. Thus, the entanglement entropy is nothing but the optimal averaged entanglement that we can distill out of Ψ by LOCC in the asymptotic limit, where we have a large number of copies.

One may consider the opposite process, called entanglement dilution (Fig. 4.1(b)). Given n maximally entangled states (of qubits), and by applying the optimal LOCC, how many copies, m, of the state Ψ can we obtain? The ratio $\bar{D} = n/m$ (on average) in the limit $n \to \infty$ turns out to coincide again with $E(\Psi)$, again giving a physical meaning to the latter. In fact, in this limit, states $\Psi_{1,2}$ can be converted into each other with a yield $E(\Psi_2)/E(\Psi_1)$. This implies that, at least when we consider an scenario where we dispose of many copies of a state and are allowed to perform LOCC, there is just a single measure of entanglement, namely the entropy of entanglement.

We finish this section by mentioning other quantities that are usually employed to quantify entanglement. One is the concurrence [14], which is the square of the determinant of the matrix c in (4.1) and is nothing but the product of the Schmidt coefficients. Another one is the fidelity with a maximally entangled state, i.e.,

$$F(\Psi) = \max|\langle\Phi^+|(U_A \otimes V_B)|\Psi\rangle|^2, \tag{4.11}$$

where the maximization is with respect to the unitary operators U and V. This measures in a sense how close we are to a maximally entangled state; U and V just correspond to a basis change.

4.2.2 Multipartite systems

Entanglement in multipartite systems becomes more complicated than in bipartite ones. First of all, one can have that certain objects are entangled to others, but not to all of them. Second, the quantification becomes more subtle since it is not known whether a property like the interconvertibility of states by distillation and dilution exists.

We say that a state Ψ of systems $A, B, \ldots, Z$, is a *product state* if there exists $|\varphi_X\rangle \in H_X$, such that $|\Psi\rangle = \otimes_X |\varphi_X\rangle_X$. Otherwise, we say that we have an entangled state. Still, it may happen that some of the systems are disentangled. To characterize the entanglement, we consider all possible partitions of the systems, and for each of them we apply the above definition. Thus, entanglement is characterized in terms of a partition that indicates which systems are entangled among themselves. For instance, for three parties we can have: (i) they are in a product state; (ii) only A and B are entangled; (iii) only A and C; (iv) only B and C; (v) all are entangled. These cases are mutually disjoint. An example of case (i) is the state $|0,0,0\rangle$, of case (ii) $|\Phi^+\rangle_{AB} \otimes |0\rangle_C$, and of case (iii) the states [15; 16]

$$\begin{aligned} |W\rangle &= |0,0,1\rangle + |0,1,0\rangle + |1,0,0\rangle \\ |GHZ\rangle &= |0,0,0\rangle + |1,1,1\rangle \end{aligned} \tag{4.12}$$

(we have omitted the normalization).

These two last examples illustrate the difficulty of quantifying entanglement in many-body quantum systems. It is not clear which of those states is 'more' entangled. In the context of quantum information, it depends on the application we have in mind.

For some, one of them is more useful, whereas for others, it is the other one. Furthermore, in this case it is not possible to convert the state $|W\rangle$ into the state $|GHZ\rangle$ by LOCC in the asymptotic limit without losing copies (equivalently, entanglement), and thus we cannot assign a quantity like the entanglement entropy to them [17]. One could hope to be able to convert n copies of any state Ψ into m_1 copies of $|W\rangle$ and m_2 copies of $|GHZ\rangle$, and then back in the limit $n \to \infty$, in which case one could define two measures of entanglement, given by the ratios $m_{1,2}/n$. However, this is also impossible. One may then try to include other representative states beyond those two and thus define more measures of entanglement. However, it is not even clear that a finite set of representative states exists. Thus, one cannot follow the procedure we reviewed in Section 4.2.1 to assign a meaningful entanglement measure to multipartite states.

What we can still do is consider bipartite partitions, in which we consider the entropy of entanglement of two disjoint sets of subsystems. This, in fact, has interesting applications in the context of many-body physics at zero temperature and will be analysed in more detail in Section 4.4.

Another approach is to look at fidelities with respect to certain particular states. For instance, similarly to Eq. (4.11), one can define the fidelity with respect to a GHZ, a W state, or products of Bell states. Yet another possibility is to define measures that give figures of merit in specific applications. For instance, in the context of quantum repeaters, one usually defines the *localizable entanglement* [18] as follows. Consider that we perform measurements on all particles except for two and the goal is to obtain, on average, the maximum entanglement for those two particles. Let us consider four particles, A, B, C, D, and define the localizable entanglement of Ψ with respect to particles A and D. Assume that we measure certain observables $O_{B,C}$ in B and C, respectively, and denote by $|o_i\rangle$ an eigenbasis of those operators. Then, the average entanglement we obtain in A and D if we perform the measurement will be

$$E_{AD}(O_B, O_C) = \sum_{i,j} P_{i,j} E(\phi_{i,j}), \tag{4.13}$$

where $P_{i,j}$ is the probability that we obtain the outcomes i and j in B and C, respectively, and $\phi_{i,j}$ the state of AD in such a case. This entanglement will, in general, depend on the observables we decide to measure. We thus define the localizable entanglement as the maximum with respect to all possible observables. As mentioned before, this has a specific meaning in the context of quantum repeaters, whereby the goal is to obtain as much entanglement as possible between the first and the last node by measuring in the intermediate ones.

4.3 Entanglement in many-body systems: mixed states

In this section we consider mixed states. These are described by a density operator, ρ, fulfilling $\rho = \rho^\dagger \geq 0$ (meaning that all eigenvalues are non-negative), and $\text{tr}(\rho) = 1$ (normalization condition). It can always be written as

$$\rho = \sum_i p_i |\Psi_i\rangle\langle\Psi_i|, \tag{4.14}$$

where $p_i > 0$ and fulfil $\sum_i p_i = 1$ and the normalized states Ψ_i need not be orthogonal. The interpretation of Eq. (4.14) is the following: the state ρ can be obtained by preparing the state Ψ_i with probability p_i (and then 'forgetting' which state has been prepared). If one of the $p_i = 1$, we have a pure state, which obviously fulfils $\mathrm{tr}(\rho^2) = 1$; otherwise, we say that our state is mixed. Note that for mixed states there exist many ways of writing Eq. (4.14). That is, there exist other $q_i > 0$ and $|\Phi_i\rangle$ such that ρ can be decomposed in their terms. For instance,

$$\rho = \frac{1}{2}(|0\rangle\langle 0| + |1\rangle\langle 1|) = \frac{1}{2}(|+\rangle\langle +| + |-\rangle\langle -|), \tag{4.15}$$

where $|\pm\rangle = (|0\rangle \pm |1\rangle)/\sqrt{2}$. This simply means that the same state can be prepared in different ways, either by mixing $|0\rangle$ and $|1\rangle$, or by mixing $|\pm\rangle$.

The probability of obtaining an outcome μ when performing a (generalized) measurement is given by Eq. (4.8). Thus, the density operator contains all the information we have about our system. This entails that we will not be able to distinguish by any means how we prepared a state (which decomposition we used), given that the statistics of any measurement do not depend on how we prepared the state.

In this section, we will consider the entanglement of a mixed state ρ. We will start out with the simplest case, that of two subsystems, and later on analyse the multipartite case.

4.3.1 Bipartite systems

We consider again two subsystems, A and B. We say that ρ represents a *product state* whenever we can find $\rho_{A,B}$, operators acting on $H_{A,B}$ such that $\rho = \rho_A \otimes \rho_B$. A state is *separable* [19] if ρ can be written as a mixture of product state; that is, if

$$\rho = \sum_i p_i |a_i, b_i\rangle\langle a_i, b_i|, \tag{4.16}$$

where $p_i > 0$. Otherwise, we say that ρ represents (or is) an *entangled state*.

The definition of entangled states is self-explanatory. A product state is obviously not entangled, since it does not yield correlations. A separable state cannot be entangled either, since it is a mixture of product states, which themselves are not. But separable states may contain correlations; and this is what differentiates the pure and mixed states. For example, the state

$$\rho = \frac{1}{2}(|0,0\rangle\langle 0,0| + |1,1\rangle\langle 1,1|) \tag{4.17}$$

fulfils $\langle \sigma_z^A \otimes \sigma_z^B \rangle = 1$ whereas $\langle \sigma_z^{A,B} \rangle = 0$. These correlations are, however, very trivial. If we had a classical system we could also have them. Only entangled states may display non-classical correlations.

A subtle point is that a state may look entangled even though it is separable. Let us take, for instance,

$$\sigma = \frac{1}{2}(|\Phi^+\rangle\langle\Phi^+| + |\Phi^-\rangle\langle\Phi^-|). \tag{4.18}$$

According to this formula, we can prepare σ by mixing two maximally entangled states. However, there is another way of preparing the same state σ that does not involve using entangled states at all. This immediately follows from the fact that $\sigma = \rho$ (one just has to replace the definition of the Bell states in this formula), and thus according to Eq. (4.17) one can prepare it by mixing two product states. The state σ is thus separable. This simple example illustrates the difficulty of finding out whether a state is entangled or not. We have to check all possible decompositions: only if none of them involves product states will we have an entangled state. Unfortunately, there exist infinitely many decompositions, so that this task is hopeless. Fortunately, in some special cases there are shortcuts which can give us the right answer with much less effort.

Entanglement witnesses An entanglement witness [20] is an observable that detects (witnesses) the presence of entanglement. Given an operator $W = W^\dagger$, we say that it is a witness if for all product states $|a, b\rangle$, $\langle a, b|\, W|a, b\rangle \geq 0$, but W possesses negative eigenvalues. From the definition of a separable state of Eq. (4.16), it is clear that if $\mathrm{tr}(\rho W) < 0$, then ρ must be entangled. Thus, a negative expectation value of a witness indicates the presence of entanglement. Note, however, that the converse is not necessarily true: if the expectation value of a witness is positive, this does not imply that the corresponding state is entangled. However, one can show that for any entangled state there always exists a witness that detects it. Sadly, there is no simple way of finding out such a witness, which makes the problem of detecting entanglement in a mixed state rather non-trivial.

Let us consider some examples. First, for two qubits, $W = 2 - S$, where $S = \sigma_1^A \otimes (\sigma_1^B + \sigma_2^B) + \sigma_2^A \otimes (\sigma_1^B - \sigma_2^B)$ and the sigmas are Pauli operators, is an entanglement witness. This can be shown by noting that for a product state, $|\langle S\rangle| \leq |\langle\sigma_1^B\rangle + \langle\sigma_2^B\rangle| + |\langle\sigma_1^B\rangle - \langle\sigma_2^B\rangle| \leq 2$, since $\langle\sigma\rangle \leq 1$. By choosing the sigmas as in the game in Section 4.2.1, we see that for $\rho(p) = (1-p)1/4 + p|\Phi^+\rangle\langle\Phi^+|$, $\mathrm{tr}[\rho(p)\, W] = 2\sqrt{2}p$, and thus entanglement is detected for $p > 1/\sqrt{2}$.

For our second example, we choose continuous variable systems. In this case, $H_{A,B}$ are infinite-dimensional (and isomorphic to $L^2(R)$). We consider two canonical operators, X and P, for A and another two, Y and Q, for B. They fulfil canonical commutation relations $[X, P] = [Y, Q] = i$. We define $W = S - 2$ with $S = (X - Y)^2 + (P + Q)^2$. For a product state, $\langle S\rangle = \Delta X^2 + \Delta Y^2 + \Delta P^2 + \Delta Q^2 + (\langle X\rangle - \langle Y\rangle)^2 + (\langle P\rangle + \langle Q\rangle)^2 \geq 2$, as a consequence of Heisenberg uncertainty relation $\Delta X^2 + \Delta P^2 \geq 1$ (and similarly for Y and Q). On the other hand, since $[X - Y, P + Q] = 0$, there always exist states for which both quantities are as small as we want, so that W becomes negative. For instance, two mode-squeezed states are detected by this witness.

Partial transposition Another method of detecting entanglement is via the partial transposition [21]. Given ρ, we can always write it in terms of an orthonormal basis $\{|n, m\rangle\}$, as

$$\rho = \sum_{i,j,k,l} \rho_{i,j;k,l} |i, k\rangle\langle j, l|. \tag{4.19}$$

We define its partial transpose with respect to A in the basis $\{|n\rangle\}$,

$$\rho^{T_A} = \sum_{i,j,k,l} \rho_{i,j;k,l} |j, k\rangle\langle i, l|. \tag{4.20}$$

For separable states, ρ^{T_A} is a valid density operator (Hermitian and positive), as can be directly shown from Eq. (4.16). However, for entangled states this need not be the case. Thus, if ρ^{T_A} has any negative eigenvalue, the ρ must necessarily be entangled. In particular, for two qubits it can be shown [20] that for all entangled states, the partial transpose is no longer positive semidefinite, which provides us with a very powerful tool for detecting entanglement. For higher-dimensional systems, however, there typically exist entangled states where ρ^{T_A} is positive semidefinite. In these cases, the detection of entanglement can be very complex, as one has to find an appropriate witness.

As an example, let us consider the state $\rho(p)$ defined earlier. A negative eigenvalue of $\rho(p)^{T_A}$ appears whenever $p > 1/3$, and thus this state is entangled in such a case and separable otherwise.

Entanglement measures and mutual information For mixed states, entanglement measures similar to the entanglement entropy can be defined. However, they are much harder to evaluate. For instance, we can consider the distillation procedure as before but now with mixed states [22]. That is, we may try to distill out of n copies of a state ρ the maximal number m of states Φ^+ using LOCC. The ratio m/n in the limit $n \to \infty$ is called distillable entanglement, $D(\rho)$. Analogously, we may define the process of entanglement dilution and define the entanglement cost, $E_c(\rho)$. In general, $D(\rho) < E_c(\rho)$ so that we cannot first distill and then get back the same state as before. Furthermore, there exist very few examples where these quantities can be evaluated. Another measure that can be determined in practice (at least for qubits) is the entanglement of formation [14]

$$E_F(\rho) = \min \sum_i p_i E(\Psi), \tag{4.21}$$

where the minimization is done with respect to all decompositions of ρ (cf. Eq. (4.14)). This quantity is related to the entanglement cost through $E_F(\rho^{\otimes n})/n \to E_c(\rho)$ in the limit $n \to \infty$.

Another way of measuring the entanglement is through the fidelity with a maximally entangled state, as in Eq. (4.11) but now with $F(\rho) = \max\langle\Phi^+|(U \otimes V)\rho(U^\dagger \otimes V^\dagger)|\Phi^+\rangle$.

Finally, another way of measuring entanglement is by using the definition of partial transposition. One defines the *negativity* [23] as $N(\rho) = \max\left(||\rho^{T_A}||_1 - 1, 0\right)$, where the 1-norm is given by the sum of the absolute values of the eigenvalues (one can also define in terms of the logarithm of such an expression). The negativity can be positive only if we have an entangled state; however, there exist entangled states for which it is zero. Nevertheless, it possesses certain properties that make it very useful to quantify entanglement in a simple way.

Another quantity of interest in the context of quantum information is the quantum mutual information, $I(A:B)$ [1]. This does not measure entanglement, but rather correlations. In fact, it is the finest measure of correlations, in the sense that it detects them even when correlation functions do not. It is defined through

$$I(A:B) = S_A + S_B - S_{AB}. \tag{4.22}$$

Here, S_X is the von Neumann entropy of ρ restricted to system X. Thus, $S_{AB} = -\text{tr}(\rho \log_2 \rho)$, and $S_A = -\text{tr}(\rho_A \log_2 \rho_A)$, where $\rho_A = \text{tr}_B(\rho)$ is the reduced density operator of subsystem A. The mutual information has the following properties:

$$I(A:B) \geq 0; \quad I(A:B) = 0 \Leftrightarrow \rho = \rho_A \otimes \rho_B, \tag{4.23}$$

$$\frac{1}{2}||\rho - \rho_A \otimes \rho_B||_1^2 \leq I(A:B) \leq \log_2(d)||\rho - \rho_A \otimes \rho_B||_1, \tag{4.24}$$

$$I(A:B) \leq I((aA):B) \leq I(A:B) + 2S_a. \tag{4.25}$$

The first line indicates that it is only zero for product states, i.e., when there are no correlations. The second relates it to standard correlations: whenever there are strong correlations, the mutual information is large; the converse is not entirely true, since the dimension of the Hilbert space, d, appears in the expression, which may be very large. Finally, the last line indicates that it decreases whenever we discard a subsystem (in this case a), but it cannot decrease by more than twice the entropy of such a system. The mutual information can be interpreted as the amount of qubits we must erase in order to obtain a product state [24]. The proof of all those properties (except for the second one [1]) easily follows from the strong subadditivity of the von Neumann entropy: given three subsystems, X, Y, and Z, $S_{XY} + S_{XZ} \geq S_{XYZ} + S_X$. If we choose $Y = A$, $Z = B$, and X a system that is disentangled, we immediately obtain the first property. By choosing a $X = A$, $Y = B$, and Z such that the whole state Ψ is pure and the reduced state in AB is our state ρ, we obtain $S_{AB} \geq S_A - S_B$, which can be used to prove the third property: taking $Z = a$ instead, we obtain the first inequality, whereas using $S_{AaB} \geq S_{AB} - S_a$ and $S_a + S_A \geq S_{Aa}$, one readily obtains the second one.

4.3.2 Multipartite systems

The description of multipartite entanglement must still confront several challenges for mixed states. While the definitions of product and separable states are straightforward, now we have to consider again different partitions in order to characterize this eluding property. So, given a partition of all subsystems in disjoint sets, we say that this partition is entangled if we cannot write the state as a mixture of product states along each set (but still entangled within each set; compare with multipartite pure states, and bipartite mixed states). We end up with a table in which, for each partition, we state whether the state is entangled or not. To check the entanglement for each partition, we have to find the appropriate witness, whose definition very naturally follows that for bipartite systems. Such a table contains some redundancies since, for instance, if a tripartite state is separable with respect to the partition $(A)(B)(C)$ it is automatically so for any other partition. However, not all the implications that occur for pure states concur for mixed ones. For instance, there can be tripartite states that are entangled with respect to the partitions $(AB)(C)$, $(AC)(B)$, but not for $(BC)A$ [25]. This simply means that the state may be prepared by acting together (i.e., letting them interact) on A and B, and independently on C (plus classical communication), or, alternatively, by acting together on A and C, and independently on B. But never if A is not allowed to interact with B or C.

As measures of entanglement, one can use the ones defined for mixed bipartite states properly extended to many subsystems, or those for multipartite pure states, like the localizable entanglement. However, one typically finds other measures that are more appropriate to describe specific experimental situations. In the following, we briefly review *spin squeezing* [26], which plays an important role in certain precision measurements.

Spin squeezing We consider a set of N qubits, although it is simple to extend it to d-level systems. We denote by $\vec{S} = (S_x, S_y, Sz)$ the collective spin operators

$$S_\alpha = \frac{1}{2}\sum_{n=1}^{N} \sigma_\alpha^n. \tag{4.26}$$

These operators fulfil angular momentum commutation relations $[S_\alpha, S_\beta] = 2\mathrm{i}\epsilon_{\alpha,\beta,\gamma} S_\gamma$ (with ϵ the antisymmetric tensor and where a sum over γ is understood). We define the spin squeezing parameter;

$$\xi = \frac{N\Delta S_z^2}{\langle S_x\rangle^2 + \langle S_y\rangle^2}. \tag{4.27}$$

It can be shown that whenever $\xi < 1$, the state must be entangled [27]. This quantity cannot be written in terms of an entanglement witness, since it is not linear in the state. However, one can easily define it in terms of a witness for two copies of the state.

For separable states, $\xi \geq 1$. This inequality is saturated when all the qubits are in the same pure state along the XY plane in the Bloch sphere, for instance in $|+\rangle$. Actually, the Bloch sphere can help us to get an intuitive picture of the meaning of this quantity. There, given a state, we represent the possible values of the measurement of the S_α by an ellipsoid centred at $\langle \vec{S}\rangle$, with its axis given by the corresponding variances $\Delta S_x, \Delta S_y$, and ΔS_z. For the state $|+\rangle^{\otimes N}$, we will have a circle in the YZ plane centred at the point $(N/2, 0, 0)$ and with radius $\sqrt{N/4}$, yielding $\xi = 1$. This quantity may decrease if we deform the circle into an ellipse stretched along the Y-direction and squeezed along Z. In that case, we can decrease the value of ΔS_z

while keeping $\langle S_x \rangle$ practically constant (in reality it will decrease, but since it is close to $N/2 \gg 1$ this will not affect ξ much, as long as the deformation is small). Thus, $\xi < 1$ corresponds to states that are close to pure product states (i.e., they are close to the surface of the Bloch sphere in the XY plane) but for which the variance of S_z is reduced with respect to product states (at the expense of increasing some other variance). In fact, if N is large and we are close to the Bloch sphere, we can use the formalism of Holstein and Primakoff [28] to re-express the situation in terms of two canonical variables. Let us assume that we work with states for which $\langle S_x \rangle \simeq N/2$; then we can treat it as a c-number (i.e., neglect its quantum fluctuations, which will be very small compared with its expectation value), and define two operators $X = 2S_y/N$, and $P = 2S_z/N$. The commutation relation $[X, P]$, when acting on the states we consider can be replaced by $[X, P] \simeq \mathrm{i}$, and thus we are left with two canonical operators, which fulfil the uncertainty relation $\Delta X^2 + \Delta P^2 \leq 1$. States fulfilling $\xi < 1$ are now states fulfilling $\Delta P^2 < 1/2$, i.e., the so-called squeezed states.

Particle versus mode entanglement We close this section by illustrating something already mentioned in the introduction; namely, that the property of entanglement depends on how we define our subsystems. Consider bosonic particles, each of them possessing two levels, $|0\rangle$ and $|1\rangle$. We can first consider the Hilbert space $H = H_A \otimes H_B \otimes \ldots$, where H_X defines the space for each particle. Obviously, since we have bosons, not all states in H are relevant, but just those which are symmetric under exchange of any pair of particles. On the other hand, we can consider Fock space $H_\mathrm{F} = H_0 \otimes H_1$, where H_0 contains states $|n\rangle_0$ with n particles in state $|0\rangle$ (and similar for H_1). Again, if we have a fixed number of particles, not all the states will be relevant. In the first (second) case we say that we write the states in first (second) quantization. Let us consider two states written in both languages:

$$\begin{aligned} |\psi_1\rangle &= \frac{1}{\sqrt{2}}(|0\rangle_A \otimes |1\rangle_B + |1\rangle_A \otimes |0\rangle_B) = |1\rangle_0 \otimes |1\rangle_1, \\ |\psi_2\rangle &= \frac{1}{\sqrt{2}}(|0\rangle_A + |1\rangle_A) = \frac{1}{\sqrt{2}}(|1\rangle_0 \otimes |0\rangle_1 + |0\rangle_0 \otimes |1\rangle_1), \end{aligned} \tag{4.28}$$

The first state is entangled in the first quantization but a product in the second, while the second one is the other way round. In fact, in the second one there is only one particle, but we can still have entanglement if we express the states in terms of modes and thus divide our space into that of the first and that of the second mode. This illustrates that, to speak properly about entanglement, we have to specify the subsystems that we are considering.

4.4 Entanglement and area laws

So far we have been dealing with many-body states and their entanglement properties, regardless of whether they are in equilibrium or not. We did not specify any interactions, or temperature, but just defined some of the tools that are required to study entanglement for any state, in equilibrium or not. In this section, we concentrate on many-body states in thermal equilibrium that interact with short-range interactions in lattices. This corresponds to some of the most interesting situations in experiments with cold atoms and other condensed-matter systems. The requirement that we concentrate on lattices is in order to avoid some of the mathematical inconveniences of working in the continuum, although some of the ideas reviewed here may be extended to the continuum simply by taking the lattice constant a to zero.

Entanglement in many-body systems has been and is an active area of theoretical research in the last years. Of particular interest has been the behaviour of such a property along phase transitions. We will first very briefly review some of those results, and then concentrate on a property, the area law, that seems to be fulfilled by all systems in equilibrium, as we will explain, and thus strongly characterizes many-body quantum states that appear in nature. We will consider systems at both zero and finite temperature. In the first case, we will use the entanglement entropy and a bipartition of our system. In the latter, stronger statements can be made, where we will use the quantum mutual information.

We consider a spin system on a lattice in d spatial dimensions. Some of the ideas can be extended to bosonic and fermionic systems; nevertheless, for the sake of simplicity we will use the spin language and comment on the possible extensions to fermions. Spins interact according to a Hamiltonian, H, which we will assume only involves few spins (typically only two, i.e., we will have two-body interactions) close to each other. In other words, we will consider finite-range interactions only. We will also be interested in the thermodynamic limit where the number of spins $N \to \infty$. We consider thermal equilibrium states at temperature T, i.e., described by the density operator

$$\rho_T = \frac{\mathrm{e}^{-H/T}}{Z}, \tag{4.29}$$

where $Z = \mathrm{tr}(\mathrm{e}^{-H/T})$ is the partition function (and we have used units with $k_{\mathrm{B}} = 1$). At zero temperature, the state ρ reduces to a projector onto the ground subspace, i.e., that fulfilling $H|\Psi_i\rangle = E_0|\Psi_i\rangle$, where E_0 is the ground-state energy. If the ground state is not degenerate (which we will assume on many occasions), we will just use a pure state $|\Psi_0\rangle$ to denote the ground state.

4.4.1 Entanglement and phase transitions

In thermal equilibrium, the state of Eq. (4.29) will depend on external parameters driving the Hamiltonian, which we will denote with a single letter B, as well as on the temperature. B could be an external magnetic field, interaction strengths, etc. Thus, we can write $\rho(B, T)$. As we change one of those external parameters, the state may change its properties abruptly in the thermodynamic limit. Strictly speaking, some of the observables may depend non-analytically on some of those parameters. At the point of non-analyticity, we say that the system undergoes a phase transition. Particularly interesting transitions may occur at $T = 0$ when we change the external parameter B. In that case, we speak about a *quantum phase transition*, since quantum fluctuations are now responsible for the transition.

If we study the reduced state of two particles, σ, as we undergo a (quantum) phase transition, we may experience an abrupt behaviour in its entanglement [29; 30]. The reason can be easily understood as follows. Imagine that we use a measure of entanglement (like the ones introduced for bipartite mixed states, Section 4.3.1), say $E[\sigma(B)]$. Then, if a correlation function presents a non-analyticity with respect to B, this will also show up as a non-analyticity in the entanglement. The reason is that σ is an analytic function of all correlation functions, and thus whenever the latter is not analytic, then neither is σ and thus E. For example (schematically),

$$\frac{\mathrm{d}E[\sigma(B)]}{\mathrm{d}B} = \frac{\mathrm{d}E}{\delta\sigma}\frac{\delta\sigma}{\mathrm{d}B} \tag{4.30}$$

can become discontinuous whenever $\mathrm{d}\sigma/\mathrm{d}B$ is, or even when the measure of entanglement is not analytic. In other words, if E is not analytic at some point then it is because of the presence of a quantum phase transition, or because the definition of the entanglement measure is not analytic at that point. Thus, in some sense, looking at the entanglement of two subsystems does not really give more information about the phase transition than looking at two-body correlations. One has to look at the whole many-body system.

There exist two methods that look at the whole many-body state to study phase transitions. The first is based on looking at quantities like $F(B_0) = \lim_{\epsilon\to 0} |\langle\Psi(B_0+\epsilon)|\Psi(B_0)\rangle|$, properly re-scaled, since they display special features whenever there is a quantum phase transition [31; 32]. The other method consists of using the localizable entanglement introduced in Section 4.2.2. There, for instance, hidden orders can be identified [33].

4.4.2 Area laws at zero temperature

In this section, we will consider $T = 0$ and a pure ground state of H. We consider a connected region A of the lattice with a smooth boundary and the complementary region B and concentrate on the entropy of entanglement, E_A, between those two regions and its growth as we make region A larger and larger. According to Eq. (4.7), this is given by the von Neumann entropy of the reduced density operator corresponding to region A, ρ_A. In general, since the entropy is an extensive quantity, one would expect that it scales with the number of spins in region A. In fact, this is the case if we take a random state on the lattice (according to the appropriate measure). However, for ground states of Hamiltonians, as we are considering here, this seems not to be the case. Instead, the entanglement scales with the number of particles at the border of region A, $N_{\partial A}$. In a sense, the entanglement scales not with the volume of region A but with the area; thus the name area law [34].

Let us consider a simple example, the Majumdar–Ghosh [35] Hamiltonian

$$H = \sum_i \left(\vec{S}^i \cdot \vec{S}^{i+1} + \frac{1}{2}\vec{S}^i \cdot \vec{S}^{i+2}\right), \tag{4.31}$$

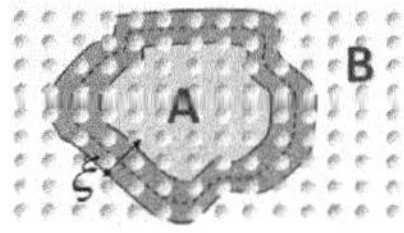

Figure 4.2 Only particles that are at a distance smaller than the correlation length, ξ, are correlated with (or entangled to) the particles in B. Thus, only they contribute to the entanglement entropy and therefore the area law, $E_A \leq \xi N_{\partial A}$.

where the spins are qubits and N is even. Each term in the sum can be written as a projector onto the subspace of two qubits with total spin 1 plus an irrelevant constant. Thus, the ground state is a dimerized state, where singlet states $|\Psi^-\rangle$ are formed between nearest neighbours. In fact, it is doubly degenerate: one can have dimers formed between spins $2i-1$ and $2i$ or between $2i$ and $2i+1$. If we choose a set of neighbouring spins as region A, the entropy of entanglement is bounded by 2, since at most two singlets will contribute to it. As we see, this limit does not increase with the size of A, and the border of A has a fixed number of spins (two), which does not increase either.

The intuition behind the area law is displayed in Figure 4.2. Note that for continuous systems, the entropy is generally unbounded, since $N_{\delta A} \to \infty$ when we take the lattice constant a to zero. Nevertheless, by scaling it properly, an area law also arises.

In the last few years, the area law has been studied in a variety of systems. Here is a summary of some of the results [4; 5]. In one-dimensional chains, it has been proven that all gapped systems fulfil the area law [36]. For gapless systems, which can be described in terms of a conformal field theory, the area law is slightly violated: $E_A \leq \frac{c}{3}\log_2(\ell)$, where $\ell = L/a$ is the number of spins in region A, and c the central charge of the conformal theory [37; 38]. Interestingly enough, this is a universal law, since the coefficient in front of the logarithmic violation is a constant independent of the lattice constant. Many other models have been studied both analytically and numerically, and no violation (beyond the logarithmic one) has been found, at least for all reasonable models (i.e., homogeneous, finite range interactions, etc.).

In higher dimensions, the situation is not so clear. It has been conjectured, and it seems to be corroborated by all examples studied so far, that the area law is fulfilled for gapped systems. For critical systems, there is at most a logarithmic correction [39; 40], even though for some of them one can show that an area law is not violated [41]. In two dimensions, we can thus conjecture $E_A = c_1\ell\log\ell + c_2\ell + c_3\log^m\ell + \gamma + \ldots$, where $\ell = L/a$ and L the length of the boundary, $c_1 \neq 0$ for certain critical systems (such as free fermions), and $c_3 \neq 0$ for the rest. All coefficients c_i are now not universal, since they change if we change the lattice constant. However, γ can indeed be universal and, in fact, it can be connected to the existence of topological properties [42; 43].

As an illustration, we calculate the entanglement entropy for a set of N free bosons in one dimension. We consider R lattice sites, and a Hamiltonian for free Bosons with periodic boundary conditions,

$$H = -t\sum_{r=1}^{R}\left(a_r^\dagger a_{r+1} + a_{r+1}^\dagger a_r\right), \tag{4.32}$$

where a_r are annihilation operators with standard commutation relations $[a_r, a_s^\dagger] = \delta_{r,s}$ and with $a_{R+1} = a_1$. The ground state is given by

$$|\Psi_0\rangle = \frac{1}{\sqrt{N!}}\left(\frac{1}{\sqrt{R}}\sum_{r=1}^{R} a_r\right)^N |\mathrm{vac}\rangle, \tag{4.33}$$

with $|\mathrm{vac}\rangle$ the state with 0 bosons. We now separate the R sites into the first $L \ll R$ and the rest, and are interested in the mode entanglement (see Section 4.3.2) between these two subsystems. We thus separate the whole Hilbert space into the tensor product of the Fock space corresponding to the first L modes, and the rest. In this representation, we can write

$$|\Psi_0\rangle = \sum_{n=0}^{N}\sqrt{\binom{N}{n}}\left[\frac{L}{R}\right]^{n/2}\left[\frac{R-L}{R}\right]^{(N-n)/2}|n, N-n\rangle. \tag{4.34}$$

From this expression, it is simple to calculate the reduced density operator for the first subsystem, $\rho = \sum_{n=0}^{N} p(n)|n\rangle\langle n|$. For $N \gg 1$, we have

$$E_L \simeq -\int_{-\infty}^{\infty} p(n)\log_2[p(n)]\mathrm{d}n, \tag{4.35}$$

with

$$p(n) \simeq \frac{1}{\sqrt{2\pi\sigma^2}}\mathrm{e}^{(n-n_0)^2/2\sigma^2}, \tag{4.36}$$

$n_0 = NL/R$, and $\sigma^2 = NL(R-L)/R^2$. Performing the integral, we obtain $E_A = \log_2\sigma + \text{const.}$, which in the limit $N, R \to \infty$ with N/R constant scales as $1/2\log_2 L$.

4.4.3 Area laws at finite temperature

For finite temperature, we can no longer use the entropy to derive an area law since: (i) it does not measure correlations any more; (ii) it is an extensive quantity at any finite T. However, by using the quantum mutual information introduced in Section 4.3.1, we will be able rigorously to derive [44] an area law valid for arbitrarily finite temperature $T \neq 0$, as long as we have finite-range interactions. Note that at zero temperature the mutual information reduces (up to a factor of two) to the entanglement entropy, and thus it provides us with a suitable generalization of such a quantity. Nevertheless, our limit will diverge at zero temperature, as should be expected from the fact that for critical systems one obtains logarithmic corrections to the area law, as explained in the previous section.

We will consider nearest-neighbour interactions, which can always be achieved under finite-range interactions by redefining a spin that accumulates those of region of size equal to the interaction length. We write the Hamiltonian as

$$H = \sum_{i,j} h_{i,j}, \tag{4.37}$$

where $h_{i,j}$ is a Hamiltonian acting on nearest-neighbour spins, and which we assume to be bounded above $||h_{i,j}||_1 < h$ by a constant. Given a region A, as before, and its complement, we want to show that

$$I(A:B) \leq N_{\delta A} \frac{h}{T} \log_2(\mathrm{e}). \tag{4.38}$$

To show this, we will use the fact that the free energy $F(\rho) = \langle H \rangle_\rho - TS(\rho)/\log_2(\mathrm{e})$, where ρ is any valid density operator and S the von Neumann entropy, is minimized by the Gibbs state of Eq. (4.29). In particular, $F(\rho_T) \leq F(\rho_A \otimes \rho_B)$, where $\rho_{A,B}$ are the reduced density operators of regions A and B, respectively. Using the definition of the free energy, we obtain

$$I(A:B) = S_A + S_B - S_{AB} \leq \frac{(\langle H \rangle_{\rho_T} - \langle H \rangle_{\rho_A \otimes \rho_B}}{T} \log_2(\mathrm{e}). \tag{4.39}$$

By noting that the expectation value of each $h_{i,j}$ with ρ_T and $\rho_A \otimes \rho_B$ coincides whenever i and j are both in region A or B, we arrive at the desired expression, Eq. (4.38).

We remark that this area law is generally valid for any dimension and short-range Hamiltonian. In fact, one can also admit sufficiently fast decaying terms in the proof. Note also that for $T \to 0$, the bound diverges, as it should, according to our discussion of critical systems in the previous section.

4.4.4 Area law and correlation length

Section 4.4.2 gave an intuitive picture of the area law in terms of the correlation length (see Figure 4.2). In this section, I will make this connection more precise, again using the quantum mutual information [44]. This result will be independent of the temperature.

As mentioned earlier, $I(A:B)$ measures correlations between regions A and B. Thus, as those regions get further away, the mutual information must decay. Indeed, one of the properties of the mutual information, Eq. (4.25), expressed exactly this fact. If we fix region A and separate it increasingly from region B (see Figure 4.3),

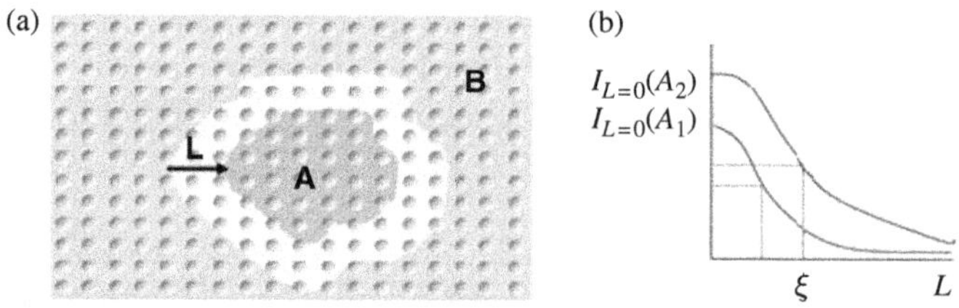

Figure 4.3 (a) We choose two regions A and B separated by a distance L. (b) The mutual information decay as a function of L for each choice of A.

the mutual information will be reduced by a factor of two at some point, $\xi_A \in [1, \infty]$. We define the correlation length ξ_I as the maximum of all ξ_A with respect to all regions A fulfilling these requirements. Note that this quantity may be infinite, as, for instance, in critical systems at zero temperature. Using Eq. (4.25) and the fact that in between regions A and B there are of the order of $N_{\partial A} L$ sites, each of them with a maximal entropy equal to $\log_2(2s+1)$ (s is the total spin), we have

$$I_{L=0}(A) \leq I_{L=\xi_I}(A) + 2N_{\partial A}\xi_I \log_2(2s+1). \tag{4.40}$$

Now using $I_{L=\xi_I}(A) \leq I_{L=0}(A)/2$, we finally arrive at the area law

$$I_{L=0}(A) \leq 4N_{\partial A}\xi_I \log_2(2s+1), \tag{4.41}$$

which gives a finite limit for finite correlation length.

We remark that the upper bound diverges for critical systems, as it should, given the discussion in Section 4.4.2. Note that the converse may not be true: if the correlation length is infinite, we may still have an area law.

4.4.5 Detecting the area law

How can we measure the area law? This seems to be difficult since measuring entropy is not an easy task, given the fact that it does not correspond to any expectation value of a physical observable. However, we may aim at measuring Renyi entropies instead, which are defined as $S_\alpha(\rho) = \log_2(\text{tr}\rho^\alpha)/(1-\alpha)$. In fact, for integer α this corresponds to measuring quantities such as $\text{tr}(\rho^n)$, which can be carried out if several copies of the system are at our disposal. The von Neumann entropy can then be determined by analytic continuation. In any case, one can use the Renyi entropies instead of the von Neumann entropy in most of our previous definitions, and expect similar behaviours.

The quantity $\text{tr}(\rho^2) = \text{tr}[(\rho \otimes \rho)\, T^{\otimes N}]$, where T is the swap operator between each site and its copy partner, i.e.,

$$T = \sum_{n,m} |n, m\rangle\langle m, n|. \tag{4.42}$$

Thus, we just have to measure the observable T on each particle and its copy, multiply the results, and then average over many measurements. For qubits, $T = \vec{S}_1 \cdot \vec{S}_2$ up to a constant, and thus could be measured by letting the qubits interact for a while, according to the Heisenberg Hamiltonian, and then measuring both of them. In a similar way, one can measure the nth Renyi entropy by using n copies. Note that one would expect that the outcome of the measurement would become exponentially small with N, something which should happen for random states. However, in the presence of an area law, this decay with N will be softened and, in one spatial dimension, may allow one to carry out experiments with subsystems containing a large number of sites.

4.5 Tensor network states

Many-body quantum systems are very hard to describe, since the number of parameters required to specify a state scales exponentially with the number of lattice sites. The reason is that the Hilbert space of the whole system is the tensor product of those corresponding to each lattice site, and thus the dimension of that space displays the exponential scaling. In the previous section, we have seen that there is a common property of thermal states in lattice systems with short-range interactions, namely the area law. This fact can guide us in finding an efficient language to describe many-body quantum systems in which the number of parameters only scales polynomially with the number of sites. Under certain conditions, this is indeed possible in terms of so-called tensor network states. In this section, we consider how the area law leads to such descriptions and briefly review some of them.

4.5.1 Area law and projected entangled-pair states

Let us consider the Majumdar–Ghosh model of Eq. (4.31), which clearly illustrates how the area law arises. Let us consider a dimerized ground states, as displayed in Figure 4.4(a). There are two spins in each site, each of them maximally entangled with the neighbouring spins. The entanglement entropy of any region A with the rest, B, equals 2, since only the entangled states at the border of the region contribute to the entanglement, each of them with one unit. If the spin Hilbert space would have dimension D instead of two, then we would have $2\log_2(D)$. Now, imagine that we map each pair of spins at each node into a single spin $s = (d-1)/2$ (d is the dimension of the corresponding space), as indicated in Figure 4.4(b). That is, we apply $P : H_D \otimes H_D \to H_d$, where H_x is a Hilbert space of dimension x. It can easily be shown, that this map may only reduce the entanglement entropy and thus we will have $E_A \leq 2\log_2(D)$, an area law. We can do exactly the same in any spatial dimension. For instance, in a square lattice, we just start with four spins at each node and use a map $P : H_D^{\otimes 4} \to H_d$. Thus, the states are generated by projecting auxiliary entangled pairs onto physical spins. The states so produced are called projected entangled-pair states (PEPS) [45] and obviously fulfil the area law for fixed D. The PEPS are completely

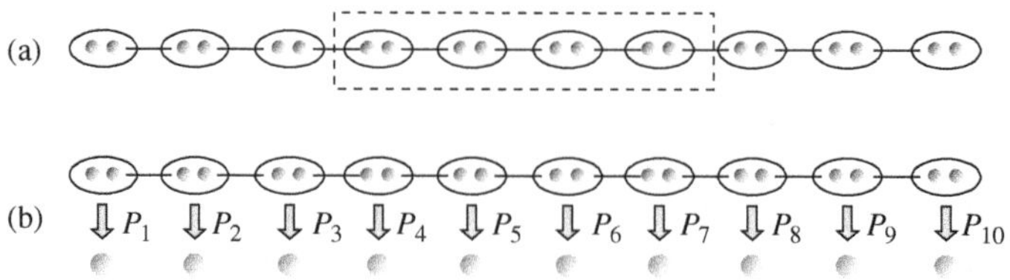

Figure 4.4 (a) Dimerized state: in each node we have two spins, which are maximally entangled with the neighbouring spins. This is represented by a line. (b) Matrix product state construction. Starting from the dimerized state, we map the state of two spins in each site to a single spin. The maps P completely characterize the state.

characterized by the maps P that we have to apply to each node, since the auxiliary states are in maximally entangled states

$$|\Phi^+\rangle = \sum_{n=1}^{D} |n, n\rangle. \tag{4.43}$$

Thus, the number of parameters required to specify a PEPS is NdD^z, where z is the coordination number.

Let us consider a simple example. Let us consider $D = 2$, one spatial dimension, and apply the map $P = |0\rangle\langle 0, 0| + |1\rangle\langle 1, 1|$ on each site. The resulting state is a GHZ state, Eq. (4.12), of N sites.

If we want to approximate the ground state of Hamiltonians with short-range interaction in terms of PEPS, like the ones considered in the previous section, we may have to increase D with N. However, since we fulfil the area law (and may only have logarithmic violations), this increase is expected to be polynomial with N, rendering an efficient description of the ground states [46]. In fact, Hastings has proven [47] that in any spatial dimension and for any finite temperature, PEPS can be used to approximate the Gibbs state efficiently, as long as the interactions are short range.

The PEPS are not restricted to spin systems or pure states. In fact, one can define fermionic PEPS [48], or mixed PEPS [49; 50], and one can extend them even to the continuous limit [51] or to the case where the dimension $D \to \infty$ [52]. Apart from that, different algorithms have been proposed and implemented in order to solve particular Hamiltonians using those families of states variationally (see, e.g., the review [6]).

4.5.2 One dimension: matrix product states

In 1D, this construction gives rise to the so-called matrix product states (MPS) [53; 54]. By writing the map $P_n = \sum A^i_{\alpha,\beta}[n]|i\rangle\langle\alpha, \beta|$, we obtain

$$|\Psi\rangle_N = \sum_{n_1,\ldots,n_N=1}^{d} \operatorname{tr}\left(A^{n_1}[1]A^{n_2}[2]\ldots A^{n_N}[N]\right) |n_1, n_2, \ldots, n_N\rangle. \tag{4.44}$$

Here, $A^n[N]$ represents a $D \times D$ matrix. For translationally invariant systems, we may choose all the matrix independent on the sites, i.e., $A^n[M] = A^n[M'] = A^n$.

Every MPS is invariant under the exchange of $A^n_M \to X_{M-1}A^n_M X^{-1}_M$, where the X are non-singular matrices, as can be checked by direct inspection of Eq. (4.44). This gives us the possibility of choosing a gauge, and thus imposing conditions on the matrices A that simplify the further calculations, or give a physical meaning. In our case, we can impose

$$\sum_{n=1}^{d} A^n[M]^\dagger A^n[M] = 1 \tag{4.45}$$

as a gauge condition that implements such a choice.

Given an MPS, one can easily determine the expectation value of the product of the observables. This task reduces to a calculation of the trace of product of matrices, something that can be implemented very easily. In fact, one can show that all connected correlation functions decay exponentially, and the correlation length may take arbitrary values. One could expect that, given the construction, only near-neighbour correlation exists. However, the projections may be understood as partial teleportations [55], which can give rise to very long range correlations.

The MPS lie at the realm of the density matrix renormalization group method (DMRG) [56], which is widely used to determine the ground state in 1D spin chains. In fact, this method can be understood as a variational calculation with respect to the matrices $A^n[M]$. The success of this method lies in the fact that MPS can efficiently approximate the ground states of 1D chains, as explained here.

4.6 Conclusions

In this chapter, we have analysed several entanglement properties of many-body quantum systems. In the first two sections, we have reviewed the basic concepts regarding entanglement of bipartite and multipartite systems, both for pure (Section 4.2) and mixed (Section 4.3) states, and related them to more standard properties (like correlations) used in quantum many-body theories. As we have seen, whereas entanglement for bipartite systems is well established, for multipartite ones there exist many possibilities of defining entanglement measures. It may turn out that one state is more entangled according to one measure, but not according to a different one. Apart from that, measures of entanglement are very hard to determine in general. Nevertheless, we have highlighted two measures of entanglement or correlations that are simple to determine and have a clear physical meaning. The first one, the entropy of entanglement, applies to bipartite pure states and is given by the von Neumann entropy of the reduced density operator of one of the subsystems. The second one, the quantum mutual information, measures the correlations between the two subsystems and for pure states it reduces to the entropy of entanglement. One can apply these two measures to multispin systems in lattices by separating all the spins into two disjoint regions, which are then considered as a bipartite system. In the particular case where we deal with ground or thermal states of short-range interaction Hamiltonians, the application of those measurements gives rise to area laws, which have been the subject of Section 4.4. These laws state that the quantum mutual information between a region A and its complementary region scales with the number of spins at the boundary of A (and not, as one would expect, with the total number of spins in A). We have also mentioned that possible violations of this law appear for certain critical systems. The violation is, however, very mild and displays universal properties (independent of the lattice constant, as expected for scale-invariant states). This is very peculiar, and provides us with a signature of the many-body quantum states that appear in nature in thermal equilibrium. In fact, the area law can guide us in finding efficient descriptions of thermal equilibrium states, where the number of parameters does not grow exponentially with the volume of the system. In Section 4.5.1, we have made use of this fact and introduced certain tensor network states, the projected entangled-pair states, and explained some of their properties. In particular, in 1D they reduced to the matrix product states, which play a very important role in numerical methods widely used in condensed matter physics.

References

1. Nielsen, MA, and Chuang, IL (2000). *Quantum Computation and Quantum Information*, Cambridge University Press, Cambridge.
2. Horodecki, R, Horodecki, P, Horodecki, M, and Horodecki, K (2009). *Rev. Mod. Phys.* **81**, 865.
3. Amico, L, Fazio, R, Osterloh, A, and Vedral, V (2008). *Rev. Mod. Phys.* **80**, 517.
4. Special issue Calabrese, P, Cardy, J, and Doyon, B (eds) (2009). Entanglement entropy in extended quantum systems, *J. Phys. A: Math. Theor.* **42**, N (40).
5. Eisert, J, Cramer, M, and Plenio, MB (2010). *Rev. Mod. Phys.* **82**, 277.
6. Verstraete, F, Cirac, JI, and Murg, V (2008). *Adv. Phys.* **57**, 143.
7. Cirac, JI, and Verstraete, F (2009). *J. Phys. A: Math. Theor.* **42**, 504004.
8. Brassard, G, Broadbent, A, and Tapp, A (2005). *Found. Phys.* **35**, 1877.
9. Bell, JS (1964). *Physics*, **1**, 195.
10. Horn, RA, and Johnson, CR (1985), *Matrix Analysis*, Cambridge University Press, Cambridge.
11. Schumacher, B (1995). *Phys. Rev. A* **51**, 2738.
12. Bennett, CH, DiVincenzo, DP, Fuchs, CA, *et al.* (1999). *Phys. Rev. A* **59**, 1070.
13. Lo, H-K, and Popescu, S (2001). *Phys. Rev. A* **63**, 022301.
14. Wootters, WK (1998). *Phys. Rev. Lett.* **80**, 2245.
15. Dur, W, Vidal, G, and Cirac, JI (2000). *Phys. Rev. A* **62**, 062314.
16. Greenberger, DM, Horne, MA, and Zeilinger, A (1989). *Quantum Theory, and Conceptions of the Universe*, Kluwer Academic, Dordrecht.
17. Bennett, CH, Popescu, S, Rohrlich, D, Smolin, JA, and Thapliyal, AV (2001). *Phys. Rev. A* **63**, 012307.
18. Verstraete, F, Popp, M, and Cirac, JI (2004). *Phys. Rev. Lett.* **92**, 027901.
19. Werner, RF (1989). *Phys. Rev. A* **40**, 4277.
20. Horodecki, M, Horodecki, P, and Horodecki, R (1996). *Phys. Lett. A* **223**, 1.
21. Peres, A (1996). *Phys. Rev. Lett.* **77**, 1413.
22. Bennett, CH, DiVincenzo, DP, Smolin, JA, and Wootters, WK (1996). *Phys. Rev. A* **54**, 3824.
23. Vidal, G, and Werner, RF (2002). *Phys. Rev. A* **65**, 032314.
24. Groisman, B, Popescu, S, and Winter, A (2005). *Phys. Rev. A*, **72**, 032317.
25. Dur, W, Cirac, JI, and Tarrach, R (1999). *Phys. Rev. Lett.* **83**, 3562.
26. Wineland, DJ, Bollinger, JJ, Itano, WM, Moore, FL, and Heinzen, DJ (1992). *Phys. Rev. A* **46**, R6797.
27. Sorensen, A, Duan, L-M, Cirac, JI, and Zoller, P (2001). *Nature* **409**, 63.
28. Holstein, T, and Primakoff, H (1940). *Phys. Rev.* **58**, 1098.
29. Osborne, TJ, and Nielsen, MA (2002). *Quantum Information Processing*, **1**, 45.

30. Osterloh, A, Amico, L, Falci, G, and Fazio, R (2002). *Nature* **416**, 608.
31. Campos-Venuti, L, and Zanardi, P (2007). *Phys. Rev. Lett.* **99**, 095701.
32. You, WL, Li, YW, and Gu, SJ (2007). *Phys. Rev. E* **76**, 022101.
33. Verstraete, F, Martin-Delgado, MA, and Cirac, JI (2004). *Phys. Rev. Lett.* **92**, 087201.
34. Srednicki, M (1993). *Phys. Rev. Lett.* **71**, 666.
35. Majumdar, CJ, and Ghosh, D (1969). *J. Math. Phys.* **10**, 1388.
36. Hastings, MB (2007). *J. Stat. Phys.* **2007**, P08024.
37. Vidal, G, Latorre, JI, Rico, E, and Kitaev, A (2003). *Phys. Rev. Lett.* **90**, 227902.
38. Calabrese, P, and Cardy, J (2004). *J. Stat. Mech.—Theory E.* **0406**, P002.
39. Wolf, MM (2006). *Phys. Rev. Lett.* **96**, 010404.
40. Gioev, D, and Klich, I (2006). *Phys. Rev. Lett.* **96**, 100503.
41. Verstraete, F, Wolf, MM, Perez-Garcia, D, and Cirac, JI (2006). *Phys. Rev. Lett.* **96**, 220601.
42. Kitaev, A, and Preskill, J (2006). *Phys. Rev. Lett.* **96**, 110404.
43. Levin, M, and Wen X-G (2006). *Phys. Rev. Lett.* **96**, 110405.
44. Wolf, MM, Verstraete, F, Hastings, MB, and Cirac, JI (2008). *Phys. Rev. Lett.* **100**, 070502.
45. Verstraete, F, and Cirac, JI (2004). arXiv:cond-mat/0407066.
46. Verstraete, F, and Cirac, JI (2006). *Phys. Rev. B* **73**, 094423.
47. Hastings, MB (2007). *Phys. Rev. B* **76**, 035114.
48. Kraus, CV, Schuch, N, Verstraete, F, and Cirac, JI (2010). *Phys. Rev. A* **81**, 052338.
49. Verstraete, F, Garcìa-Ripoll, JJ, and Cirac, JI (2004). *Phys. Rev. Lett.* **93**, 207204.
50. Zwolak, M, and Vidal, G (2004). *Phys. Rev. Lett.* **93**, 207205.
51. Verstraete, F, and Cirac, JI (2010). *Phys. Rev. Lett.* **104**, 190405.
52. Cirac, JI, and Sierra, G (2010). *Phys. Rev. B* **81**, 104431.
53. Fannes, M, Nachtergaele, B, and Werner, RF (1992). *Comm. Math. Phys.* **144**, 443.
54. Klümper, A, Schadschneider, A, and Zittartz, J (1993). *Europhys. Lett.* **24**, 293.
55. Verstraete, F, and Cirac JI (2004). *Phys. Rev. A* **70**, 060302.
56. White, SR (1992). *Phys. Rev. Lett.* **69**, 2863.

5

Quantum Hall states of ultracold atomic gases

Nigel R. Cooper

T.C.M. Group, Cavendish Laboratory, University of Cambridge,
J.J. Thomson Avenue, Cambridge CB3 0HE, United Kingdom

Chapter Contents

5.1 Introduction

5.1.1 Scope of this chapter

One of the most remarkable characteristics of a Bose–Einstein condensate is its response to rotation. As was first understood in the context of superfluid helium-4 (Donnelly, 1991), a Bose–Einstein condensate does not rotate in the manner of a conventional fluid, which undergoes rigid body rotation. Rather, the rotation leads to the formation of an array of quantized vortex lines. Quantized vortices also appear in the superfluid states of helium-3 and in type-II superconductors in applied magnetic fields,[1] which may be viewed as condensates of pairs of fermionic atoms or electrons.

The achievement of Bose–Einstein condensation in ultracold atomic gases—formed either from the condensation of atomic bosons (Cornell and Wieman, 2002; Ketterle, 2002), or of pairs of atomic fermions (Bloch *et al.*, 2008)—opens up a wide range of new features in the physics of quantized vortices and vortex arrays, allowing access to parameter regimes unlike those accessible in the helium superfluids or type-II superconductors. This can lead to novel properties of the rotating ground states, including the possibility of exotic strongly correlated phases. In recent years, there have been advances in experimental capabilities and in the theoretical understanding of rotating ultracold atomic gases in these unconventional regimes.

The aim of this chapter is to review the theoretical understanding of Bose and Fermi superfluids under rotation (or in an effective magnetic field) while placing this in the context of the current experimental status. I shall focus on theoretical predictions of the novel phases that can appear. One goal will be to establish the connections with the physics of the fractional quantum Hall effect (FQHE) of electrons in semiconductors. As this chapter is intended to be tutorial in nature, I shall concentrate on the qualitative features and simple model calculations to present the essential ideas in as physically transparent a way as possible. Further details, including numerical studies, are described in an extended review article (Cooper, 2008). For a broader discussion of the physics of rotating atomic Bose gases, the reader is also referred to the excellent reviews of (Bloch *et al.*, 2008) and (Fetter, 2009).

I begin with a brief review of the properties of Bose–Einstein condensates subjected to rotation.

5.1.2 Rotating Bose–Einstein condensates

Let us start by considering a simple Bose–Einstein condensate (BEC), formed from a gas of identical bosonic particles of mass M with mean number density $\bar{n}$, and in equilibrium at a temperature T. For a non-interacting gas, the phase transition to a BEC occurs below the critical temperature

$$T_c = \frac{2\pi\hbar^2}{Mk_B}\left(\frac{\bar{n}}{\zeta\left(\frac{3}{2}\right)}\right)^{2/3}. \tag{5.1}$$

[1] The connection between a rotating neutral fluid and a charged fluid in a magnetic field will be clarified later.

In the condensed phase, the thermal de Broglie wavelength, $\lambda_T \sim \hbar/\sqrt{Mk_B T}$ is larger than the mean particle spacing $\bar{a} \sim \bar{n}^{-1/3}$. For the low temperatures needed to form a BEC in typical ultracold atomic gases, the thermal de Broglie wavelength, $\lambda_T \gtrsim \bar{a} \simeq 0.1\text{–}1\,\mu\text{m}$, is much larger than the range of the interatomic potential. The two-particle scattering is, therefore, dominated by s-wave scattering, with a scattering length a_s that is typically of the order of a few nanometres (for ^{87}Rb, $a_s \simeq 5$nm). Thus, an atomic Bose gas is typically weakly interacting, in the sense that $\bar{n}\,a_s^3 \ll 1$, so it is well described as an ideal Bose gas with very small condensate depletion (this is in contrast to superfluid helium, for which the strong interactions cause significant condensate depletion). That said, the interactions are non-zero and are important for many physical properties. In particular, the weak repulsive interactions allow the BEC to behave as a superfluid, with a non-zero critical velocity. Atomic BECs therefore exhibit superfluidity, first explored in the studies of ^{4}He.

Quantized vortex line It was noted by Onsager and Feynman (Donnelly, 1991) that superfluid helium cannot rotate as a conventional fluid. A conventional fluid rotating at angular frequency $\boldsymbol{\Omega}$ has the velocity field of rigid body rotation

$$\boldsymbol{v} = \boldsymbol{\Omega} \times \boldsymbol{r}, \tag{5.2}$$

for which the 'vorticity' of the flow, $\boldsymbol{\nabla} \times \boldsymbol{v}$, is uniform,

$$\boldsymbol{\nabla} \times \boldsymbol{v} = 2\boldsymbol{\Omega}\,. \tag{5.3}$$

If, as is believed to be the case, the superfluid is described by a superfluid wavefunction $\psi_s = \sqrt{n_s}\mathrm{e}^{\mathrm{i}\phi(\boldsymbol{r})}$, then the superfluid velocity is

$$\boldsymbol{v}_s = \frac{\hbar}{M}\boldsymbol{\nabla}\phi\,. \tag{5.4}$$

Hence, the fluid vorticity apparently vanishes:

$$\boldsymbol{\nabla} \times \boldsymbol{v}_s = \frac{\hbar}{M}\boldsymbol{\nabla} \times \boldsymbol{\nabla}\phi = 0\,. \tag{5.5}$$

The last equality follows from the identity that the 'curl of the gradient' of a smooth function vanishes. This overlooks the possibility that the phase ϕ might have line-like singularities (point-like in 2D), around which ϕ changes by an integer multiple of 2π. These are the *quantized vortex lines*. Integrating the vorticity over a 2D surface containing such a singularity gives

$$\int \boldsymbol{\nabla} \times \boldsymbol{v}_s \cdot \mathrm{d}\boldsymbol{S} = \oint \boldsymbol{v}_s \cdot \mathrm{d}\boldsymbol{l} = \frac{h}{M} \times \text{integer}, \tag{5.6}$$

indicating a delta-function contribution to the fluid vorticity on the vortex line. The 'circulation' of the vortex, defined as

$$\kappa \equiv \oint \boldsymbol{v}_s \cdot \mathrm{d}\boldsymbol{l} = \frac{h}{M} \times \text{integer}, \tag{5.7}$$

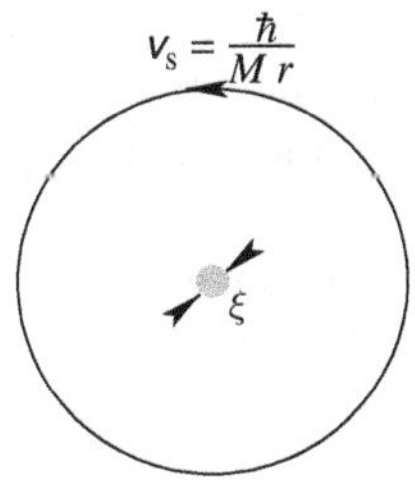

Figure 5.1 Schematic diagram of the velocity field around a quantized vortex line with one quantum of circulation, $\kappa = \frac{h}{M}$. The superfluid density falls to zero within the vortex core, which has a scale set by the healing length ξ.

is therefore quantized in units of $\frac{h}{M}$. This leads to a characteristic velocity profile, with an azimuthal velocity $|\boldsymbol{v}_s| \sim 1/r$ that diverges as $r \to 0$, see Figure 5.1. To avoid the infinite kinetic energy associated with this divergent velocity, in the core of the vortex the superfluid density n_s tends to zero, over a length scale of the *healing length*, ξ.

The healing length is an important characteristic of a superfluid. Within the Gross–Pitaevskii mean-field theory for an interacting Bose condensate, the energy is

$$E - \mu N = \int \left[\frac{\hbar^2}{2M} |\boldsymbol{\nabla}\psi|^2 + \frac{1}{2} g |\psi|^4 - \mu |\psi|^2 \right] \mathrm{d}^3\boldsymbol{r}, \tag{5.8}$$

where ψ is the condensate wavefunction. I have taken contact interactions, with strength

$$g = \frac{4\pi\hbar^2 a_s}{M} \tag{5.9}$$

chosen to reproduce the s-wave scattering length a_s. In many situations, Eq. (5.8) is an accurate microscopic description of an atomic BEC, owing to the small condensate depletion $\bar{n} a_s^3 \ll 1$. For (strongly interacting) superfluid helium, Eq. (5.8) can be interpreted as the Ginzburg–Landau free energy of the superfluid order parameter, replacing ψ by ψ_s and interpreting the coefficients as phenomenological parameters. For a uniform fluid, minimization with respect to $|\psi|^2$ sets the chemical potential to $\mu = g|\psi|^2$. There is, therefore, one characteristic length scale of the equilibrium fluid, set by

$$\frac{\hbar^2}{2M\xi^2} = g|\psi|^2 = \mu \quad \Rightarrow \quad \xi = \sqrt{\frac{\hbar^2}{2M\mu}} = \frac{1}{\sqrt{8\pi\bar{n}\,a_s}}. \tag{5.10}$$

In the case of superfluid helium, for which interactions are strong, the healing length is very short, $\xi \sim 0.8\,\text{Å}$. Thus, the vortex cores have a size of the order of the interparticle spacing. For an atomic gas, the healing length is typically of order $\xi \simeq 0.5\,\mu\text{m}$.

Vortex lattice If superfluid helium is forced to rotate, for example, by cooling liquid helium in a rotating vessel from above the superfluid transition (where the conventional fluid rotates as a rigid body) into the superfluid phase, the superfluid establishes an array of singly quantized vortex lines.

For dilute vortices, spaced by distances $a_{\mathrm{v}} \gg \xi$, the arrangement of vortices is dominated by the influence of the kinetic energy of the superfluid flow (Fetter, 1965). This arrangement is conveniently analysed by an analogy with 2D electrostatics, writing

$$\boldsymbol{v}_{\mathrm{s}} = \frac{\hbar}{M}\boldsymbol{\nabla}\phi \equiv -\frac{\hbar}{M}\hat{\boldsymbol{z}} \times \boldsymbol{\nabla}\Phi, \tag{5.11}$$

which expresses the velocity field in terms of the function Φ, which is a smooth function of position away from the vortex cores. In this parameterization, the condition for steady state flow, $\boldsymbol{\nabla} \cdot \boldsymbol{v}_{\mathrm{s}} = 0$, is automatically satisfied, while the constraint of irrotational flow $\boldsymbol{\nabla} \times \boldsymbol{v}_{\mathrm{s}} = 0$ implies Laplace's equation $\nabla^2\Phi = 0$ away from the vortex cores. To complete the electrostatic analogy, note that a singly quantized vortex in the 2D fluid acts as a charge of strength 2π in the 'electric field', $-\boldsymbol{\nabla}\Phi$, as may be seen from

$$\oint \boldsymbol{\nabla}\phi \cdot \mathrm{d}\boldsymbol{l} = 2\pi \quad \Rightarrow \quad \oint -\hat{\boldsymbol{z}} \times \boldsymbol{\nabla}\Phi \cdot \mathrm{d}\boldsymbol{l} = \oint [-\boldsymbol{\nabla}\Phi] \cdot (\mathrm{d}\boldsymbol{l} \times \hat{\boldsymbol{z}}) = 2\pi, \tag{5.12}$$

i.e., the outward flux of the electric field $-\boldsymbol{\nabla}\Phi$ is 2π. Hence, in the presence of vortices of unit strength, the field Φ satisfies Poisson's equation

$$-\nabla^2\Phi = 2\pi \sum_i \delta^{(2)}(\boldsymbol{r} - \boldsymbol{R}_i), \tag{5.13}$$

where $\boldsymbol{R}_i$ are the positions of the vortices in the plane perpendicular to the rotation axis. Thus, Φ is the 2D electrostatic potential for a set of charges of strength 2π at the locations of the vortices.

For a superfluid forced to rotate with angular velocity $\boldsymbol{\Omega}$, one should find the flow that minimizes the kinetic energy (per unit length $\mathcal{L}$) *in the rotating frame*:

$$\frac{K}{\mathcal{L}} = \frac{1}{2}\rho_{\mathrm{s}} \int |\boldsymbol{v}_{\mathrm{s}} - \boldsymbol{\Omega} \times \boldsymbol{r}|^2 \, \mathrm{d}^2\boldsymbol{r} \tag{5.14}$$

$$= \frac{1}{2}\frac{\hbar^2}{M^2}\rho_{\mathrm{s}} \int \left| -\hat{\boldsymbol{z}} \times \boldsymbol{\nabla}\Phi - \frac{M}{\hbar}\Omega\hat{\boldsymbol{z}} \times \boldsymbol{r} \right|^2 \mathrm{d}^2\boldsymbol{r} \tag{5.15}$$

$$= \frac{1}{2}\frac{\hbar^2}{M^2}\rho_{\mathrm{s}} \int \left| -\boldsymbol{\nabla}\Phi - \frac{M\Omega}{\hbar}\boldsymbol{r} \right|^2 \mathrm{d}^2\boldsymbol{r}, \tag{5.16}$$

where ρ_{s} is the mass density of the superfluid. This is the electrostatic field energy for the sum of the field $-\boldsymbol{\nabla}\Phi$ and a background field $-M\Omega\boldsymbol{r}/\hbar$.

From Eq. (5.16) one can deduce some useful consequences:

(i) Since Φ is the potential for a set of discrete charges of size 2π, it cannot perfectly compensate the background charge density $\boldsymbol{\nabla}\cdot(-M\Omega\boldsymbol{r}/\hbar) = -2M\Omega/\hbar$. However, the condition of electrical neutrality on average fixes the mean vortex density to

$$n_\mathrm{v} = \frac{2M\Omega}{h} \,. \tag{5.17}$$

This is Feynman's result for the mean density of quantized vortices in a rotating superfluid (Donnelly, 1991).

(ii) The vortices, being 2π point charges in 2D, experience logarithmic repulsion, leading to effective vortex–vortex interactions

$$\frac{\rho_s \hbar^2}{M^2}\sum_{i<j} -\frac{(2\pi)^2}{2\pi}\ln\left(|\boldsymbol{R}_i - \boldsymbol{R}_j|/\xi\right), \tag{5.18}$$

where the healing length cuts off the divergence in kinetic energy close to the vortex cores.[2] The ground-state configuration of classical particles interacting with logarithmic repulsion is a triangular lattice. One therefore expects the vortices to arrange in a triangular lattice at low temperature. The observation of vortex lattice arrays in superfluid helium is hampered by the very small size of the vortex core $\xi \sim 0.8\,\text{Å}$. However, in remarkable experiments (Yarmchuk *et al.*, 1979), images of small arrays of vortices were obtained.

Vortex lattices in atomic Bose condensates Vortex lattices may be generated in atomic BECs by confining the gas in a magnetic trap of cylindrical symmetry, and stirring the gas with a rotating deformation (Bloch *et al.*, 2008).[3] Once the rotating deformation has spun the condensate up to a large angular momentum, the deformation can be removed; the gas then comes to equilibrium but now with a large angular momentum, which is conserved, owing to the cylindrical symmetry. Images of the particle density, taken following expansion of the cloud, show large ordered arrays of vortex lines, which appear as regions at which the density is suppressed (see, for example, (Madison *et al.*, 2000; Abo-Shaeer *et al.*, 2001)).

Atomic gases offer possibilities for studying the properties of quantized vortices in regimes that are inaccessible in superfluid helium. In particular, atomic Bose gases can be driven into a regime of high vortex density, where the separation between the vortices a_v becomes less than the healing length ξ.[4] In Section 5.2, I explain that $a_\mathrm{v} \lesssim \xi$ leads to a restriction to single-particle states in the lowest Landau level. These Landau levels are characteristic of the motion of a charged particle in a uniform magnetic field. I shall explain in Section 5.3 how this spectrum can be formed by the used of optically

[2] The total energy also contains single-vortex contributions, arising from the kinetic energy of the superfluid flow around individual vortices and their interaction with 'image' vortices (Fetter, 1965).

[3] The dynamics of vortex lattice formation is an interesting subject, involving surface wave instabilities and turbulent flow. I shall discuss only the equilibrium states.

[4] For superfluid ^{4}He, where the zero temperature healing length is $\xi \sim 0.8\,\text{Å}$, this would require an unrealistically high rotation rate of $\Omega \gtrsim 10^{13}\,\mathrm{rads}^{-1}$!

induced gauge potentials (instead of by rotation). In the remaining sections, I shall describe the consequences of this underlying Landau-level spectrum on the properties of interacting gases of bosons (Section 5.4) and fermions (Section 5.5). The main goal will be to describe some of the interesting strongly correlated phases that can form, and to give an indication of their consequences for experiment.

5.2 Rapid rotation

The central theme that underpins all of the ideas that I shall describe is the appearance of Landau levels in the single-particle energy spectrum of the atomic gas. In this section, I show how this emerges for a harmonically trapped gas under conditions of rapid rotation. I further explain why a weakly interacting atomic Bose gas is restricted to the set of quasi-2D lowest-Landau-level states.

The Hamiltonian for a single particle in a harmonic trap is

$$H^{(1)} = \frac{|\boldsymbol{p}|^2}{2M} + \frac{1}{2}M\omega_\perp^2(x^2+y^2) + \frac{1}{2}M\omega_\parallel^2 z^2 \,. \tag{5.19}$$

The harmonic trap, with natural frequencies $\omega_\perp$ and $\omega_\parallel$, is axially symmetric, so the angular momentum around the z-axis is conserved. We can therefore discuss the spectrum in terms of eigenstates of fixed angular momentum. However, to motivate the form of the single-particle states, it is convenient to work not in terms of fixed angular momentum, but in terms of the conjugate variable, which is the rotation rate, Ω. This is an exactly equivalent formulation. However, it provides a more intuitive view of the essential physics.

We consider the system in a frame of reference rotating about the z-axis with angular momentum $\boldsymbol{\Omega} = \Omega\hat{z}$. In this frame, the Hamiltonian is (Landau and Lifshitz, 1981)

$$H_\Omega = H - \boldsymbol{\Omega}\cdot\boldsymbol{L}\,. \tag{5.20}$$

The Hamiltonian can be written in a suggestive way:

$$H_\Omega^{(1)} = \frac{|\boldsymbol{p}|^2}{2M} + \frac{1}{2}M\omega_\perp^2(x^2+y^2) + \frac{1}{2}M\omega_\parallel^2 z^2 - \boldsymbol{\Omega}\cdot\boldsymbol{r}\times\boldsymbol{p} \tag{5.21}$$

$$= \frac{|\boldsymbol{p} - M\boldsymbol{\Omega}\times\boldsymbol{r}|^2}{2M} + \frac{1}{2}M(\omega_\perp^2 - \Omega^2)(x^2+y^2) + \frac{1}{2}M\omega_\parallel^2 z^2. \tag{5.22}$$

The kinetic term in this Hamiltonian is equivalent to that of a particle of charge q^* experiencing a magnetic field $\boldsymbol{B}^*$ with

$$q^*\boldsymbol{B}^* = 2M\boldsymbol{\Omega}\,. \tag{5.23}$$

This connection shows that the Coriolis force in the rotating frame plays the same role as the Lorentz force on a charged particle in a uniform magnetic field. The radial confining potential is reduced, owing to the centrifugal forces.

Our interest will be in the regimes in which the rotation is sufficiently rapid that $\Omega \simeq \omega_\perp$, so that the residual trapping term is a weak effect. Throughout these notes, I shall assume that $\Omega = \omega_\perp$; any residual trapping will be included within the local density approximation.

In this limit, the energy spectrum for motion in the 2D plane perpendicular to the rotation axis becomes that of a charged particle in a uniform magnetic field, which is a Landau-level spectrum. The total energy is

$$E_{n,m,n_\parallel} = \hbar\omega_{\rm c}(n + 1/2) + \hbar\omega_\parallel(n_\parallel + 1/2), \tag{5.24}$$

where $n = 0, 1, 2, \ldots$ is the Landau level index, and $m = 0, \pm 1, \pm 2, \ldots$ is the projection of angular momentum along the rotation axis (for $\Omega = \omega_\perp$ the energy is independent of m). We have defined the cyclotron energy:

$$\hbar\omega_{\rm c} \equiv \hbar\frac{q^* B^*}{M} = 2\hbar\Omega = 2\hbar\omega_\perp. \tag{5.25}$$

This energy scale will play a central role throughout this chapter. Finally, $n_\parallel = 0, 1, 2 \ldots$ is the oscillator index for motion parallel to the rotation axis.

If the chemical potential μ and temperature are small compared with the level spacing $\hbar\omega_\parallel$, then only $n_\parallel = 0$ is occupied and the particles are in a quasi-2D regime. The density of states is illustrated in Figure 5.2. It is a set of degenerate Landau levels. The degeneracy scales with the area of the system, with a density (per unit area) that is set by the density of magnetic flux quanta

$$n_\phi \equiv \frac{q^* B^*}{h} = \frac{2M\Omega}{h} = \frac{2M\omega_\perp}{h}. \tag{5.26}$$

We recognize this density of degenerate states (per unit area) as the *vortex* density for the Bose gas, Eq. (5.17). In view of this equivalence, I shall use the terms 'vortex density' and 'flux density' interchangeably throughout these notes, and it should be

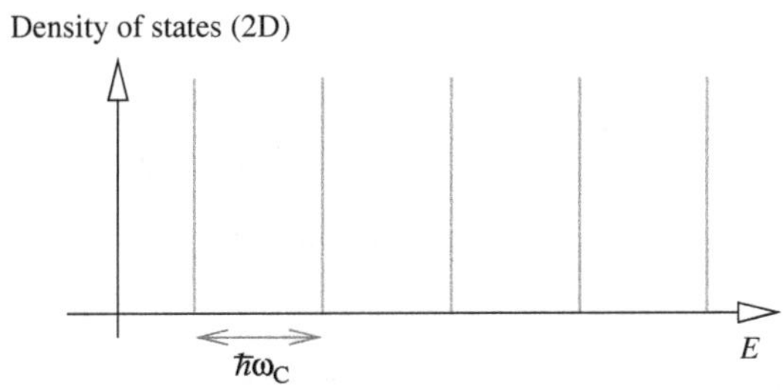

Figure 5.2 Schematic diagram of the density of states for the two-dimensional charged particle in a uniform magnetic field: it is a set of highly degenerate Landau levels, spaced by the cyclotron energy $\hbar\omega_{\rm c}$, each with a degeneracy of n_ϕ states per unit area.

remembered that $n_{\rm v} = n_\phi$. For conventional atomic BECs in a magnetic trap, the trapping frequencies are typically on the order of $\omega_\perp \simeq 2\pi \times 100$Hz. Hence, the maximum vortex/flux density (when $\Omega = \omega_\perp$) is typically

$$n_\phi^{\rm max} = \frac{2M\omega_\perp}{h} \simeq 3 \times 10^{11}{\rm m}^{-2}\,. \tag{5.27}$$

In later work on Bose gases, we shall focus on situations in which the chemical potential, set by the mean interaction energy $g\bar{n}$, is small compared with the single-particle level spacings

$$g\bar{n} \ll \hbar\omega_\parallel,\ \hbar\omega_{\rm c} = 2\hbar\omega_\perp\,. \tag{5.28}$$

These conditions are equivalent to the conditions that the healing length is large compared with the intervortex spacing, $\xi \gg a_\perp$ (for a large vortex array, where $\Omega \simeq \omega_\perp$), and to the subband thickness in the z-direction, $\xi \gg a_\parallel$. Although Eq. (5.28) is typically not satisfied for a non-rotating gas, for a rapidly rotating gas the centrifugal forces spread the cloud out, the density falls, and the system tends towards this weakly interacting regime.

Under the conditions of Eq. (5.28), the single-particle states are restricted to (quasi)-2D and to the lowest Landau level (LLL) (Wilkin *et al.*, 1998). The wavefunctions for these LLL states can be written

$$\psi_m(\boldsymbol{r}) \propto (x + {\rm i}y)^m\, {\rm e}^{-({\rm x}^2+{\rm y}^2)/2{\rm a}_\perp^2}\, {\rm e}^{-{\rm z}^2/2{\rm a}_\parallel^2}, \tag{5.29}$$

where

$$a_{\perp,\parallel} \equiv \sqrt{\frac{\hbar}{M\omega_{\perp,\parallel}}} \tag{5.30}$$

are the trap lengths in the radial and axial directions. To make connections with the fractional quantum Hall effect as clear as possible, I introduce the complex representation

$$\zeta \equiv \frac{x + {\rm i}y}{\ell} \qquad \ell \equiv \sqrt{\frac{\hbar}{M\omega_{\rm c}}} = \frac{a_\perp}{\sqrt{2}}, \tag{5.31}$$

where ℓ is the conventional magnetic length. The 2D LLL basis states are then

$$\psi_m(\zeta) = \frac{1}{\sqrt{2\pi 2^m m!}}\zeta^m \times \left[\frac{{\rm e}^{-|\zeta|^2/4}}{\ell}\frac{1}{(\pi a_\parallel^2)^{1/4}}{\rm e}^{-z^2/2a_\parallel^2}\right]. \tag{5.32}$$

In the following, for simplicity I shall omit the ubiquitous (bracketed) exponential terms in the wavefunctions, focusing only on the prefactors that are polynomial in ζ.

An alternative way to derive the restriction to the 2D LLL for weak interactions is to work in the laboratory frame, and consider the total angular momentum as the control parameter. Even including (rotationally invariant) interactions between the particles, the *total* angular momentum of the many-particle system is conserved. In this frame, one can ask: what is the ground state of the many particle system at total angular momentum L? For the weak interactions of Eq. (5.28), to determine the ground-state wavefunction one must first minimize the kinetic and potential energies. The spectrum of the single-particle Hamiltonian of Eq. (5.19) may be written

$$E = \hbar\omega_\perp(2n_\perp + |m| + 1) + \hbar\omega_\parallel(n_\parallel + 1/2), \tag{5.33}$$

where $m = 0, \pm 1, \pm 2, \ldots$ is the angular momentum (in units of $\hbar$) about the z-axis and $n_\perp, n_\parallel \geq 0$ are the radial and axial quantum numbers. The lowest-energy states have $n_\perp = n_\parallel = 0$. For fixed total angular momentum (in units of $\hbar$),

$$L = \sum_{i=1}^{N} m_i\,; \tag{5.34}$$

the lowest energy states are also those for which $m_i \geq 0$ (Wilkin *et al.*, 1998). The single-particle states with $n_\perp = n_\parallel = 0$ and $m_i \geq 0$ are the 2D LLL states of Eq. (5.29). A collection of N particles with total angular momentum, Eq. (5.34), restricted to the 2D LLL has total single-particle energy

$$E = \sum_{i=1}^{N} \left[\hbar\omega_\perp(m_i + 1/2) + \hbar\omega_\parallel/2\right] = \left(\hbar\omega_\perp + \frac{1}{2}\hbar\omega_\parallel\right) N + \hbar\omega_\perp L\,. \tag{5.35}$$

To see why one requires $m_i \geq 0$, consider moving particle $i = 1$ from a state $m_1 \geq 0$ to a state $m_1{}'$. To conserve the total angular momentum L, one also needs to add $m_1 - m_1{}'$ units of angular momentum to the other particles; keeping these particles in states with $m \geq 0$ leads to an overall change in energy

$$\Delta E = \hbar\omega_\perp\left[(-|m_1| + |m_1{}'|) + (m_1 - m_1{}')\right] = \hbar\omega_\perp\left[|m_1{}'| - m_1{}'\right],$$

which is an energy *increase* if $m_1{}' < 0$.

Determining the ground state of a rapidly rotating atomic gas poses a problem that is very closely related to that which appears in the FQHE. For a system of N particles with a given total angular momentum, L, one must distribute the particles within the 2D LLL orbitals such that the angular momentum, Eq. (5.34), is fixed. The total kinetic and potential energy of these many-particle states, Eq. (5.35), is the same. Therefore, at the single-particle level (i.e., neglecting interactions), there is a very high degeneracy. The true ground state is determined from within this degenerate set of states by the action of the interactions. This is a fundamentally non-perturbative problem, as in the FQHE.

5.3 Optically induced gauge fields

A very interesting recent development has been the use of optical fields to generate synthetic gauge fields. Theoretical proposals have indicated several ways in which this can be achieved: for Abelian and non-Abelian fields, and either in the continuum or in the lattice. Here, I shall focus on a method that has been used by the NIST group to generate an Abelian gauge field (and thereby simulate a uniform synthetic magnetic field) in the continuum. The intention is to provide some intuition on the origin of this effect, and to allow a simple interpretation of the strength of the synthetic gauge field. This presentation is incomplete, certainly from a historical point of view, so the interested reader is referred to the recent review (Dalibard *et al.*, 2011).

5.3.1 Continuum: momentum space

I start by describing a simplified version of the scheme used in the experimental work of the NIST group to imprint a gauge field (Lin *et al.*, 2009a). This involves the use of two-photon transitions to dress the three hyperfine states in the $F = 1$ manifold of ^{87}Rb. However, for simplicity, I shall consider a two-level setting, which illustrates the essential physics.

The two levels, labelled $|1\rangle$ and $|2\rangle$, are coupled by a two-photon transition in a geometry in which the two beams are counterpropagating along the x-direction, as in Figure 5.3(b). Thus, the two-photon transition from state $|1\rangle$ to state $|2\rangle$ involves the absorption of a photon of wavevector $k_1\hat{\boldsymbol{x}}$ and the (stimulated) emission of a photon of wavevector $-k_2\hat{\boldsymbol{x}}$. By momentum conservation, the atom undergoes a change in wavevector (in the x-direction) of

$$\Delta k \equiv k_1 + k_2 \simeq 2k_{\mathrm{r}}\,, \tag{5.36}$$

where $k_1 \simeq k_2 = k_{\mathrm{r}}$ with k_{r} the recoil momentum. Classifying the atomic states also by their wavevector k_x, we see that the state $|1, k_x\rangle$ is coupled to the state $|2, k_x + \Delta k\rangle$.

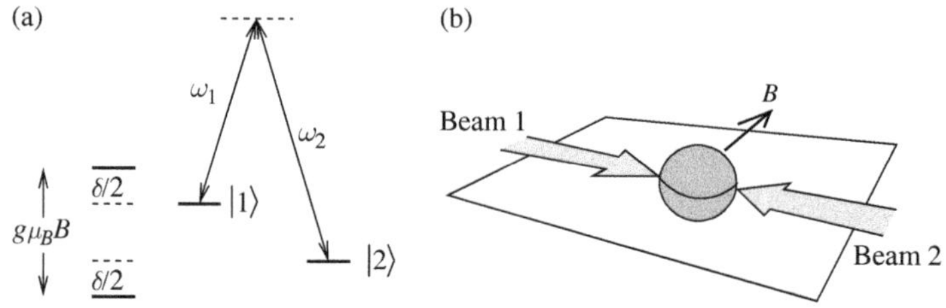

Figure 5.3 Illustration of a method to imprint an optically induced gauge potential on a two-level system. States $|1\rangle$ and $|2\rangle$ are coupled by a two-photon transition, with detuning δ, using counterpropagating beams.

Within the rotating wave approximation, the dressed levels and their energies follow from the eigenvectors and eigenvalues of

$$H_{k_x} = \begin{pmatrix} \frac{\hbar^2}{2M}k_x^2 + \hbar\delta/2 & \hbar\Omega_{\rm R}/2 \\ \hbar\Omega_{\rm R}/2 & \frac{\hbar^2}{2M}(k_x+\Delta k)^2 - \hbar\delta/2 \end{pmatrix}, \tag{5.37}$$

where $\Omega_{\rm R}$ is the two-photon Rabi frequency and $\delta = g\mu_B B/\hbar - (\omega_2 - \omega_1)$ is the detuning from resonance, set by the Zeeman energy in the (real) magnetic field B. In the absence of light, $\Omega_{\rm R} = 0$, the solutions are the uncoupled parabolas indicated by the dotted lines in Figure 5.4. With the light on, $\Omega_{\rm R} \neq 0$, they anticross, leading to the optically dressed states with k_x-dependence shown as solid lines in Figure 5.4. For sufficiently large $\Omega_{\rm R}$, the lower dressed energy band has a simple minimum, close to which it can be approximated by

$$E_0(k_x) \simeq \epsilon_0 + \frac{(\hbar k_x - q^* A_x^*)^2}{2M^*} + \ldots \tag{5.38}$$

The offset in the zero of the momentum $q^* A_x^*$ represents a synthetic vector potential in the x-direction. The value of this offset, which is a function of $\delta/\Omega_{\rm R}$, can be varied between $q^* A_x^* = -\Delta k$ ($\delta/\Omega_R \gg 1$) and $q^* A_x^* = 0$ ($\delta/\Omega_{\rm R} \ll -1$).

The creation of a uniform synthetic vector potential was reported in (Lin *et al.*, 2009b). A uniform vector potential can have no dynamical effect on the system, since

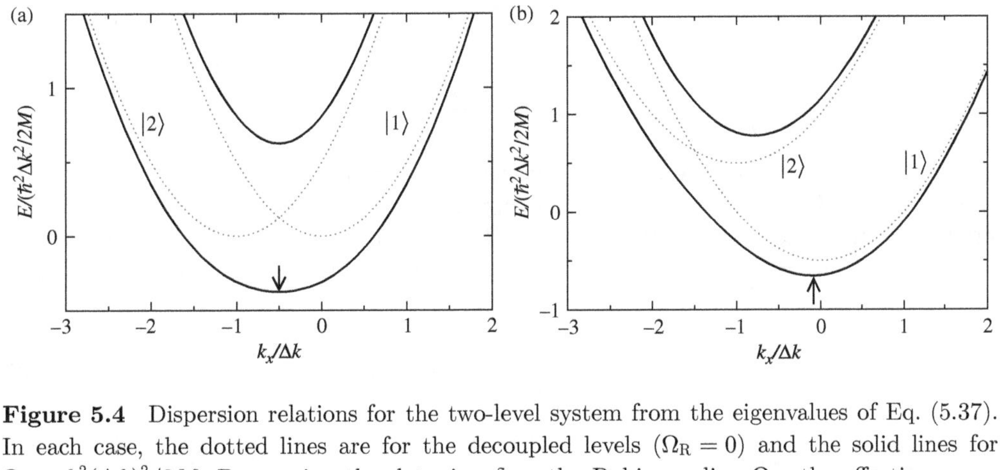

Figure 5.4 Dispersion relations for the two-level system from the eigenvalues of Eq. (5.37). In each case, the dotted lines are for the decoupled levels ($\Omega_{\rm R} = 0$) and the solid lines for $\Omega_{\rm R} = \hbar^2(\Delta k)^2/2M$. By varying the detuning δ or the Rabi coupling $\Omega_{\rm R}$, the effective gauge potential $q^* A_x^*$, defined by the position of the minimum in the dispersion relation for the lowest energy dressed state (marked by the arrows), can be varied: (a) $\delta = 0$ for which $q^* A_x^* = -\Delta k/2$ (by symmetry); (b) $\delta = -\hbar^2(\Delta k)^2/2M$, for which $q^* A_x^* \simeq 0$.

it is curl-free ($\nabla \times \boldsymbol{A} = 0$) so can be removed by a gauge transformation. All physical properties of the trapped gas are gauge-invariant. However, the effect of the gauge potential can be detected in experimental measurements that go beyond this effective Hamiltonian and that 'expose' the inner structure of the dressed state. In particular, the expansion images are gauge-dependent, and measure the *canonical* momentum distribution (Möller and Cooper, 2010).

To have non-trivial effects on the trapped condensate, one must generate a vector potential with non-zero curl. This can be achieved by making $\delta/\Omega_{\mathrm{R}}$ spatially dependent in the y-direction. The approach of the NIST group is to apply a gradient of the (real) magnetic field, which causes the Zeeman splitting of the hyperfine levels to depend on y, and therefore the detuning $\delta(y)$. Then, there is an effective synthetic magnetic field with

$$q^* B^* = -\frac{\mathrm{d}[q^* A_x^*]}{\mathrm{d}y} \propto \frac{\mathrm{d}\delta}{\mathrm{d}y}. \tag{5.39}$$

By increasing the field gradient, the NIST group was able to generate condensates containing quantized vortices (Lin *et al.*, 2009a).

It is instructive to determine the largest (uniform) flux density that can be constructed in this geometry. The flux density

$$n_\phi = \frac{q^* B^*}{h} = -\frac{1}{h}\frac{\mathrm{d}[q^* A_x^*]}{\mathrm{d}y} \tag{5.40}$$

is maximized by maximizing the gradient of $q^* A_x^*$. For simplicity, I consider a system that is uniform along the x-direction, with a fixed overall width w_y in the y-direction. The maximum change in $q^* A_x^*$ is from $-\Delta k$ to 0. The maximum (uniform) flux across the width w_y is achieved by setting $q^* A_x^* = -\Delta k$ on $y = -w_y/2$ and $q^* A_x^* = 0$ on $y = w_y/2$, giving

$$n_\phi^{\mathrm{max}} = \frac{\Delta k}{2\pi w_y}. \tag{5.41}$$

This formula applies for the two-level system we have discussed. More generally, if the hyperfine manifold has spin F, such that there are $2F+1$ coupled states, the maximum change in wavevector is $\Delta k(2F)$ (since $2F$ two-photon transitions are required to change from $m_F = -F$ to $m_F = +F$). Then the maximum change in $q^* A_x^*$ is $\Delta k(2F)$ and

$$n_\phi^{\mathrm{max}} = \frac{(2F)\Delta k}{2\pi w_y}. \tag{5.42}$$

Since $\Delta k \sim 4\pi/\lambda_{\mathrm{r}}$ where λ_{r} is an optical wavelength, the mean vortex separation $a_{\mathrm{v}} = 1/\sqrt{\pi n_\phi} \sim \sqrt{\lambda_{\mathrm{r}} w_y / F}$, i.e., the geometric mean of the optical wavelength and the transverse trap size. For a cloud of dimension $w_y \simeq 10\,\mu\mathrm{m}$ and assuming light with $\lambda_{\mathrm{r}} = 800\,\mathrm{nm}$ coupling three hyperfine levels ($F = 1$), Eq. (5.42) leads to a maximum flux density of $5 \times 10^{11}\,\mathrm{m}^{-2}$. This flux density is comparable to the flux densities achieved in rotating gases in typical magnetic traps, Eq. (5.27).

5.3.2 Continuum: real space

A more general viewpoint on the origin of the gauge potential can be obtained by understanding the optical dressing in *real* space. To discuss this, I follow the notation in (Bloch *et al.*, 2008).

We start from a 'Born–Oppenheimer' approximation, and consider an atom localized in space at a position $\boldsymbol{r}$. The atom is dressed by the local optical fields at this point. I denote the local dressed states by $|n_{\boldsymbol{r}}\rangle$ and their energies $E_n(\boldsymbol{r})$, with $n = 0, 1, 2, \ldots, 2F$. A general state of the atom can be expanded in this local basis

$$|\psi(\boldsymbol{r})\rangle = \sum_n \psi_n(\boldsymbol{r})|n_{\boldsymbol{r}}\rangle. \tag{5.43}$$

The full Hamiltonian for the atom is

$$\hat{H} = \frac{\hat{\boldsymbol{p}}^2}{2M} + \sum_n E_n(\boldsymbol{r})|n_{\boldsymbol{r}}\rangle\langle n_{\boldsymbol{r}}|, \tag{5.44}$$

where the kinetic energy term accounts for effects beyond the Born–Oppenheimer limit. Provided the motion of the particle is slow (adiabatic), in the sense that the typical kinetic energy $\langle \boldsymbol{p}^2\rangle/2M$ is small compared with the energy spacings $E_{n+1} - E_n$, then to a good approximation one can project the Hamiltonian onto states labelled by n alone (ignoring mixing between different dressed states to lowest order in the kinetic energy). The effective Hamiltonian for the nth state, obtained from $\hat{H}_n\psi_n = \langle n_{\boldsymbol{r}}|\hat{H}\psi_n|n_{\boldsymbol{r}}\rangle$, is

$$\hat{H}_n = \frac{(\hat{\boldsymbol{p}} - q^*\boldsymbol{A}^*)^2}{2M} + V_n(\boldsymbol{r}), \tag{5.45}$$

where

$$q^*\boldsymbol{A}^* = \mathrm{i}\hbar\langle n_{\boldsymbol{r}}|\boldsymbol{\nabla} n_{\boldsymbol{r}}\rangle, \tag{5.46}$$

$$V_n(\boldsymbol{r}) = E_n(\boldsymbol{r}) + \frac{\hbar^2}{2M}\left(\langle \boldsymbol{\nabla} n_{\boldsymbol{r}}|\boldsymbol{\nabla} n_{\boldsymbol{r}}\rangle - |\langle n_{\boldsymbol{r}}|\boldsymbol{\nabla} n_{\boldsymbol{r}}\rangle|^2\right). \tag{5.47}$$

These additional terms are readily derived by considering the action of the momentum operator $\hat{\boldsymbol{p}}$ on the local basis state $|n_{\boldsymbol{r}}\rangle$.

Note the appearance of a gauge field. This gauge field is nothing but the Berry connection arising from the (spatial) parallel transport of the local state $|n_{\boldsymbol{r}}\rangle$. The Berry phase accumulated as the atom moves in the xy plane plays the role of the Aharonov–Bohm phase of a charged particle moving in an effective magnetic field.

To establish the equivalence of the Aharonov–Bohm phase of the synthetic gauge field and the Berry phase, let us explore the properties for the two-component system $F = 1/2$. In this case, we can parameterize the local dressed state as

$$|n_{\boldsymbol{r}}\rangle = \begin{pmatrix} \cos(\theta/2) \\ \sin(\theta/2)\mathrm{e}^{-\mathrm{i}\phi} \end{pmatrix}, \tag{5.48}$$

where (θ, ϕ) are the polar and azimuthal angles representing the state of the two-component system on the Bloch sphere. A direct calculation leads to

$$q^* \boldsymbol{A}^* = \hbar \sin^2 \theta/2 \boldsymbol{\nabla}\phi, \tag{5.49}$$

from which one finds

$$q^* \boldsymbol{B}^* \equiv \boldsymbol{\nabla} \times (q^* \boldsymbol{A}^*) = \frac{\hbar}{2} \sin\theta \boldsymbol{\nabla}\theta \times \boldsymbol{\nabla}\phi\,. \tag{5.50}$$

Consider the Aharonov–Bohm phase that is picked up by an atom that traces out the closed path enclosing an area A, as in Figure 5.5. This phase is

$$\frac{1}{\hbar}\oint q^* \boldsymbol{A}^* \cdot \mathrm{d}\boldsymbol{l} = \frac{1}{\hbar}\int_{\text{area } A} \boldsymbol{\nabla} \times (q^* \boldsymbol{A}^*) \cdot \hat{\boldsymbol{z}}\, \mathrm{d}^2 r = \frac{1}{2}\int_{\text{solid angle } \Omega} \sin\theta\, \mathrm{d}\theta\, d\phi, \tag{5.51}$$

which is 1/2 the solid angle Ω that the area A maps to on the surface of the Bloch sphere.[5] This value, $\Omega/2$, is nothing but the well-known Berry phase for a two-level system ($F = 1/2$) that is adiabatically transported around a closed path on the Bloch sphere (Shapere and Wilczek, 1989).

This provides a simple geometric interpretation of the flux density of the synthetic gauge field. An area containing one flux quantum gives an Aharonov–Bohm phase change of 2π. For the two-level system, this area maps to a solid angle of $\Omega = 4\pi$, which is the total surface of the sphere. Thus, the number of flux quanta in a region A is equal to the number of times that the Bloch vector wraps over the sphere in the solid angle Ω derived from A. The generalization to a system with $2F + 1$ levels is straightforward: the Berry phase is $\Omega \times F$, so *the number of flux quanta in a region A is $2F$ times the number of times that the Bloch vector wraps over the sphere* in the solid angle Ω derived from A.

Let us apply this geometrical interpretation to the situation described in the previous section. Recall that we considered a system that is extended in the x-direction,

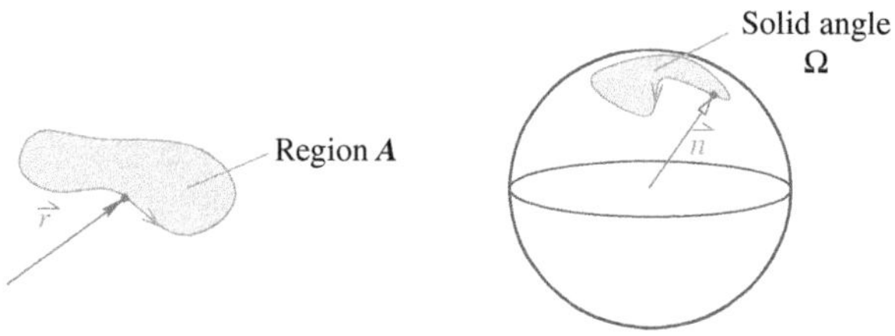

Figure 5.5 In the adiabatic limit, as a particle moves around a closed path in position space (left), the Bloch vector describing the dressed state traces out a closed path on the unit sphere (right). The area A maps to the solid angle Ω. The Berry phase accumulated due to the motion of the Bloch vector appears as an Aharonov–Bohm phase for the motion in real space.

[5]The mathematical relation follows from the fact that $\sin\theta\nabla\theta \times \nabla\phi$ is the Jacobian for the coordinate transformation from (x, y) to (θ, ϕ).

and having a finite width $-w_y/2 \leq y \leq w_y/2$ in the y-direction. To maximize the flux density, we required $\delta/\Omega \gg 1$ on $y = -w_y/2$ and $\delta/\Omega \ll -1$ on $y = w_y/2$. These conditions fix $\theta(-w_y/2) = \pi$ and $\theta(w_y/2) = 0$. In real space, the counterpropagating lasers (in the rotating wave approximation) give rise to a Rabi coupling $\Omega_{\rm R}(x) = \Omega_{\rm R}e^{i\Delta kx}$, such that the phase of the lowest-energy dressed state is $\phi = \Delta kx$. The spatial pattern of the Bloch vector of the lowest-energy dressed state is shown in Figure 5.6.

Note that over the rectangular area of size $(2\pi/\Delta k) \times w_y$, the Bloch vector wraps once over the surface of the sphere. Thus, for a $2F + 1$-level system, this rectangular area contains $2F$ flux quanta. The maximum (average) flux density is therefore

$$n_\phi^{\max} = \frac{(2F)\Delta k}{2\pi w_y}, \tag{5.52}$$

which reproduces the result of Eq. (5.42) derived previously in a momentum space description.

The derivation in real-space makes it clear that, in addition to the gauge potential, there are scalar potentials.[6] Indeed, the spatially varying light-shift on the atoms acts as a potential $V(y)$. Assuming, for simplicity, that $|\Omega_{\rm R}|$ is constant in space, introducing a detuning that varies linearly in space, $\delta = \delta' y$, to generate a non-zero local effective flux density $n_\phi \propto \delta'$ will, in addition, lead to a light-shift that gives an *antitrapping* potential term of the form $V(y) \propto -(\delta')^2 y^2$. Since the coefficient of this antitrapping term scales with the square of the flux density, it plays a similar role to the centrifugal potential in the rotating gases. At large flux densities, this antitrapping potential can be large. Any practical implementation to realize a uniform magnetic field requires this potential to be balanced out by other potentials.

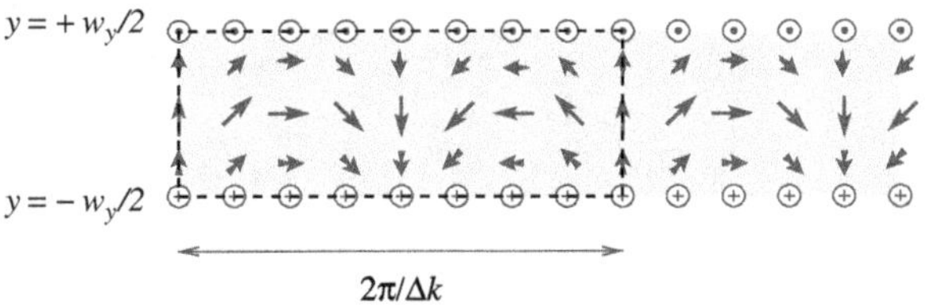

Figure 5.6 Schematic diagram of spatial orientation of the Bloch vector. The sample is extended in the horizontal (x) direction, with width w_y in the vertical (y) direction. On $y = w_y/2$, $\delta/\Omega \ll -1$, so the local dressed state is $|1\rangle$ ($\theta = 0$). On $y = -w_y/2$, $\delta/\Omega \gg 1$, so the local dressed state is $|2\rangle$ ($\theta = \pi$). Between these limits, the Bloch vector rotates from $\theta = 0$ to $\theta = \pi$ via a path with azimuthal angle $\phi = \Delta kx$, set by the local phase of the optical fields. The dashed line marks an area within which the Bloch vector wraps once over the sphere. For a spin-F system this area contains $2F$ flux quanta. Note that the relative spatial scales in x and y are very much distorted. For the system to be in a 2D regime with $w_y \gg 1/\sqrt{n_\phi}$ requires $w_y \gg 1/\Delta k$.

[6] In the momentum-space description, the same conclusion arises from noting that ϵ_0 varies with δ, and hence acquires a y-dependence when $d\delta/dy \neq 0$.

5.3.3 Lattice

This description in real space can be used to understand some general features of proposals to generate synthetic gauge potentials on a lattice. The ideas presented in (Jaksch and Zoller, 2003; Gerbier and Dalibard, 2010) essentially amount to forming a periodic array of the system in Figure 5.6, repeated many times in the y-direction. I replace the distance w_y by the lattice constant a_y, since the overall width of the system can now be many times this scale a_y. An optical lattice is imposed, with lattice constants a_x and a_y, in such a way that the atoms are either in state $|1\rangle$ or in state $|2\rangle$ (and alternating between these two cases in subsequent horizontal rows). See Figure 5.7. Tunnelling along the x-direction preserves the internal state of the atom. Tunnelling between nearest-neighbour lattice sites in the y-direction requires a change of state from $|1\rangle$ to $|2\rangle$ or vice versa. This tunnelling is photon-assisted (a two-photon transition in (Jaksch and Zoller, 2003) or a single-photon process in (Gerbier and Dalibard, 2010)). The optical field stimulating this hopping has a phase variation $\phi(x) = \Delta kx$, along the x-direction, meaning that the tunnelling matrix element in the y-direction from state $|1\rangle$ to $|2\rangle$ acquires a phase $\phi(x)$. This phase has precisely the same origin as in the continuum case, and is denoted in Figure 5.7 by the in-plane vectors.

While tunnelling between lattice sites is not an adiabatic process, the previous geometric interpretation of the accumulated phase on encircling a plaquette still applies: the phase acquired on hopping around any closed plaquette can be determined by the path taken on the Bloch sphere, with the phase of the hopping ϕ fixing the angle of the great circle that connects the poles. Using this, it is then straightforward to see

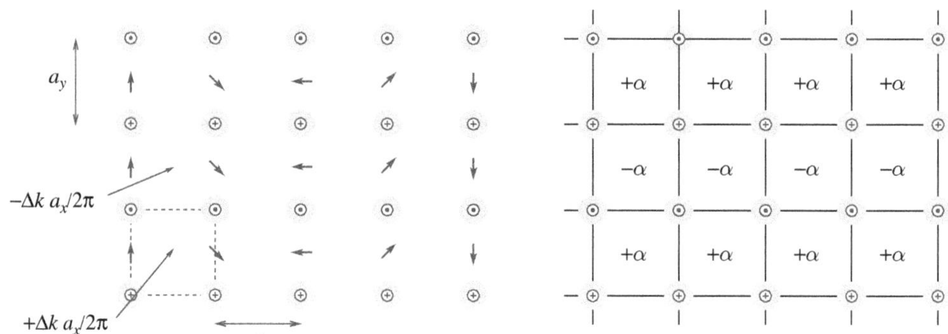

Figure 5.7 (a) Schematic diagram of a proposed geometry to imprint gauge fields on a lattice. The atoms are trapped at the sites of the lattice, alternating in state $|1\rangle$ ($\theta = 0$) or in state $|2\rangle$ ($\theta = \pi$), along the y-direction. Tunnelling along the x-direction involves no change of internal state. Tunnelling in the y-direction requires the atom to be flipped between the two internal states. This tunnelling is optically induced by a laser field, with a phase $\phi = \Delta kx$. The number of flux quanta piercing the plaquette shown by the dashed line is $\alpha = \Delta k a_x/(2\pi)$. The row of plaquettes above (e.g., above the dashed plaquette) are pierced by $-\alpha$ flux quanta. (b) Overall, the system has a staggered flux arrangement.

that for hopping in an anticlockwise direction around a plaquette, as shown by the dashed line in Figure 5.7, the net phase accumulation is $\Delta k a_x$. Thus, this plaquette encloses $\alpha \equiv \Delta k a_x/(2\pi)$ flux quanta. Owing to the constant phase accumulation in the x-direction, all plaquettes in the same row contain the same number of flux α. However, hopping in an anticlockwise direction around a plaquette in the *adjacent* row (above the plaquette, marked by the dashed line in Figure 5.7) gives a path on the Bloch sphere that is exactly reversed, so the phase accumulated is $-\Delta k a_x$ and the plaquette encloses a flux $-\alpha$. Thus, the optically induced hopping leads to a flux that alternates between α and $-\alpha$ on consecutive horizontal rows. The ground states of interacting Bose gases on lattices experiencing these alternating fluxes have recently been discussed in (Möller and Cooper, 2010).

References (Jaksch and Zoller, 2003; Mueller, 2004; Gerbier and Dalibard, 2010) introduce inventive ways of rectifying these fluxes such that a uniform magnetic field can be generated (albeit at the cost of additional potentials and lasers). I do not describe these methods here, but refer the reader to the original papers and the reviews of (Bloch *et al.*, 2008; Dalibard *et al.*, 2011). With these methods in place, the gauge field that can be generated on the lattice can have a flux per plaquette, $|\alpha|$, which is up to its maximum value of 1/2. (Gauge invariance means that the physical properties are unchanged under $\alpha \to \alpha + 1$, so we can restrict to $-1/2 \leq \alpha < 1/2$.) Thus the maximum flux density is

$$n_\phi^{\max} = \frac{1}{2 a_x a_y}. \tag{5.53}$$

Since the a_x and a_y are on the scale of (half) an optical wavelength π/k_r, one has $n_\phi^{\max} \simeq 1/\lambda_\mathrm{r}^2$. This flux density is significantly larger than the flux density achievable in the continuum scheme described in Eq. (5.41). For $a_x = a_y = 400\,\mathrm{nm}$, the flux density is $n_\phi^{\max} = 3 \times 10^{12}\,\mathrm{m}^{-2}$, which is an order of magnitude larger than the flux densities achieved in the continuum scheme described previously.[7]

5.4 Bose gases

In the previous sections, I have established how rapid rotation, or an optically induced gauge field, can give rise to an effective uniform magnetic field. It was explained that for a weakly interacting Bose gas, satisfying Eq. (5.28), the particles are restricted to single-particle states in the 2D LLL, Eq. (5.29). The many particle states within the 2D LLL are highly degenerate. The true ground state is determined from within this degenerate set of many particle states by the action of the interactions

$$\hat{H}_\mathrm{I} = \sum_{i<j} V\left(\boldsymbol{r}_i - \boldsymbol{r}_j\right). \tag{5.54}$$

[7] While preparing this course the author devised a way to adapt the continuum approach to generate flux densities of $n_\phi^{\max} \sim (2F)/\lambda_\mathrm{r}^2$ (Cooper, 2011).

For the most part, I shall choose contact interactions

$$V(\boldsymbol{r}) = g\delta^{(3)}(\boldsymbol{r}) \tag{5.55}$$

with $g = \frac{4\pi\hbar^2 a_s}{M}$. However, I shall also consider the effects of dipolar interactions.

5.4.1 Gross–Pitaevskii mean-field theory

The Gross–Pitaevskii mean-field theory for an interacting Bose gas amounts to the assumption that the many-body state is a pure condensate, in which all particles occupy the same single particle state $\psi(\boldsymbol{r})$

$$\Psi(\{\boldsymbol{r}_i\}) = \prod_{i=1}^{N} \psi(\boldsymbol{r}_i)\,. \tag{5.56}$$

As an approximation to the ground state, the (normalized) condensate wavefunction ψ is chosen to minimize the expectation of the Hamiltonian. (I shall discuss the limits of applicability of this ansatz later.)

For a rapidly rotating Bose gas the single-particle orbitals consist only of the states in Eq. (5.29). Thus, the condensate wavefunction can be expanded in terms of these states

$$\psi(\zeta, z) = \sum_{m\geq 0} c_m \psi_m \tag{5.57}$$

for $\sum_m |c_m|^2 = 1$. Noting that the condensate wavefunction, Eq. (5.57), is a polynomial in ζ, it may be expressed in terms of its zeros as

$$\psi(\zeta, z) = A \prod_{\alpha=1}^{m_{\max}} (\zeta - \zeta_\alpha)\,, \tag{5.58}$$

where $m_{\max}$ is introduced as a cut-off in the degree of the polynomial of Eq. (5.57), and A is a normalization factor. These zeros are the complex positions of the quantized vortices. Thus the LLL Gross–Pitaevskii wavefunction is fully described by the positions of the vortices.

The mean-field ground state is obtained by choosing the coefficients c_m to minimize the expectation value of the interaction energy, which for contact interactions, Eq. (5.55), is

$$\frac{1}{2} g N^2 \int |\psi(\boldsymbol{r})|^4 \, \mathrm{d}^3\boldsymbol{r}.$$

For a system of finite size, one must minimize the energy with a constraint on the average angular momentum

$$L = N \sum_m m|c_m|^2 , \tag{5.59}$$

leading to a mean-field ground state that depends only on L/N. The minimization for contact interactions can readily be implemented numerically. For large L/N the number of vortices grows as $N_\mathrm{v} \simeq 3L/N$, and the system exhibits a triangular vortex lattice (weakly distorted by the confinement), see Figure 5.8.

The appearance of this triangular vortex lattice is very closely related to the appearance of a triangular lattice of flux lines in type-II superconductors close to the upper critical field. In that case, the order parameter is determined by the Ginzburg–Landau energy functional (per unit length along the field direction)

$$\frac{F - F_n}{L} \sim \int \left[\frac{1}{2M} |(-\mathrm{i}\hbar\boldsymbol{\nabla} + 2e\boldsymbol{A})\psi_s|^2 + \alpha|\psi_s|^2 + \frac{\beta}{2}|\psi_s|^4 \right] \mathrm{d}^2\boldsymbol{r}. \tag{5.60}$$

Close to $H_{\mathrm{c}2}$ the order parameter is small, so to a first approximation one can neglect the quartic term and minimize the quadratic terms. These describe electron pairs (hence charge $-2e$) in a uniform magnetic field: the lowest-energy states are the (degenerate) lowest Landau level orbitals. As explained by Abrikosov (1957), the ground state is selected as the linear combination of the lowest Landau level orbitals that minimizes the quartic term. This is mathematically equivalent to the problem described here, at least in the uniform case (i.e., close to the centre of a trap containing many vortices); the ground state in both cases is a *triangular* vortex lattice.

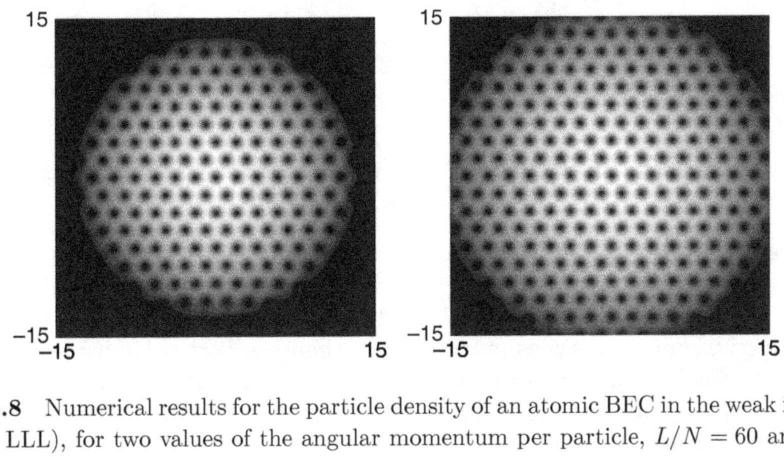

Figure 5.8 Numerical results for the particle density of an atomic BEC in the weak interaction limit (2D LLL), for two values of the angular momentum per particle, $L/N = 60$ and 90. The lattice structure is determined only by the interactions, which are here chosen to be contact interactions, Eq. (5.55). Close to the centre of the trap the vortices form a regular triangular lattice. Reproduced with permission from (Cooper *et al.*, 2004).

For both $\xi \ll a_\perp$ (dilute vortices) and $\xi \gg a_\perp$ (dense vortices) the mean-field ground state is a triangular vortex lattice. As $\xi/a_\perp$ varies between these two limits, there is a smooth crossover (Fischer and Baym, 2003). This crossover has been observed in experiments, which have achieved a chemical potential $\mu \sim g\bar{n}$ that is less than the effective cyclotron splitting $2\hbar\omega_\perp$. The apparent core size of the vortices shows the behaviour expected on the crossover into the LLL regime (Coddington *et al.*, 2004). In the regime of strong interactions $\xi \gtrsim a_\mathrm{v}$, the area of the density depletion close to a vortex core is found to be proportional to ξ^2. For the weak interaction regime $\xi \lesssim a_\mathrm{v}$, this area is found to be proportional to $a_\perp^2$; this is consistent with the expectation that in this regime the single particle states are restricted to the LLL wavefunctions, and the only length scale is the mean vortex spacing $a_\mathrm{v} \simeq a_\perp$.

Given that very different physics controls the energetics in the two limits—for strong interactions, $\xi \lesssim a_\perp$, the vortex lattice is determined by the kinetic energy; for weak interactions, $\xi \gtrsim a_\perp$, the vortex lattice is determined entirely by the interactions—one should view the smooth crossover as a coincidence. As I now show, changing the form of the interaction potential can lead to large changes in the vortex lattice structure for $\xi \gtrsim a_\perp$.

Dipolar interactions The interesting possibilities that are afforded by the ability to trap and cool atoms (or molecules) with large magnetic or electric dipole moments are explained in detail in Chapter 6.

From the point of view of this presentation, the long-range nature of the dipolar interaction offers the possibility of changing the qualitative form of the interaction potential for bosonic atoms that are in the 2D LLL regime. Specifically, I consider a situation in which the dipole moments of the atoms are aligned along the rotation axis $\hat{z}$. Then the interaction potential is

$$V(\boldsymbol{r}) = \frac{4\pi\hbar^2 a_\mathrm{s}}{M}\,\delta^3(\boldsymbol{r}) + C_\mathrm{d}\frac{1-3(z/r)^2}{r^3}. \tag{5.61}$$

Gross–Pitaevskii mean-field theory in the 2D LLL now requires one to minimize the expectation value of the interaction energy

$$\frac{1}{2}\int\!\!\int |\psi(\boldsymbol{r})|^2\, V(\boldsymbol{r}-\boldsymbol{r}')|\psi(\boldsymbol{r}')|^2\, \mathrm{d}^3\boldsymbol{r}\, \mathrm{d}^3\boldsymbol{r}' \tag{5.62}$$

for $\psi(\boldsymbol{r})$ in the lowest Landau level and for the potential given in Eq. (5.61). Although this is a simple generalization of the Abrikosov problem, it is one that does not naturally appear in that context, where the microscopic physics determining the superconductivity acts on scales much smaller than the vortex lattice period.

The mean-field ground states have been found numerically for the case of a translationally invariant vortex lattice (Cooper *et al.*, 2005a). The results are shown in Figure 5.9, as a function of

$$\alpha \equiv \frac{V_2}{V_0}, \tag{5.63}$$

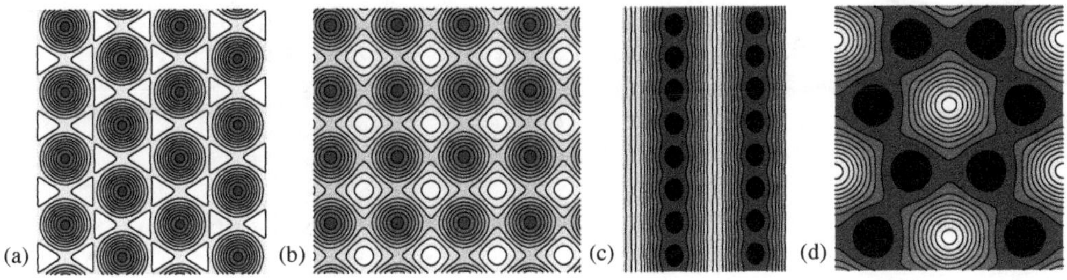

Figure 5.9 Mean-field ground states of a rotating BEC in the 2D LLL, for particles interacting by both contact and dipolar interactions. The relative size of these is controlled by $\alpha = V_2/V_0$. The structure of the vortex lattice varies from (a) triangular ($0 \leq \alpha \leq 0.20$) through (b) square ($0.20 \leq \alpha \leq 0.24$) and (c) 'stripe crystal', with simple rectangular unit cell, ($0.24 \leq \alpha \leq 0.60$) to 'bubble crystal' phases ($\alpha \geq 0.60$), the simplest of which is shown in (d). Reproduced with permission from (Cooper *et al.*, 2005a).

where

$$V_0 = \sqrt{\frac{2}{\pi}}\frac{\hbar^2 a_s}{Ma_\perp^2 a_\parallel} + \sqrt{\frac{2}{\pi}}\frac{C_d}{a_\perp^2 a_\parallel} - \sqrt{\frac{\pi}{2}}\frac{C_d}{a_\perp^3}, \tag{5.64}$$

$$V_{m>0} = \sqrt{\frac{\pi}{2}}\frac{(2m-3)!!}{m!\,2^m}\frac{C_d}{a_\perp^3}. \tag{5.65}$$

are the Haldane pseudo-potentials for the interaction of Eq. (5.61) in the limit $a_\parallel/a_\perp \ll 1$.

The results show that the mean-field ground state is extremely sensitive to long-range repulsion, passing through a series of vortex lattice phases as α increases. The contact interaction aims to make $|\psi(\boldsymbol{r})|^2$ as uniform as possible, while the long-range repulsion causes $|\psi(\boldsymbol{r})|^2$ to cluster, leading to crystalline phases of clusters of particles at large α. These are referred to as 'bubble crystal' phases, in analogy with the terminology used for structures of similar symmetry in 2D electron gases at high Landau levels (Koulakov *et al.*, 1996; Moessner and Chalker, 1996).

5.4.2 The filling factor

The Gross–Pitaevskii mean-field theory described previously is an approximate theory, based on the assumption that the ground state is a simple condensate. It is important to ask under what conditions its results are reliable.

To answer this, it is helpful to take advantage of the connection between cold atoms in the 2D LLL and the FQHE of electrons. In the FQHE of electrons, it is known that the ground states are characterized by the electron *filling factor*

$$\nu_e \equiv n_e \frac{h}{eB}, \tag{5.66}$$

where n_e is the 2D number density of electrons. From the mapping of the rotating atomic gas of Eq. (5.23), the analogous quantity is (Cooper *et al.*, 2001)

$$\nu \equiv n_{2d} \frac{h}{q^* B^*} = n_{2d} \frac{h}{2M\Omega}, \tag{5.67}$$

which, from Eq. (5.17), can be written in terms of the vortex density as

$$\nu = \frac{n_{2d}}{n_v}, \tag{5.68}$$

or as

$$\nu = \frac{N}{N_v}, \tag{5.69}$$

if one considers the particles to be uniformly distributed over an area containing N_v vortices. This result follows from the equivalence, discussed in Section 5.2, of vortex density n_v and flux density n_ϕ in a large system.

What is the phase diagram of a rapidly rotating atomic gas as a function of the filling factor ν?

Evidence has been given (Cooper *et al.*, 2001) that, for ν large (but finite) the mean-field theory is accurate—at least by correctly predicting a triangular vortex lattice (even though there may be quantitative corrections at finite ν). However, for small ν, quantum fluctuations are large, and can cause mean-field theory to fail. There are several ways in which to understand why the filling factor determines the degree of quantum fluctuations. One approach, which I find very instructive, is to determine the quantum fluctuations of the vortices directly.

5.4.3 Quantum fluctuations of vortices

Consider the dynamics of a single vortex line in a 2D fluid (i.e., a straight vortex line). The classical dynamics of a 2D vortex, at a position X and Y in an external potential $V(X, Y)$, follows from the standard Magnus force dynamics of a classical fluid,

$$-\rho_s \kappa \dot{Y} + F_X^{\mathrm{ext}} = 0 \tag{5.70}$$

$$+\rho_s \kappa \dot{X} + F_Y^{\mathrm{ext}} = 0, \tag{5.71}$$

where ρ_s is the mass density (per unit area) of the fluid and κ the circulation of the vortex. The only amendment for a quantized vortex in a superfluid is that the circulation is quantized, $\kappa = h/M$, so one may write

$$\rho_s \kappa = (n_{2d} M) \frac{h}{M} = n_{2d} h\,. \tag{5.72}$$

A Lagrangian that reproduces this classical dynamics is[8]

$$L = n_{2\mathrm{d}} h\, \dot{X}\, Y - V(X, Y), \qquad \left[\vec{F}^{\mathrm{ext}} = -\vec{\nabla} V\right]. \tag{5.73}$$

Constructing the momentum conjugate to the particle coordinate X and applying canonical quantization leads to

$$\Pi_X \equiv \frac{\partial L}{\partial \dot{X}} = n_{2\mathrm{d}} h\ Y \tag{5.74}$$

$$[\hat{X}, \hat{\Pi}_X] = \mathrm{i}\hbar \Rightarrow [\hat{X}, \hat{Y}] = \frac{\mathrm{i}}{2\pi n_{2\mathrm{d}}}. \tag{5.75}$$

The X and Y coordinates[9] are conjugate, and obey the generalized uncertainty relation

$$\Delta X \Delta Y \geq \frac{1}{4\pi n_{2\mathrm{d}}}, \tag{5.76}$$

which implies

$$\Delta X^2 + \Delta Y^2 \geq \frac{1}{2\pi n_{2\mathrm{d}}}. \tag{5.77}$$

The result of Eq. (5.77) makes physical sense: one cannot locate the vortex line to a distance less than the mean 2D separation between the particles. It is interesting to note that this result has an entirely classically interpretation, but emerges from a *quantum* calculation, owing to the cancellation of Planck's constant in the circulation with Planck's constant in the commutator.

The importance of the filling factor becomes clear if one applies a form of Lindemann criterion and asserts that the vortex lattice will become unstable to quantum fluctuations if the rms fluctuation is larger than some multiple α_{L} of the vortex spacing

$$\sqrt{\Delta X^2 + \Delta Y^2} = \frac{1}{\sqrt{2\pi n_{2\mathrm{d}}}} \geq \alpha_{\mathrm{L}} \times a_{\mathrm{v}} = \alpha_{\mathrm{L}} \sqrt{\frac{2}{\sqrt{3} n_{\mathrm{v}}}}, \tag{5.78}$$

$$\nu \equiv \frac{n_{2\mathrm{d}}}{n_{\mathrm{v}}} \geq \nu_{\mathrm{c}} = \frac{\sqrt{3}}{4\pi \alpha_{\mathrm{L}}^2}. \tag{5.79}$$

For a typical value for the Lindemann parameter, $\alpha_{\mathrm{L}}^2 \simeq 0.02$ (Rozhkov and Stroud, 1996), one finds $\nu_{\mathrm{c}} \simeq 7$.[10] Calculations of the rms vortex fluctuations that allow for

[8]This is easily checked by constructing the Euler–Lagrange equations.

[9]These are the guiding-centre coordinates of a particle in a single Landau level.

[10]For a single vortex, the quantum fluctuations of the guiding centre coordinate and of the cyclotron motion give equal contributions. Combining the fluctuations of both these degrees of freedom, one finds that $\Delta x^2 + \Delta y^2 \geq \frac{1}{\pi n_{2\mathrm{d}}}$—that is, *twice* the value of Eq. (5.77) from the guiding centre fluctuations alone. Using $\Delta x^2 + \Delta y^2$ in the Lindemann criterion leads to a critical value that is twice as large, $\nu_{\mathrm{c}} \simeq 14$.

the collective modes of a full vortex lattice (Sinova *et al.*, 2002; Baym, 2004) give very nearly the same results as this one-vortex result, and therefore the same ν_c if the same Lindemann parameter is used.

While instructive, these considerations are hardly predictive, depending very sensitively on α_L, which itself is estimated from the thermal melting of 3D crystals! For this reason it is useful to have a direct determination of the transition. This has been studied in large-scale exact diagonalization studies (Cooper *et al.*, 2001). The strategy is to work on a system with periodic boundary conditions (the torus geometry), which is consistent with the formation of a vortex lattice. The signal of crystallization is the collapse to very low energies (above the ground-state energy) of a set of excitations at momenta that are reciprocal lattice vectors of the vortex lattice. By looking for the emergence of broken translational symmetry, it was found that there is a transition to the triangular vortex lattice at $\nu_c \simeq 6$.

5.4.4 Strongly correlated regime

For $\nu < \nu_c$, the vortex lattice phase is unstable to quantum fluctuations and is replaced by a series of strongly correlated phases. These phases are best understood at small filling factors, far from the transition ν_c. As ν approaches ν_c, our understanding becomes much poorer.

Laughlin state $\nu = 1/2$ The prototype strongly correlated phase is the (bosonic) Laughlin state (Laughlin, 1983), which has the wavefunction

$$\Psi_L(\{\zeta_i\}) \propto \prod_{i<j}^{N} (\zeta_i - \zeta_j)^2 . \tag{5.80}$$

This is the *exact* ground state for contact repulsion, Eq. (5.55), at filling factor $\nu = 1/2$. For $\nu < 1/2$, the ground state of the contact interaction becomes degenerate, with a subspace spanned by quasi-hole excitations of Eq. (5.80).[11]

Although the particle density of this state is uniform, there are still vortices in the system. However, unlike the vortex lattice phase, Eq. (5.58), these vortices are not localized in space (translational symmetry is not broken). Rather, the vortices are *bound to the particles* (Girvin and MacDonald, 1987; Read, 1989). This can be seen by noting that in the state of Eq. (5.80), the wavefunction for the particle i changes phase by $2 \times 2\pi$ around the position of every other particle $\zeta_{j \neq i}$. Each particle experiences two vortices bound to the position of every other particle, giving $N_v = 2(N-1)$ vortices in total. Thus, Eq. (5.80) describes a phase that has filling factor $\nu = N/N_v = 1/2$ in the thermodynamic limit.

The Laughlin state for bosons shows all of the characteristics familiar from the conventional FQHE (Prange and Girvin, 1990): it is an incompressible fluid, with gapped collective excitations in the bulk (see Figure 5.10) and gapless edge modes; the

[11] There are no incompressible states at ν for the contact interaction. In particular, the other bosonic Laughlin states at $\nu = 1/4, 1/8 \ldots$ are compressible.

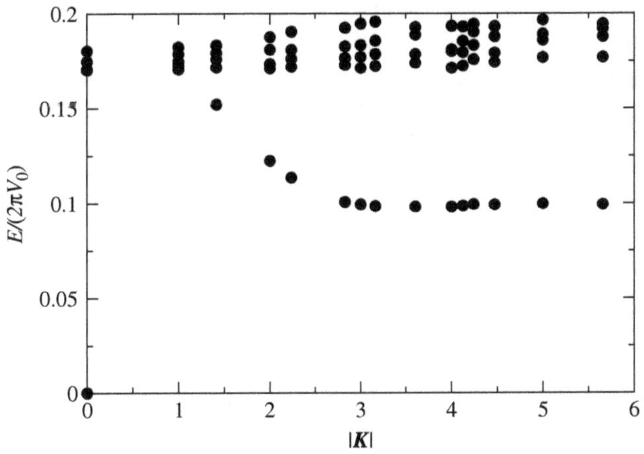

Figure 5.10 Low energy spectrum for the $\nu = 1/2$ Laughlin state of rotating bosons with contact interactions, obtained by exact diagonalization studies on a square (torus) geometry for $N = 8$ and $N_v = 16$. The states are labelled by the magnitude of the conserved wavevector on this periodic geometry. ($\boldsymbol{K}$ is measured in units of $2\pi/L$ where L is the side of the (square) torus. Energy is in units of $V_0 = \sqrt{\frac{2}{\pi}} \frac{\hbar^2 a_s}{M a_\parallel a_\perp^2}$.)

particle-like excitations have fractional particle number and fractional statistics (they are 'Abelian anyons'). Incompressibility is evident from exact diagonalization studies of the excitation spectrum, which show a gapped collective mode at all wavevectors.

Composite Fermion Ansatz At certain higher filling factors, $\nu > 1/2$, the strongly correlated states are accurately described by a 'composite fermion' ansatz. Here, since the underlying particles are bosons, one can form composite fermions by binding one vortex to the location of each particle. This may be achieved within the Jain construction (Jain, 1989) by writing the many-particle wavefunction as (Cooper and Wilkin, 1999)

$$\Psi(\{\vec{r}_k\}) = \hat{P}_{\text{LLL}} \prod_{i<j} (\zeta_i - \zeta_j)\, \psi_{\text{CF}}(\{\vec{r}_k\})\,. \tag{5.81}$$

The Jastrow factor $\prod_{i<j}(\zeta_i - \zeta_j)$ causes any particle i to experience a single vortex at the location of each other particle $j \neq i$. Since this factor is completely antisymmetric in particle exchange, to obtain a bosonic wavefunction the function ψ_{CF} must also be antisymmetric: this is the wavefunction of the composite fermions. In general terms, one can appreciate why this is a useful variational state: the Jastrow factor suppresses the amplitude for two particles to approach each other, as does the composite fermion wavefunction. Taken alone, these two factors would give a wavefunction that is a zero-energy eigenstate of the contact interactions. However, for $\nu > 1/2$, this construction requires ψ_{CF} to include basis states that are not in the lowest Landau level. In general, one must then project the single particle states into the LLL, as represented by the operator $\hat{P}_{\rm LLL}$. For $\nu > 1/2$ the projected wavefunction does not vanish as two particles approach each other, so it is a state with a non-zero contact interaction energy. Nevertheless, it has been found from numerical studies that the resulting trial states accurately describe the exact ground states of the two-body contact repulsion.

Treating the composite fermions as *non-interacting* particles and completely filling p Landau levels, one is led to the bosonic version of the Jain sequence

$$\nu = \frac{p}{p \pm 1} \,. \tag{5.82}$$

The states constructed in this way have large overlaps with exact ground states in the disc geometry (Cooper and Wilkin, 1999) and account well for the bulk phases and their excitations in the spherical geometry at $\nu = 1/2, 2/3$ and (with less accuracy) at $\nu = 3/4$ (Regnault and Jolicoeur, 2003).

Moore–Read 'Pfaffian' state One of the most interesting aspects of the physics of rapidly rotating Bose gases is the prediction of the appearance of *non-Abelian* phases: incompressible phases whose quasi-particle excitations obey non-Abelian exchange statistics. The existence of exotic exchange statistics of this kind is a very remarkable theoretical prediction. There is intense interest in finding experimental evidence for non-Abelian exchange statistics. This is driven largely by the fundamental interest in exploring such exotic properties of matter. However, it is also further motivated by suggestions that these systems might one day be used for so-called topologically protected quantum computing. The interested reader is referred to (Nayak *et al.*, 2008) for more details.

Applying the composite fermion construction at filling factor $\nu = 1$ for the bosons leads to the conclusion that the composite fermions feel no residual vortices. (One vortex is bound into each composite fermion, so at $\nu = 1$ there are no vortices remaining.) Treating the composite fermions as non-interacting particles, one would be drawn to the conclusion that the system is a *compressible* Fermi liquid of composite fermions: one would replace $\Psi_{\rm CF}$ by the Slater determinant of plane waves filling a Fermi sea. However, there is another possibility. Imagine that the composite fermions *do* have some interactions. If these interactions have an attractive component, they can lead to an instability to a paired superfluid phase. The Moore–Read 'Pfaffian' state describes such a situation. This state has the wavefunction (for bosons at $\nu = 1$)[12]

$$\Psi_{\rm MR}(\{\vec{r}_k\}) = \prod_{i<j=1}^{N} (\zeta_i - \zeta_j)\, {\rm Pf}\left[\frac{1}{\zeta_k - \zeta_l}\right], \tag{5.83}$$

[12] The wavefunction is written for a number of particles N that is even. An analogous state exists for odd N.

where Pf $[A_{kl}]$ denotes the 'Pfaffian' of the antisymmetric matrix A_{kl}, which can be viewed as the square root of the determinant.[13] The Pfaffian is the (first-quantized) wavefunction for a set of one-component fermions in a BCS paired state. Equation (5.83) describes a state in which composite fermions form a $p_x + ip_y$ BCS pairing state (Read and Green, 2000).

Exact diagonalization studies of contact-interacting bosons at $\nu = 1$ have shown very convincing evidence for the appearance of the Moore–Read 'Pfaffian' state (Moore and Read, 1991). Numerical evidence has been reported in several geometries (Wilkin and Gunn, 2000; Cooper *et al.*, 2001; Regnault and Jolicoeur, 2003). Large overlaps of the exact wavefunctions with the model state of Eq. (5.83) were found. A robust characterization of topological phases is provided by the ground-state degeneracy on a periodic geometry (which has the topology of a torus). For the Moore–Read state, the expected degeneracy appears clearly in the spectrum even for relatively small system sizes, as shown in Figure 5.11.

Read–Rezayi states A related family of states, which also have non-Abelian quasi-particle excitations was proposed by Read and Rezayi (Read and Rezayi, 1999). In the same way that the Moore–Read 'Pfaffian' can be viewed as containing pairs of composite particles fermions, the Read–Rezayi 'Parafermion' states, characterized by

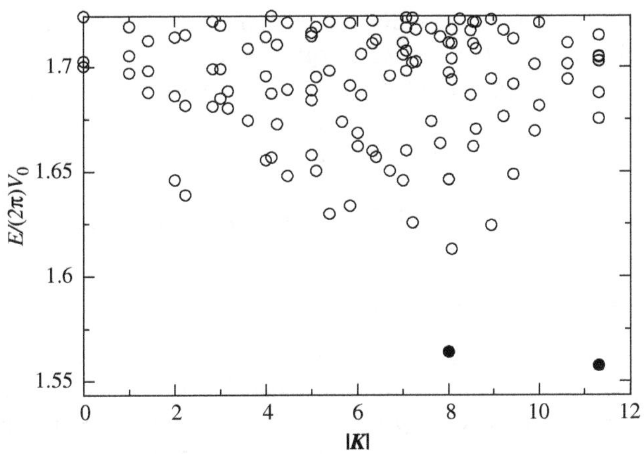

Figure 5.11 Low-energy spectrum for the $\nu = 1$ state with contact interactions, obtained by exact diagonalization studies on a torus with square geometry for $N = 16$ and $N_v = 16$. The states are labelled by the magnitude of the conserved wavevector on this periodic geometry. The appearance of low-energy states at $\boldsymbol{K} = (8, 8)$, $\boldsymbol{K} = (8, 0)$, and $\boldsymbol{K} = (0, 8)$ (indistinguishable in this plot, by symmetry), is consistent with the expected three-fold degeneracy of the Moore–Read state. ($\boldsymbol{K}$ is measured in units of $2\pi/L$, where L is the side of the (square) torus.)

[13]See (Moore and Read, 1991) or (Nayak *et al.*, 2008) for an explicit formula.

an integer k, can be viewed as containing clusters of k composite particles. For bosons, these states appear at filling factors $\nu = k/2$.[14]

Strong evidence for the appearance of the Read–Rezayi states ($k \geq 3$) at $\nu = k/2$ for bosons with contact repulsion, Eq. (5.55), has been found in numerical calculations on the periodic system (torus geometry) (Cooper *et al.*, 2001). While wavefunction overlaps are large, at least for k that is not too large, the expected ground-state degeneracy is less clearly resolved than for the Moore–Read state (Rezayi *et al.*, 2005). However, the addition of a small non-local interaction leads to a very clear ground-state degeneracy (Rezayi *et al.*, 2005), and leads to very strong evidence that the ground state is indeed the phase described by the $k = 3$ Read–Rezayi state.

The numerical results described in this section provide by far the most convincing theoretical evidence for the existence of non-Abelian phases of matter, described by the Moore–Read and Read–Rezayi states, in a physically realistic model. In the FQHE of electrons, where these states were first proposed, there are complicating factors and the numerical evidence is much less clear. It would be a major achievement if experiments on ultracold atomic gases could access and explore the properties of these very exotic phases of matter.

Smectic states The range of filling factors $2 \lesssim \nu \lesssim 6$ is difficult to study theoretically, the system being close to the transition to the vortex lattice phase. Exact diagonalization studies show large finite-size effects. At $\nu = 2$, these numerical results show strong evidence for the appearance of a 'smectic' state (Cooper and Rezayi, 2007)—that is, a stripe state with broken translational order in only one direction. This can be viewed as a phase in which the vortex lattice is quantum melted by fluctuations of lines of vortices. While this state appears to describe the ground state in the system sizes that can be studied numerically ($N_{\mathrm{v}} \leq 12$), it is possible that finite-size effects may favour this over competing phases (vortex lattice and $k = 4$ Read–Rezayi). Estimates suggest that the finite-size effects in experiments on cold gases (N_{v} is, of course, finite!) will mean that the smectic phases are likely to be important in practice.

5.4.5 Experimental considerations

To date, the highest vortex densities in rotating atomic Bose gases in conventional magnetic traps have been achieved in Eric Cornell's group at JILA. In these experiments, the mean interaction energy $g\bar{n}$ is sufficiently small that the gases are in or close to the 2D LLL regime, $g\bar{n} \lesssim 2\hbar\omega_\perp, \hbar\omega_\parallel$. However, the smallest filling factor reported is $\nu \simeq 500$ (Schweikhard *et al.*, 2004a). This is deep in the regime $\nu > \nu_{\mathrm{c}}$, where one expects the ground state to be a vortex lattice, and indeed this is what is observed.

Clearly further special efforts are required to reach the regime $\nu \lesssim 10$, where one expects quantum fluctuations of vortices to destabilize the lattice. There appear to be two main difficulties in reaching this regime:

Weak interactions In the 2D LLL, the typical interaction energy that stabilizes the strongly correlated phases is small. For a rapidly rotating gas, the maximum vortex

[14] The sequence of states defined by Read–Rezayi reduces to the Laughlin state at $k = 1$ and to the Moore–Read Pfaffian at $k = 2$. The new states are $k \geq 3$, which are all non-Abelian.

density of Eq. (5.17), $n_v = 2M\omega_\perp/h$, so at filling factor ν the maximum 2D particle density is $n_{2d} = \nu n_v = \nu 2M\omega_\perp/h$ and so the maximum 3D density is of order $\bar{n} \sim \nu 2M\omega_\perp/(ha_\parallel)$. The typical interaction scale is then

$$V_\mathrm{I} \sim \frac{\hbar^2 a_\mathrm{s}}{M}\frac{\nu}{a_\perp^2 a_\parallel} \sim \nu \times \frac{a_\mathrm{s}}{a_\parallel}\hbar\omega_\perp\,, \tag{5.84}$$

In the experiments of the JILA group (Schweikhard *et al.*, 2004b), the parameters were such that $V_\mathrm{I} \sim \nu \times 0.1\,\mathrm{Hz}$. The smallest filling factor achieved was $\nu \simeq 500$, corresponding to $V_\mathrm{I} \simeq 50\,\mathrm{Hz}$. Reducing this to $\nu = 1$ would lead to an interaction energy that is far below the lowest temperatures currently achievable.

While these small energy scales are discouraging, Eq. (5.84) indicates ways to overcome this difficulty: increasing a_s, decreasing $a_\parallel$, or increasing $\omega_\perp$. Recent experiments have produced interesting results on arrays of rotating traps, located on the sites of an optical lattice (Gemelke *et al.*, 2010), with very large trap frequencies $(\omega_\perp, \omega_\parallel) = 2\pi \times (2.1\,\mathrm{kHz}, 28\,\mathrm{kHz})$. As compared with the JILA experiments described before, the interaction scale Eq. (5.84) is increased by a factor of 18 000. The interaction scale at $\nu \simeq 1$ is then sizeable, $V_\mathrm{I} \sim 2\,\mathrm{kHz}$. Although these small traps only contain small numbers of particles ($\bar{N} \sim 5$), numerical studies show that this may be enough to see some emerging properties of the Laughlin correlations.[15] Indeed, the experimental results (Gemelke *et al.*, 2010) show evidence for the formation of local correlations.

Heating An atomic gas that has come to equilibrium at a rotation rate Ω is at rest in the rotating frame of reference. In the laboratory frame, the fluid at radius $r_\perp$ is moving at velocity of $\Omega r_\perp$. As a result, the rotating gas experiences any non-axisymmetric perturbation that is static in the laboratory frame as a time-dependent potential in the rotating frame. This can lead to excitation or heating. Simple estimates (Cooper, 2008) show that the quantum Hall phases are susceptible to heating by static perturbations of this kind.

The use of optical dressing to simulate a synthetic magnetic field has the advantage that the gas remains at rest in the laboratory frame. This removes the relative motion of the gas with respect to possible static perturbations, eliminating this source of heating. As described already, the current scheme that has been implemented in NIST (Lin *et al.*, 2009a) is limited to a flux density that is comparable to that for a rotating gas, so the energy scale of Eq. (5.84) is still small. However, the gas being at rest in the laboratory frame should facilitate the use of optical lattices to increase the 2D confinement and decrease $a_\parallel$. Moreover, the recently proposed 'optical flux lattices' (Cooper, 2011) will allow the flux density to be increased by two orders of magnitude beyond that in the NIST scheme. Schemes to generate a gauge field directly on an optical lattice could achieve a small filling factor $\nu \sim 1$, with a large energy scale, set by the Hubbard U or (for the hardcore limit $U \gg J$) the hopping J, which can be a few kHz.

[15] For contact repulsive interactions, the Laughlin state is the exact ground state at $L = N(N-1)$, even for small numbers of particles.

5.4.6 Experimental signatures of strongly correlated phases

Experimental detection of the properties of cold atomic gases is typically performed following the removal of the trapping potentials and expansion of the gas. An important aspect of the physics of rapidly rotating atomic gases is that, provided the interactions are weak enough that the gas is in the LLL limit, Eq. (5.28), the expanded gas provides a direct map of the wavefunction prior to expansion: the only effect of expansion is a rotation and rescaling of the coordinates (Read and Cooper, 2003). The use of expansion as a 'wavefunction microscope' makes probing the rotating gas particularly straightforward.

The most dramatic feature of the strongly correlated phases predicted by theory is that they are incompressible. As a result, the density distribution in a trapped system will show the characteristic 'wedding cake' structure (Cooper *et al.*, 2005b), with regions in which the 2D particle density is pinned to values $\nu^* n_\mathrm{v}$, where ν^* are the filling factors at which the ground state is incompressible. This density distribution differs markedly from the expected density distribution of the vortex lattice phase, which is compressible and hence follows a Thomas–Fermi distribution (Watanabe *et al.*, 2004; Cooper *et al.*, 2004). An extreme case is the Laughlin state, $\nu = 1/2$, for which the average particle density is constant out to the radius $R \sim 2\sqrt{N}\ell$.

Other experimental signatures, which can probe the quantum Hall phases, are summarized in (Cooper, 2008).

5.5 Fermi gases

The possibility of cooling Fermi gases to quantum degeneracy $T < T_\mathrm{F}$ (DeMarco and Jin, 1999), to achieve regimes of strongly correlated quantum phases of two-component Fermi gases (Bloch *et al.*, 2008), raises interesting questions concerning the properties of degenerate Fermi gases under rotation.

5.5.1 Non-interacting Fermi gas

I first discuss how a non-interacting Fermi gas in an axisymmetric trap is affected by the Landau level structure arising from rapid rotation at $\Omega \simeq \omega_\perp$ or a synthetic gauge field. The spectrum, Eq. (5.24), involves the Landau-level spectrum, with cyclotron frequency ω_c, and confinement along the z-direction, with trapping frequency $\omega_\parallel$.

The quasi-2D regime arises when the Fermi energy $\epsilon_\mathrm{F} \lesssim \hbar\omega_\perp$, such that only the lowest subband, $n_\parallel = 0$, is occupied. The density of states then consists of a series of degenerate Landau levels, separated by energy $\hbar\omega_\mathrm{c}$, as sketched in Figure 5.2. Owing to the gaps in the spectrum, the Fermi gas shows a series of incompressible states whenever the Fermi energy lies in one of these energy gaps. This corresponds to the cases that the filling factor ν is an integer. Including the trapping potential within the local density approximation, one expects to see a density distribution in a trap with a 'wedding cake' structure. The density plateaus occur at the cases $\nu =$ integer where the system is incompressible. This is sketched in Figure 5.12(a). The density distribution has been explored in more detail in (Ho and Ciobanu, 2000).

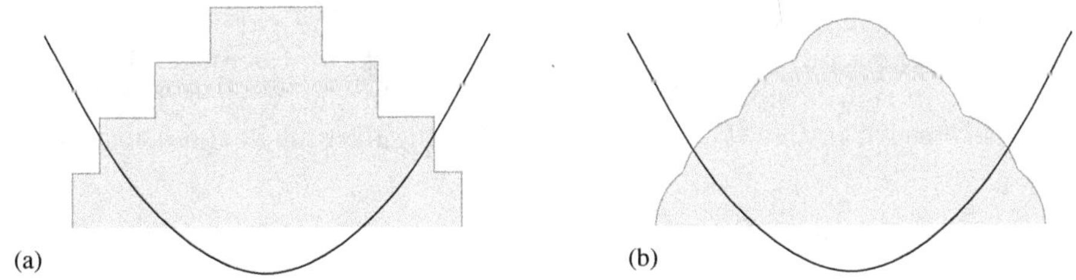

Figure 5.12 Schematic illustrations of the expected density distribution of non-interacting Fermi gas in a strong gauge potential (i.e., small number of Landau levels occupied). (a) In the quasi-2D regime, the system is incompressible for $\nu =$ integer leading to plateaus in the density. (b) In the 3D regime, the system is compressible so no plateaus arise. However, cusps in the density appear at positions where each LL depopulates and there is a peak in the (local) density of states, see Figure 5.13.

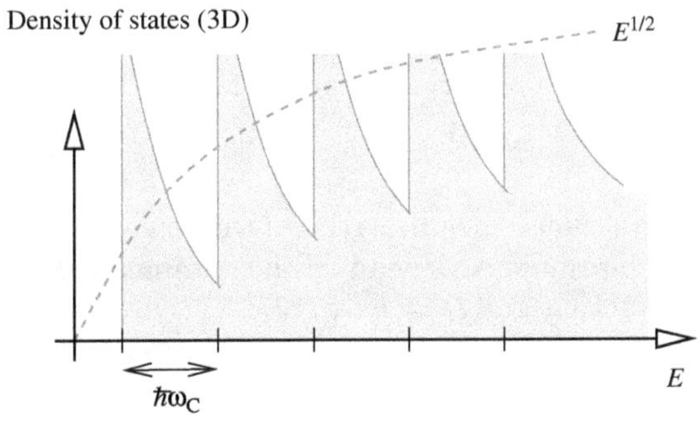

Figure 5.13 Schematic diagram of the density of states for a three-dimensional charged particle in a uniform magnetic field, with cyclotron energy $\hbar\omega_c$.

The 3D regime arises when $\omega_\| \ll \epsilon_F$ so that many subbands $n_\|$ are occupied. This can be relevant for very anisotropic traps $\omega_\| \ll \omega_c = 2\omega_\perp$. Here, for each Landau-level state, the motion along the rotation axis can be viewed as a quasi-1D gas, and the energy is more conveniently written as

$$E^{\omega_\perp}_{n,k_\|} = 2\hbar\omega_c(n + 1/2) + \frac{\hbar^2 k_\|^2}{2M}, \tag{5.85}$$

where $k_\|$ is the wavevector along the rotation axis. The density of states is shown in Figure 5.13. There are no energy gaps, so the system does not show incompressibility. However, there is a $1/\sqrt{E}$ divergence in the density of states at the energy of each Landau level, associated with the quasi-1D motion. This leads to a cusp in the density profile when a Landau level depopulates, as is illustrated in Figure 5.12(b).

Thus, the Landau-level quantization should lead to clear signatures in the density distribution of a trapped non-interacting Fermi gas. The experimental signatures depend on the aspect ratio of the trap: they are strongest for the quasi-2D limit, where there is strict incompressibility leading to plateaus in the density distribution; but features also survive in the 3D case owing to the singularities in the density of states.

5.5.2 Two-component Fermi gas

In the following, I shall explore the effects of rapid rotation on the physics of a strongly interacting two-component Fermi gas. Degenerate gases of this type may be formed either from two atomic species, or from two hyperfine states of the same fermionic atom (Grimm, 2008; Ketterle and Zwierlein, 2008). I shall denote the two components as spin-up and spin-down. For simplicity, I assume that the two components have equal densities and equal masses M.

Strong interspecies interactions can be induced by a Feshbach interaction, at which the s-wave scattering length a_s passes through a divergence,

$$a_s(B) \propto \frac{1}{B_{\rm res} - B}. \tag{5.86}$$

For a detailed discussion of the physics of the resonant scattering, the reader is referred to (Bloch *et al.*, 2008).

For short-range interactions, the properties of the (balanced) two-component Fermi gas close to the Feshbach resonance depend on the parameter $k_F a_s$. (I denote the Fermi energy of the non-interacting gas $\epsilon_F = \hbar^2 k_F^2/(2M)$.) The nature of the ground state can be understood in the limiting cases far from the resonance, when $k_F|a_s|$ is small. When a_s is small and positive, there exists a bound state even for one pair of atoms. The ground state of a many-particle system consists of tightly bound pairs of atoms of the two species; the resulting small bosonic molecules behave as a weakly interacting Bose gas and form a Bose–Einstein condensate at low temperatures. This is the BEC side of the resonance. When a_s is small and negative, the weak attractive interaction causes a BCS instability of the Fermi surface, leading to a BCS paired superfluid. There is an energy gap that, in this BCS limit of weak interaction, is exponentially small

$$\Delta \sim \epsilon_F \exp\left(-\frac{\pi}{2k_F|a_s|}\right). \tag{5.87}$$

This is the 'BCS' side of the resonance. As the resonance is swept, such that $1/k_F a_s$ varies from the BCS to BEC regime, the ground state evolves smoothly. This 'BEC to BCS crossover' has been confirmed in the experimental studies of several groups (Bloch *et al.*, 2008).

What is the effect of rapid rotation on the superfluid phase that appears in the BCS–BEC crossover?

The effects of moderate rotation are well understood. At all points across the crossover, the low-energy phase is a condensate, and responds as a superfluid of pairs of fermions. Rotation causes the formation of an array of vortices with density, $n_\text{v}^\text{pair} = 2(2M)\Omega/h$, which is the Feynman result of Eq. (5.17) with the mass, $2M$, of the pair of particles. The formation of a lattice of quantized vortices has been demonstrated in experiments described in (Zwierlein *et al.*, 2005). The results reveal the superfluid nature of the system across the whole range of the crossover regime, and are consistent with vortices having a quantum of circulation of $h/(2M)$ associated with condensation of pairs of fermions, of mass $2M$.

Effects of rapid rotation One expects the effects of rapid rotation to be rather different on the BEC and BCS sides of the resonance.

Far on the BEC side, the effects of rapid rotation are presumably similar to those discussed above for rapidly rotating bosons. In the quasi-2D regime, the mean-field vortex lattice will survive up to vortex densities at which the filling factor, Eq. (5.68), is less than the critical value $\nu_\text{c} \simeq 6$, at which point there is a transition to strongly correlated states. In the 3D regime (Cooper *et al.*, 2005b), the analogous transition to the strongly correlated regime occurs when the 3D density of molecules $\bar{n}_\text{m}$ satisfies $\bar{n}_\text{m}\xi_\text{m}/n_\text{v}^\text{pair} \lesssim \nu_\text{c}$, where ξ_m is the healing length for the molecules.

Far on the BCS side, the effects of rotation are similar to the orbital effects of a magnetic field on type-II superconductors. For conventional superconductors, the upper critical field is well described by the semiclassical approximation of Gorkov (Gorkov, 1959a; Gorkov, 1959b; Helfand and Werthamer, 1966). Translating to the quantities of interest to us, this predicts the transition to a normal state when

$$\hbar\Omega \gtrsim 4.24\Delta^2/\epsilon_\text{F}\,. \tag{5.88}$$

One can understand this semiclassical result in terms of a balance of energies. On the one hand, the condensation energy per particle on forming the superfluid is of order $\Delta^2/\epsilon_\text{F}$. This should be compared with the kinetic energy of the superfluid flow. Since the superfluid cannot rotate as a rigid body, in the rotating frame of reference the superfluid has a large kinetic energy. To estimate this kinetic energy, note that the vortex spacing $a_\text{v} \sim \left[\frac{2(2M)\Omega}{\hbar}\right]^{-1/2}$ sets a typical velocity $\hbar/Ma_\text{v}$, so the kinetic energy per pair is $M(\hbar/Ma_\text{v})^2 \sim \hbar\Omega$. The semiclassical formula states that once this kinetic energy is larger than the condensation energy, it does not pay to form the superfluid, and it takes less energy to have a rigidly rotating Fermi gas.

BCS regime: upper critical rotation rate To go beyond the semiclassical theory of Gorkov, one can study the BCS theory with a full Landau-level energy spectrum. For ultracold Fermi gases, this has been studied in (Zhai and Ho, 2006; Möller and Cooper, 2007). These works study mean-field theory in a situation in which the trapping potential is weak, so the system is in the uniform 3D regime,

$$\Omega \simeq \omega_\perp \gg \omega_\parallel\,, \tag{5.89}$$

with single-particle energy spectrum, Eq. (5.85). The BCS mean-field equations permit the determination of the critical temperature below which the system is unstable to the formation of a BCS superfluid as a function of the rotation rate Ω. A detailed analysis (Möller and Cooper, 2007) shows that the critical temperature $T_c(\Omega)$ *never* vanishes. In fact, as the rotation rate increases, the transition temperature becomes an *increasing* function of rotation rate. This is consistent with studies of BCS equations in the solid state setting beyond the semiclassical Gorkov approximation (Rasolt and Tešanović, 1992). The increase in transition temperature at high rotation rate is related to the enhancement of the density of states at the Fermi level, owing to the $1/\sqrt{E}$ singularities of the quasi-1D motion along the rotation axis.

The BCS mean-field theory predicts the ground state to be superfluid for all $k_F a_s$ and Ω is a statement that the BCS state has a lower (free) energy than the normal phase for any attractive interactions. Independently of the orbital motion induced by the rotation, the Fermi gas undergoes a superconducting (SC) instability at sufficiently low temperature. This does not rule out the possibility of destruction of the superfluid phase through a transition to another state. The BCS phase competes with a charge density wave (CDW) state (Möller and Cooper, 2007), in which the atomic density is spontaneously modulated along the rotation axis with a period such that the filling factor in each layer is $\nu = 2$, see Figure 5.15. This competition was analysed in detail for the case of interacting fermions in the LLL, allowing fluctuations of both CDW and SC. The dominant instability is towards the CDW state. (The transition temperature into the CDW is plotted as a dashed line in Figure 5.14. At large rotation rate Ω, this lies above that of the BCS superfluid, indicating the enhanced stability of the CDW phase.) Thus, an upper critical rotation frequency *does* exist on the BCS side of the transition; it is set by the transition into this CDW. At intermediate Ω, the CDW and SC can coexist leading to a novel form of supersolidity.

Transition from BEC to BCS These considerations are expected to be valid on the BCS side of the transition. Understanding how the results connect to the BEC side is an interesting, and unresolved, problem. One thing that is clear is that, in the rapid rotation limit, in which all fermions (or bosonic molecules) are confined to the LLL, the transition from the BCS to the BEC side must involve a phase transition.[16]

In the quasi-2D case, this can be seen by noting that states on the two sides of the transition are topologically distinct. On the BCS side, where the particles are weakly attractive atoms, one expects the ground state to be a compact droplet with $\nu_a = 2$ (a full Landau level for both atomic species); this state has two gapless edge modes, corresponding to density and spin excitations. On the BEC side, one anticipates that the ground state is a Laughlin state of bosonic molecules, with $\nu_m = 1/2$;[17] this state has only one gapless edge mode. The change in edge-mode structure shows that the states on the two sides are topologically distinct; they must be separated by a (quantum) phase transition point (Haldane and Rezayi, 2004; Möller and Cooper, 2007; Yang and Zhai, 2008). An effective theory has been derived for this phase transition (Yang and Zhai, 2008), but the nature of the transition remains unclear.

[16]This transition cannot be captured within the theory in the narrow resonance limit (Veillette *et al.*, 2006), as the bosonic molecules are non-interacting so the Laughlin state does not appear.

[17]The relation $\nu_m = \nu_a/4$ follows from the fact that there are half as many molecules as atoms, and the vortex density for a molecule is twice that for an atom (Cooper, 2004).

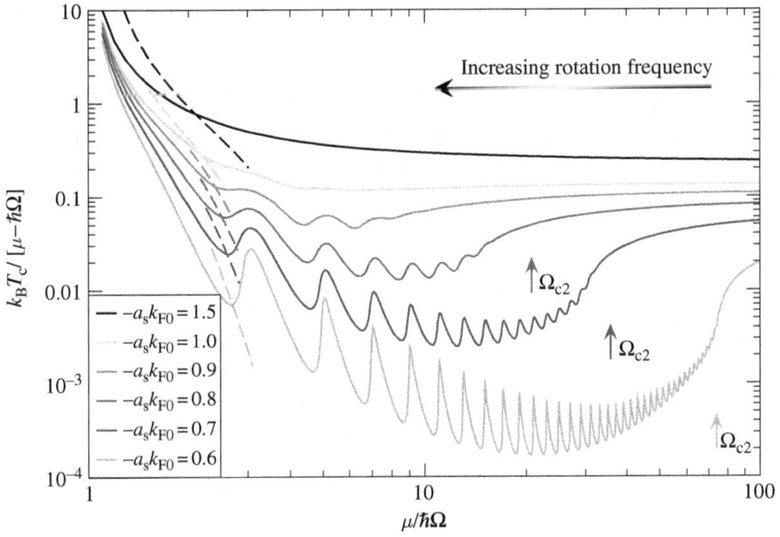

Figure 5.14 Solid lines: transition temperature for formation of a BCS superfluid state for a rotating Fermi gas in the uniform 3D limit, as a function of rotation rate Ω for different interaction strengths, $k_F a_s$. (Recall that the cyclotron frequency is $\omega_c = 2\Omega$, Eq. (5.25).) For weak interactions, as rotation rate increases at fixed chemical potential μ (right to left on the graph), the transition temperature falls abruptly at the semiclassical upper critical rotation rate. However, it remains non-zero and rises as Ω/μ decreases. For strong interactions, the transition temperature becomes a monotonically increasing function of Ω. Dashed lines: transition temperature for formation of a CDW state. From (Möller and Cooper, 2007).

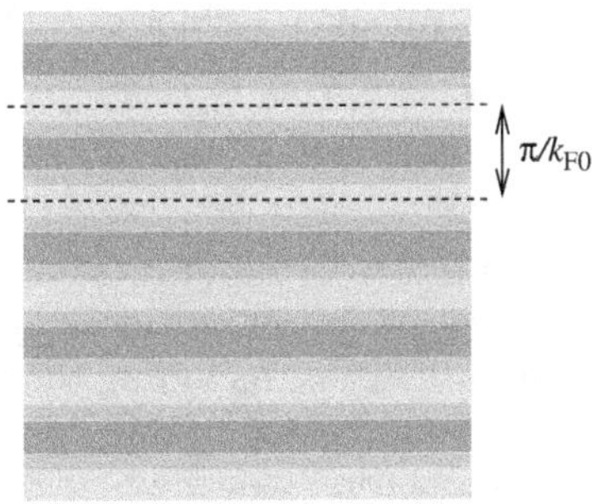

Figure 5.15 Schematic diagram of the charge density wave state that sets the upper critical rotation rate for destruction of fermionic superfluidity. The wavevector of the density wave lies along the rotation axis. Its period is set by the particle density, and is such that the atomic filling factor of each layer $\nu_a = n_a/(2\pi\ell^2) = 2$. It is a fully gapped state.

The existence of a phase transition in the 3D case (Möller and Cooper, 2007) follows directly from the 2D-argument combined with the observation that the 3D system forms a CDW of layers of $\nu_a = 2$ states at high rotation rate. If the CDW order survives with the same period on passing to the BEC side, each layer must undergo the topological phase transition; the only alternative is that the CDW does not survive (or changes period), which signals a phase transition of another kind.

5.5.3 Experimental status

The formation of a lattice of quantized vortices in the BCS–BEC crossover regime of a two-component Fermi gas has been demonstrated in experiments at MIT (Zwierlein *et al.*, 2005). From the point of view of the regimes that I have focused on in these notes, these experiments were performed at relatively low rotation rate (and high density), with $\epsilon_F/(\hbar\omega_c) \gg 1$. Experiments at higher flux density, or lower particle densities, will probe an interesting regime, $\epsilon_F/(\hbar\omega_c) \lesssim 10$, where the semiclassical approximation used in typical solid-state superconductors is inapplicable, and where qualitatively new effects can emerge.

The strongly correlated regime of most interest arises for $\epsilon_F/(\hbar\omega_c) \sim 1$ with strong interactions (close to resonance). Note that it may be easier to enter the strongly correlated regime for fermions than for bosons (Antezza *et al.*, 2007): the interaction energy, which stabilizes the strongly correlated phases, is much larger for fermions on resonance than it is (typically) for bosons. Indeed, if the conditions of chemical potential and temperature achieved in the experiments of JILA for a weakly interacting Bose gas (which reached the LLL at $\nu \sim 500$) could be reproduced for a Fermi gas, this gas would be in the strongly correlated regime ($\nu \sim 1$).

An obvious first target for future experimentation is to demonstrate the Landau level structure in a non-interacting gas, as evidenced, for example, in the density profiles of Figure 5.12. This is a robust feature of Fermi gases, and the essential ingredient of all physics of strongly correlated phases. For gauge fields imposed on a lattice, the spectrum is the richer (fractal) form of the 'Hofstadter spectrum', signatures of which may realistically be sought in experiment (Gerbier and Dalibard, 2010).

5.6 Summary

Ultracold atomic gases under conditions of rapid rotation, or experiencing a synthetic magnetic field, show a Landau level structure in their single-particle energy spectrum. This can lead to very interesting properties, including the appearance of novel strongly correlated phases. There are close connections to the physics of the fractional quantum Hall effect of electrons in semiconductors. However, ultracold gases could access very different physical situations and regimes. They would allow the first studies of the fractional quantum Hall states of *bosons*. There is convincing theoretical evidence that novel strongly correlated phases of bosonic matter can appear at small filling factors. These studies provide clear theoretical evidence that exotic non-Abelian phases can exist for physically realistic models. The response of two-component *Fermi* gases in the BCS–BEC crossover regime to rapid rotation (or strong synthetic gauge field) also presents many interesting features. Such systems have the potential to probe, for the first time, the interplay between superconducting pairing and the quantum Hall effect.

Recent developments in experimental capabilities to induce rapid rotation or synthetic gauge fields offer the hope that these new situations may soon be achieved in experiment. Our understanding of strongly correlated systems remains very incomplete, so there is every chance that experiments in these very interesting regimes will discover effects that are unanticipated theoretically.

Acknowledgements

I would like to thank my collaborators in the work described in these notes: Mike Gunn, Stavros Komineas, Gunnar Möller, Nick Read, Ed Rezayi, Steve Simon, and Nicola Wilkin.

References

Abo-Shaeer, JR, Raman, C, Vogels, JM, and Ketterle, W (2001). Observation of vortex lattices in Bose–Einstein condensates. *Science*, **292**, 476.

Abrikosov, AA (1957). On the magnetic properties of superconductors of the second group. *Zh. Exsp. Teor. Fiz.*, **32**, 1442.

Antezza, M, Cozzini, M, and Stringari, S (2007). Breathing modes of a fast rotating Fermi gas. *Phys. Rev. A*, **75**, 053609.

Baym, G (2004). Vortex lattices in rapidly rotating Bose–Einstein condensates: modes and correlation functions. *Phys. Rev. A*, **69**, 043618.

Bloch, I, Dalibard, J, and Zwerger, W (2008). Many-body physics with ultracold gases. *Rev. Mod. Phys.*, **80**, 885.

Coddington, I, Haljan, PC, Engels, P, Schweikhard, V, Tung, S, and Cornell, EA (2004). Experimental studies of equilibrium vortex properties in a Bose-condensed gas. *Phys. Rev. A*, **70**, 063607.

Cooper, NR (2004). Exact ground states of rotating Bose gases close to a Feshbach resonance. *Phys. Rev. Lett.*, **92**, 220405.

Cooper, NR (2008). Rapidly rotating atomic gases. *Adv. Phys.*, **57**, 539.

Cooper, NR (2011). Optical flux lattices for ultracold atomic gases. *Phys. Rev. Lett.*, **106**, 175301.

Cooper, NR and Wilkin, NK (1999). Composite fermion description of rotating Bose–Einstein condensates. *Phys. Rev. B*, **60**, R16279.

Cooper, NR and Rezayi, EH (2007). Competing compressible and incompressible phases in rotating atomic Bose gases at filling factor $\nu = 2$. *Phys. Rev. B*, **75**, 013627.

Cooper, NR, Wilkin, NK, and Gunn, JMF (2001). Quantum phases of vortices in rotating Bose–Einstein condensates. *Phys. Rev. Lett.*, **87**, 120405.

Cooper, NR, Komineas, S, and Read, N (2004). Vortex lattices in the lowest Landau level for confined Bose–Einstein condensates. *Phys. Rev. B*, **70**, 033604.

Cooper, NR, Rezayi, EH, and Simon, SH (2005a). Vortex lattices in rotating atomic Bose gases with dipolar interactions. *Phys. Rev. Lett.*, **95**, 200402.

Cooper, NR, van Lankvelt, FJM, Reijnders, JW, and Schoutens, K (2005b). Quantum Hall states of atomic Bose gases: density profiles in single-layer and multilayer geometries. *Phys. Rev. A*, **72**, 063622.

Cornell, EA and Wieman, CE (2002). Nobel lecture: Bose–Einstein condensation in a dilute gas, the first 70 years and some recent experiments. *Rev. Mod. Phys.*, **74**, 875.

Dalibard, J, Gerbier, F, Juzeliūnas, G, and Öhberg, P (2011). Artificial gauge potentials for neutral atoms. *Rev. Mod. Phys.*, **83**, 1523.

DeMarco, B and Jin, DS (1999). Onset of Fermi degeneracy in a trapped atomic gas. *Science*, **285**, 1703.

Donnelly, RJ (1991). *Quantized Vortices in Helium II*. Cambridge University Press, Cambridge.

Fetter, AL (1965). Vortices in an imperfect Bose gas. i. The condensate. *Phys. Rev.*, **138**, A429.

Fetter, AL (2008). Rotating trapped Bose–Einstein condensates. *Rev. Mod. Phys.*, **81**, 647.

Fischer, UR and Baym, G (2003). Vortex states of rapidly rotating dilute Bose–Einstein condensates. *Phys. Rev. Lett.*, **90**, 140402.

Gemelke, N, Sarajlic, E, and Chu, S (2010). Rotating few-body atomic systems in the fractional quantum Hall regime. arXiv:1007.2677.

Gerbier, F and Dalibard, J (2010). Gauge fields for ultracold atoms in optical superlattices. *New J. Phys.*, **12**, 033007.

Girvin, SM and MacDonald, AH (1987). Off-diagonal long-range order, oblique confinement, and the fractional quantum Hall effect. *Phys. Rev. Lett.*, **58**, 1252.

Gorkov, P (1959a). The critical supercooling field in superconductivity theory. *Zh. Eksp. Teor. Phys.*, **37**, 833. (JETP **10**, 593 (1960)).

Gorkov, P (1959b). Microscopic derivation of the Ginzburg–Landau equations in the theory of superconductivity. *Zh. Eksp. Teor. Phys.*, **36**, 1918. (JETP **9**, 1364 (1959)).

Grimm, R (2008). Ultracold Fermi gases in the BEC–BCS crossover: a review from the Innsbruck perspective. In *Proc. of the Int. School of Physics Enrico Fermi* (ed. M Inguscio, W Ketterle, and C Salomon), IOS Press, Amsterdam.

Haldane, FDM and Rezayi, EH (2004). Workshop on "Exotic Order and Criticality in Quantum Matter", KITP, University of California, Santa Barbara; and private communication.

Helfand, E and Werthamer, NR (1966). Temperature and purity dependence of the superconducting critical field, H_{c2}. ii. *Phys. Rev.*, **147**, 288.

Ho, T-L and Ciobanu, CV (2000). Rapidly rotating Fermi gases. *Phys. Rev. Lett.*, **85**, 4648.

Jain, JK (1989). Composite-fermion approach for the fractional quantum Hall effect. *Phys. Rev. Lett.*, **63**, 199.

Jaksch, D and Zoller, P (2003). Creation of effective magnetic fields in optical lattices: the Hofstadter butterfly for cold neutral atoms. *New J. Phys.*, **5**, 56.

Ketterle, W (2002). Nobel lecture. When atoms behave as waves: Bose–Einstein condensation and the atom laser. *Rev. Mod. Phys.*, **74**, 1131.

Ketterle, W and Zwierlein, M (2008). Making, probing and understanding ultracold Fermi gases, in ultracold Fermi gases. In *Proc. of the Int. School of Physics Enrico Fermi* (ed. M Inguscio, W Ketterle, and C Salomon), IOS Press, Amsterdam.

Koulakov, AA, Fogler, MM, and Shklovskii, BI (1996). Charge density wave in two-dimensional electron liquid in weak magnetic field. *Phys. Rev. Lett.*, **76**, 499.

Landau, LD and Lifshitz, EM (1981). *Statistical Physics Part 1*. Volume 5. Butterworth Heinemann, Oxford.

Laughlin, RB (1983). Anomalous quantum Hall effect: an incompressible quantum fluid with fractionally charged excitations. *Phys. Rev. Lett.*, **50**, 1395.

Lin, Y-J, Compton, RL, Jimenez-Garcia, K, Porto, JV, and Spielman, IB (2009a). Synthetic magnetic fields for ultracold neutral atoms. *Nature*, **462**, 628.

Lin, Y-J, Compton, RL, Perry, AR, Phillips, WD, Porto, JV, and Spielman, IB (2009b). Bose–Einstein condensate in a uniform light-induced vector potential. *Phys. Rev. Lett.*, **102**, 130401.

Madison, KW, Chevy, F, Wohlleben, W, and Dalibard, J (2000). Vortex formation in a stirred Bose–Einstein condensate. *Phys. Rev. Lett.*, **84**, 806.

Moessner, R and Chalker, JT (1996). Exact results for interacting electrons in high Landau levels. *Phys. Rev. B*, **54**, 5006.

Möller, G and Cooper, NR (2007). Density waves and supersolidity in rapidly rotating atomic Fermi gases. *Phys. Rev. Lett.*, **99**, 190409.

Möller, G and Cooper, NR (2010). Condensed groundstates of frustrated Bose–Hubbard models. *Phys. Rev. A*, **82**, 063625.

Moore, G and Read, N (1991). Nonabelions in the fractional quantum Hall-effect. *Nucl. Phys. B*, **360**, 362.

Mueller, EJ (2004). Artificial electromagnetism for neutral atoms: Escher staircase and Laughlin liquids. *Phys. Rev. A*, **70**, 041603.

Nayak, C, Simon, SH, Stern, A, Freedman, M, and Sarma, Sankar D (2008). Non-Abelian anyons and topological quantum computation. *Rev. Mod. Phys.*, **80**, 1083.

Prange, RE and Girvin, SM (eds.) (1990). *The Quantum Hall Effect* 2nd edn. Springer-Verlag, Berlin.

Rasolt, M and Tešanović, Z (1992). Theoretical aspects of superconductivity in very high magnetic fields. *Rev. Mod. Phys.*, **64**, 709.

Read, N (1989). Order parameter and Ginzburg–Landau theory for the fractional quantum Hall effect. *Phys. Rev. Lett.*, **62**, 86.

Read, N and Cooper, NR (2003). Free expansion of lowest-Landau-level states of trapped atoms: a wave-function microscope. *Phys. Rev. A*, **68**, 035601.

Read, N and Green, D (2000). Paired states of fermions in two dimensions with breaking of parity and time-reversal symmetries and the fractional quantum Hall effect. *Phys. Rev. B*, **61**, 10267.

Read, N and Rezayi, EH (1999). Beyond paired quantum Hall states: parafermions and incompressible states in the first excited landau level. *Phys. Rev. B*, **59**, 8084.

Regnault, N and Jolicoeur, T (2003). Quantum Hall fractions in rotating Bose–Einstein condensates. *Phys. Rev. Lett.*, **91**, 030402.
Rezayi, EH, Read, N, and Cooper, NR (2005). Incompressible liquid state of rapidly rotating bosons at filling factor 3/2. *Phys. Rev. Lett.*, **95**, 160404.
Rozhkov, A and Stroud, D (1996). Quantum melting of a two-dimensional vortex lattice at zero temperature. *Phys. Rev. B*, **54**, R12697.
Schweikhard, V, Coddington, I, Engels, P, Mogendorff, VP, and Cornell, EA (2004a). Rapidly rotating Bose–Einstein condensates in and near the lowest Landau level. *Phys. Rev. Lett.*, **92**, 040404.
Schweikhard, V, Coddington, I, Engels, P, Tung, S, and Cornell, EA (2004b). Vortex-lattice dynamics in rotating spinor Bose–Einstein condensates. *Phys. Rev. Lett.*, **93**, 210403.
Shapere, A and Wilczek, F (1989). *Geometric Phases in Physics*. World Scientific, Singapore.
Sinova, J, Hanna, CB, and MacDonald, AH (2002). Quantum melting and absence of Bose–Einstein condensation in two-dimensional vortex matter. *Phys. Rev. Lett.*, **89**, 030403.
Veillette, MY, Sheehy, DE, Radzihovsky, L, and Gurarie, V (2006). Superfluid transition in a rotating Fermi gas with resonant interactions. *Phys. Rev. Lett.*, **97**, 250401.
Watanabe, G, Baym, G, and Pethick, CJ (2004). Landau levels and the Thomas–Fermi structure of rapidly rotating Bose–Einstein condensates. *Phys. Rev. Lett.*, **93**, 190401.
Wilkin, NK and Gunn, JMF (2000). Condensation of 'composite bosons' in a rotating BEC. *Phys. Rev. Lett.*, **84**, 6.
Wilkin, NK, Gunn, JMF, and Smith, RA (1998). Do attractive bosons condense? *Phys. Rev. Lett.*, **80**, 2265.
Yang, K and Zhai, H (2008). Quantum Hall transition near a fermion Feshbach resonance in a rotating trap. *Phys. Rev. Lett.*, **100**, 030404.
Yarmchuk, EJ, Gordon, MJV, and Packard, RE (1979). Observation of stationary vortex arrays in rotating superfluid helium. *Phys. Rev. Lett.*, **43**, 214.
Zhai, H and Ho, T-L (2006). Critical rotational frequency for superfluid fermionic gases across a Feshbach resonance. *Phys. Rev. Lett.*, **97**, 180414.
Zwierlein, MW, Abo-Shaeer, JR, Schirotzek, A, Schunck, CH, and Ketterle, W (2005). Vortices and superfluidity in a strongly interacting Fermi gas. *Nature*, **435**, 1047.

6
Theory of dipolar gases

Luis Santos

University of Hannover, Institut für Theoretische Physik, Theoretical Quantum Optics Group, Appelstrasse 2, 30167 Hannover, Germany

Chapter Contents

In this chapter, I discuss some interesting features of the physics of dipolar gases, with a particular emphasis on those phenomena that differ qualitatively from those known in non-dipolar gases. I do not intend to give a complete review of previous work on dipolar gases (for that see recent reviews (Baranov 2008; Lahaye *et al.* 2009)), and in this sense I apologize from the very beginning to those whose work is not explicitly cited here. I have tried to condense as much as possible, adapting to the reduced time of three lectures, and this has forced me to leave some very interesting topics aside, including most of the theory of dipolar Fermi gases, crystallization, and rapidly rotating dipolar gases. I apologize for that. I hope that this chapter will serve, however, as a basic introduction to the topic of dipolar gases, and in particular of why dipolar gases are so interesting, since indeed dipolar gases (most relevantly, polar molecules) may change the rules of the game!

6.1 The dipole–dipole interaction

Before discussing the physics of dipolar gases, it is of crucial importance to understand the main features of the dipole–dipole interaction (DDI). We shall have a look as well at various examples of dipolar gases.

6.1.1 Main features of the dipole–dipole interaction

For two particles with dipole moments along the unit vectors $\boldsymbol{e}_1$ and $\boldsymbol{e}_2$, whose relative position is $\boldsymbol{r}$ (see Figure 6.1), the energy due to the DDI is

$$U_{\rm dd}(\boldsymbol{r}) = \frac{C_{\rm dd}}{4\pi} \frac{(\boldsymbol{e}_1 \cdot \boldsymbol{e}_2)\, r^2 - 3\,(\boldsymbol{e}_1 \cdot \boldsymbol{r})\,(\boldsymbol{e}_2 \cdot \boldsymbol{r})}{r^5}. \tag{6.1}$$

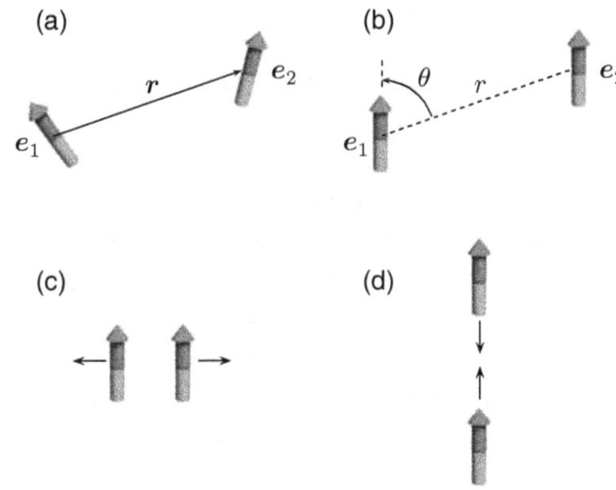

Figure 6.1 Two particles interacting via the dipole–dipole interaction. (a) Non-polarized case; (b) polarized case. (c) Two polarized dipoles side by side repel each other (black arrows). (d) Two polarized dipoles in a 'head to tail' configuration attract each other (black arrows).

The coupling constant C_{dd} is $\mu_0\mu^2$ for particles having a permanent magnetic dipole moment μ (μ_0 is the permeability of the vacuum) and d^2/ε_0 for particles having a permanent electric dipole moment d (ε_0 is the permittivity of the vacuum). For a polarized sample where all dipoles point in the same direction z (Figure 6.1b), this expression simplifies to

$$U_{dd}(\boldsymbol{r}) = \frac{C_{dd}}{4\pi}\frac{1-3\cos^2\theta}{r^3}, \tag{6.2}$$

where θ is the angle between the direction of polarization and the relative position of the particles.

This interaction must be compared with the (up to now) usual van-der-Waals-like interaction $-C_6/r^6$, which is both isotropic and short-range. On the contrary, the DDI is:

Anisotropic As θ varies between 0 and $\pi/2$, the factor $1-3\cos^2\theta$ varies between -2 and 1, and thus the DDI is repulsive for particles sitting side by side, while it is attractive for dipoles in a 'head-to-tail' configuration (see Figures 6.1(c) and (d)). For the special value $\theta_m = \arccos\left(1/\sqrt{3}\right) \simeq 54.7°$ (the 'magic-angle'), the DDI vanishes.

Long-range One must recall that in a 3D scattering, a potential $U(\boldsymbol{r})$ is long-range if it decays as $1/r^n$ with $n \leq 3$. In that case $\int U(\boldsymbol{r})\,\mathrm{d}^3r$ diverges. In this sense, the DDI is strictly long-range in three dimensions.

6.1.2 Scattering properties; pseudo-potential

The long-range anisotropic character of the DDI leads to peculiar low-energy scattering properties: not only the s-wave, but all partial waves contribute to scattering, in contrast with the case of a short-range interaction, where only s-wave scattering typically matters. Here we should recall the general theory of low-energy scattering (see, e.g. (Landau and Lifshitz 1977)), which states that for a central potential falling off at large distances as $1/r^n$, the scattering phase shifts $\delta_\ell(k)$ scale, for $k \to 0$, as $k^{2\ell+1}$ if $\ell < (n-3)/2$, and as k^{n-2} otherwise (Landau and Lifshitz 1977). For a van-der-Waals-like potential ($n=6$), only $l=0$ (s-wave) matters at low energies. In the ultracold regime, the scattering is thus fully characterized by the s-wave scattering length a. In the study of quantum gases, the true short-range interaction potential between the atoms can then be replaced by a pseudo-potential having the same scattering length, the so-called contact interaction, given by $4\pi\hbar^2 a\delta(\boldsymbol{r})/m \equiv g\delta(\boldsymbol{r})$.

The situation is very different if, in addition to the short-range interactions, there is a significant DDI. The $1/r^3$ decay at large distances implies that for all ℓ, $\delta_\ell \sim k$ at low momentum, and all partial waves contribute to the scattering amplitude. Moreover, owing to the anisotropy of the DDI, the angular momentum (i.e., l) is not conserved during scattering, and as a consequence the DDI mixes all partial waves with even (for the case of bosons) and odd (for the case of fermions) angular momenta. For dipoles polarized along z, cylindrical symmetry is, however, preserved and hence the m quantum number (related to the projection L_z of the angular momentum) is preserved.

The scattering matrix is therefore of the form $t_{l,m}^{l',m'} = t_l^{l'} \delta_{m,m'}$, where (l, m) and (l', m') characterize the incoming and outgoing channels, respectively. The problem may be rigorously treated by means of a multichannel scattering theory, but I will not discuss all the details here (see, e.g., (Marinescu and You 1998)). The main conclusion of this theory is that we may replace the actual interparticle interaction with an effective pseudo-potential of the simple form:

$$U(\boldsymbol{r}) = g\delta(\boldsymbol{r}) + U_{dd}(\boldsymbol{r}), \tag{6.3}$$

where g is defined as before, but in principle $a = a(d)$. Note that because of coupling between different scattering channels, the DDI generates a short-range contribution to the total effective potential in the s-wave channel ($l = 0$) that adds to the short-range part of the interparticle interaction. As a result, by changing the strength of the DDI, one may modify a as well. This may manifest itself quite dramatically in the appearance of scattering resonances, so-called shape resonances (see, e.g., (Marinescu and You 1998; Deb and You 2001; Bortolotti *et al.* 2006)), when a virtual state transforms into a new bound state. The previous pseudo-potential has been shown to be valid away from shape resonances (Deb and You 2001; Bortolotti *et al.* 2006).

The scattering of bosonic dipoles is hence determined by both long-range and short-range interactions, whose interplay plays, as we shall see, a crucial role in the physics of dipolar BECs. For fermionic dipoles, the s-wave channel is absent, and hence the low-energy scattering of fermionic dipoles is determined only by the long-range part. This is, of course, crucial for polarized Fermi gases, since contrary to the case of a short-range interaction, which freezes out at low temperature, the collisional cross section for identical fermions interacting via the DDI does not vanish, even at zero temperature.

6.1.3 Fourier transform of the dipole–dipole interaction

In the analysis of dipolar gases, the Fourier transform of the DDI turns out to be very helpful. This Fourier transform may be easily retrieved by re-writing:

$$U_{dd}(\boldsymbol{r}) = \frac{C_{dd}}{4\pi r^3}\left(-4\sqrt{\frac{\pi}{5}}\right) Y_{lm}(\theta), \tag{6.4}$$

where Y_{lm} are the spherical harmonics. We employ the expansion of a plane wave in spherical harmonics:

$$\mathrm{e}^{-\mathrm{i}\boldsymbol{k}\cdot\boldsymbol{r}} = 4\pi \sum_{l=0}^{\infty} i^l j_l(kr) \sum_{m=-l}^{l} Y_{lm}^*(\theta, \phi) Y_{lm}(\theta_k, \phi_k). \tag{6.5}$$

Employing the orthonormality of the spherical harmonics $\int \mathrm{d}\theta \sin\theta \mathrm{d}\phi\, Y_{lm}^* Y_{l',m'} = \delta_{ll'}\delta_{mm'}$, and $\int \mathrm{d}r j_2(kr)/r = 1/3$, we may then easily obtain

$$\tilde{U}_{dd}(\boldsymbol{k}) = \int \mathrm{d}^3 r U_{dd}(\boldsymbol{r}) \mathrm{e}^{-\mathrm{i}\boldsymbol{k}\cdot\boldsymbol{r}} = \frac{C_{dd}}{3}\left(3\cos^2\theta_k - 1\right). \tag{6.6}$$

This must be compared with the short-range interaction, which, owing to its contact character, presents no momentum dependence. This different momentum dependence of both interactions plays a crucial role in the physics of dipolar gases, and more so in the presence of trapping, as we shall see later on in this chapter.

6.1.4 Dipolar gases: from tiny to huge dipoles

Before entering into a detailed discussion of dipolar gases, we should discuss what we actually mean by a dipolar gas. A gas may be defined as dipolar if the DDI plays at least some role in its properties. This is, of course, a very broad definition, which may range from tiny to huge dipoles, since whether the DDI is important or not largely depends on other energy scales.

As mentioned, the interplay between DDI and short-range interactions is crucial in dipolar gases. We may quantify this interplay by the ratio

$$\epsilon_{dd} = C_{dd}/3g \tag{6.7}$$

between the dipole strength (given by the prefactor $C_{dd}/3$ of $\tilde{U}_{dd}$) and the short-range coupling constant g. In principle, a dominant DDI demands $\epsilon_{dd} > 1$, although dipolar effects may occur even for $\epsilon_{dd} \ll 1$, as discussed later.

In the following, I discuss some interesting details concerning atomic magnetic dipoles and polar molecules, skipping the discussion about, e.g., Rydberg gases, which may present an extraordinarily large dipole moment, although they have a short lifetime as well.

Atomic magnetic dipoles In alkali atoms, the maximum magnetic moment in the ground state is $\mu = 1\,\mu_{\rm B}$ (with $\mu_{\rm B}$ the Bohr magneton), and thus the magnetic dipolar effects are very weak. However, reducing the scattering length to $a \simeq 0$ by means of magnetic fields (Feshbach resonances) may allow observation of the effects of the magnetic DDI (Fattori *et al.* 2008; Pollack *et al.* 2009). However, this is not even necessary, because ϵ_{dd} do not always fully characterize the importance of the DDI. This is particularly true of spinor gases, which are composed of various internal Zeeman states. Spin-changing collisions, which are crucial in the magnetic properties of these gases, are typically characterized by very low energies. As a result, even a weak DDI, as that of, e.g., ^{87}Rb F $= 1$, may lead to a strong modification of the magnetic properties of spinor condensates, as recently observed at Berkeley (Vengalattore *et al.* 2008).

Some other atoms, such as chromium, erbium, europium, and dysprosium, have a large magnetic moment of several $\mu_{\rm B}$ in their ground state, and thus experience significant magnetic DDI. Among them, only chromium has been Bose-condensed to date (Griesmaier *et al.* 2005; Beaufils *et al.* 2008). Chromium has a magnetic dipole moment of $6\,\mu_{\rm B}$, and a scattering length of about $100\,a_0$ (Schmidt *et al.* 2003). This gives $\varepsilon_{\rm dd} \simeq 0.16$ (Griesmaier *et al.* 2006), which allows us to observe a perturbative effect of the dipolar interaction on the expansion dynamics of the cloud (Stuhler *et al.* 2005). The role of the DDI may be enhanced by means of Feshbach resonances, as shown in recent experiments in Stuttgart (Lahaye *et al.* 2008).

Polar molecules Polar molecules constitute a huge leap in dipole moment (this time electric). Heteronuclear molecules, like KRb, may present dipole moments larger than 0.5 debye ($1\,\mathrm{D} \simeq 3.335 \times 10^{-30}\,\mathrm{C \cdot m}$). This must be compared with chromium, which has a dipole moment equivalent to 0.05 D. Recall that the DDI is proportional to d^2, and hence polar molecules may present a DDI two orders of magnitude (or even more) larger than chromium. Dipolar effects are hence expected to be dominant in quantum gases of polar molecules ($\epsilon_{dd} \simeq 20$ for fully polarized KRb).

A polar molecule is maximally polar when placed in its lowest ro-vibrational state (the dipole moment scales asymptotically as R^{-7} with the internuclear separation R (Kotochigova *et al.* 2003)). However, although a molecule may exhibit a permanent dipole moment $\hat{\boldsymbol{d}}$ in the molecular frame, it must be oriented in the laboratory frame by an external electric field.

This orientation may be understood from a rigid rotor ('dumb-bell') model, which is a simple model of diatomic molecules, which are considered basically as two atoms joined by a rigid (weightless) rod. The Hamiltonian for a rigid rotor reads $\hat{H}_{\mathrm{rot}} = B\hat{\boldsymbol{J}}^2$, where $\hat{\boldsymbol{J}}$ is the molecule angular momentum operator (in units of $\hbar$) and B the rotational constant, linked to the equilibrium internuclear distance R and the reduced mass m_{r} by the relationship $B = \hbar^2/(2m_{\mathrm{r}}R^2)$; its typical order of magnitude is $B/h \sim 10$ GHz. The eigenstates of $\hat{H}_{\mathrm{rot}}$ (i.e., the rotational spectrum) are the angular momentum eigenstates $|J, m_J\rangle$ with energy $BJ(J+1)$, and are $2J+1$ times degenerate (Figure 6.2(a)).

This degeneracy is altered in the presence of an external field $\boldsymbol{E} = \mathcal{E}\boldsymbol{e}_z$, which leads to a Stark shift $\hat{H} = \hat{H}_{\mathrm{rot}} - d\mathcal{E}\cos\theta$, with θ the angle between z and the internuclear axis (Figure 6.2(b)). We may then obtain the average dipole moment $\langle d_z\rangle = d\langle\cos\theta\rangle$ for the ground state using the Hellmann–Feynman theorem: $\langle d_z\rangle = -\frac{\partial E_0}{\partial\mathcal{E}}$, where $E_0(\mathcal{E})$ is the ground-state energy. The dipole moment $\langle d_z\rangle$ increases linearly at small $\mathcal{E}$, saturating asymptotically for $d\mathcal{E} \gg B$ towards a value d (Figure 6.2(c)). For typical

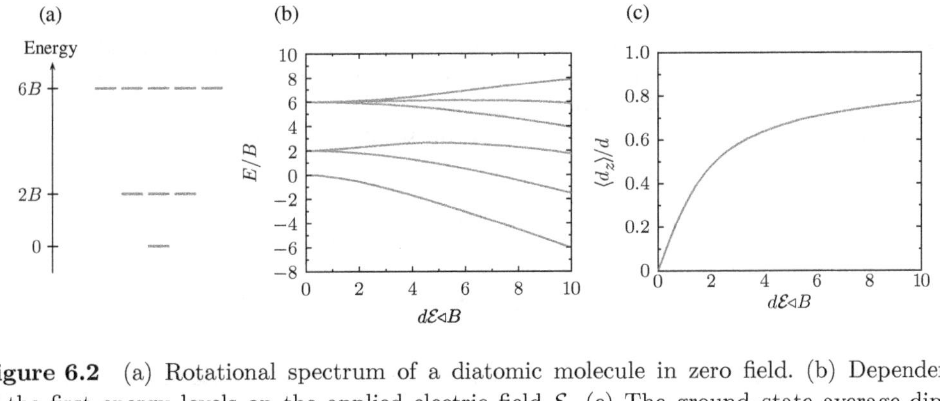

Figure 6.2 (a) Rotational spectrum of a diatomic molecule in zero field. (b) Dependence of the first energy levels on the applied electric field $\mathcal{E}$. (c) The ground-state average dipole moment $\langle d_z\rangle$ in the laboratory frame as a function of the applied field $\mathcal{E}$. Figure from (Lahaye *et al.* 2009).

values of $d \sim 1$ D and $B/h \sim 10$ GHz, the electric field strength corresponding to $d\mathcal{E} \sim B$ is of the order of 10^4 V/cm, which, from an experimental point of view, is accessible in a relatively easy way. Note, finally, that $B \sim m_r^{-1}$ and smaller m_r means larger B and hence larger $\mathcal{E}$ to polarize (e.g., LiCs is more difficult to orient than KRb). For more details on the orientation of polar molecules, see, e.g., (Micheli *et al.* 2007).

At this time, polar molecules have not been yet brought to quantum degeneracy, but this quest constitutes nowadays the focus of very active on-going efforts. Interestingly, polar molecules in the lowest ro-vibrational and hyperfine states have recently been obtained (Deiglmayr *et al.* 2008; Ni *et al.* 2008). I will not comment in detail about that since this topic is covered in Chapter 7.

6.2 Dipolar Bose–Einstein condensates

We shall have a look now at the properties of dipolar bosons, and in particular at dipolar BECs (more extensive reviews may be found in (Baranov 2008; Lahaye *et al.* 2009)). We shall focus on relatively simple scenarios, discussing some key features, including the different forms of instability, and some interesting non-linear phenomena in dipolar BECs.

6.2.1 Non-local Gross–Pitaevskii equation

Let us consider a gas of dipolar bosons. The second-quantized Hamiltonian of the system reads:

$$\hat{H} = \int d\boldsymbol{r}\hat{\psi}^\dagger(\boldsymbol{r})\left[-\frac{\hbar^2}{2m}\nabla^2 + V(\boldsymbol{r}) - \mu\right]\hat{\psi}(\boldsymbol{r}) + \frac{1}{2}\int d\boldsymbol{r}d\boldsymbol{r}'\hat{\psi}^\dagger(\boldsymbol{r})\hat{\psi}^\dagger(\boldsymbol{r}')U(\boldsymbol{r}-\boldsymbol{r}')\hat{\psi}(\boldsymbol{r}')\hat{\psi}(\boldsymbol{r}), \tag{6.8}$$

where $\hat{\psi}(\boldsymbol{r})$ and $\hat{\psi}^\dagger(\boldsymbol{r})$ are the particle annihilation and creation operators, which fulfil the usual bosonic commutation relations, $V(\boldsymbol{r})$ is the trapping potential, and μ is the chemical potential. The interaction potential $U(\boldsymbol{r})$ may be approximated by the pseudo-potential, Eq. (6.3), and then our Hamiltonian becomes:

$$\hat{H} = \int d\boldsymbol{r}\hat{\psi}^\dagger(\boldsymbol{r})\left[-\frac{\hbar^2}{2m}\nabla^2 + V(\boldsymbol{r}) - \mu + \frac{1}{2}g\hat{\psi}^\dagger(\boldsymbol{r})\hat{\psi}(\boldsymbol{r})\right]\hat{\psi}(\boldsymbol{r}) + \frac{1}{2}\int d\boldsymbol{r}d\boldsymbol{r}'\hat{\psi}^\dagger(\boldsymbol{r})\hat{\psi}^\dagger(\boldsymbol{r}')U_{dd}(\boldsymbol{r}-\boldsymbol{r}')\hat{\psi}(\boldsymbol{r}')\hat{\psi}(\boldsymbol{r}), \tag{6.9}$$

We may then easily obtain the Heisenberg equations for the dynamics of the field operators, by employing the bosonic commutation rules. Since we are interested in the case of BECs far from the critical condensation temperature, we may introduce the usual Bogoliubov approximation $\hat{\psi}(\boldsymbol{r}) \simeq \psi(\boldsymbol{r})$. In this way, we find that the Heisenberg equation transforms into the following equation for the dynamics of the condensate wavefunction:

$$\mathrm{i}\hbar\frac{\partial}{\partial t}\psi(\boldsymbol{r},t) = \left[-\frac{\hbar^2}{2m}\nabla^2 + V(\boldsymbol{r}) - \mu + g|\psi(\boldsymbol{r},\boldsymbol{t})|^2 + \frac{C_{dd}}{4\pi}\int \mathrm{d}\boldsymbol{r}'\frac{1-3\cos^2\theta}{|\boldsymbol{r}-\boldsymbol{r}'|^3}|\psi(\boldsymbol{r}',\boldsymbol{t})|^2\right]\psi(\boldsymbol{r},\boldsymbol{t}). \tag{6.10}$$

Note that this equation is a modified version of the well-known Gross–Pitaevskii equation (GPE), or, equivalently, the non-linear Schrödinger equation. In the absence of DDI, the non-linearity is given by the $g|\psi(\boldsymbol{r})|^2$ term; hence, it is a local non-linearity similar to that found in many Kerr media in non-linear optics. On the contrary, the non-linearity introduced by the DDI is non-local, i.e., the wavefunction in $\boldsymbol{r}$ depends on the wavefunction in $\boldsymbol{r}'$ through a kernel given by $U_{dd}(\boldsymbol{r}-\boldsymbol{r}')$. Interestingly, this links the physics of dipolar BECs with other non-local non-linear systems, e.g. plasmas, where the non-locality is introduced by thermal effects (Litvak *et al.* 1975), or nematic liquid crystals (Conti *et al.* 2003), where the non-local non-linearity is given by long-range intermolecular interactions.

6.2.2 Phonon instability

The DDI is attractive along some directions, and repulsive along others. In some sense, we should hence naively expect a sort of hybrid behaviour between a repulsive gas (as that of a non-dipolar gas with $a > 0$) and an attractive gas (as that of a non-dipolar gas with $a < 0$). Interestingly, this is to some extent what happens. In particular, we know that a gas with attractive interactions has a dangerous tendency to collapse.

Stability is hence an issue of obvious concern in dipolar gases. Let us try to understand this important point in more detail. We shall consider a simplified 3D homogeneous model (no trapping), and perform a standard Bogoliubov–de Gennes analysis of the stability. We consider a 3D BEC with a density n_0. Is the dipolar BEC stable? We shall quickly see that the answer may be no!

Let us return to the Hamiltonian of Eq. (6.8). We introduce the Fourier transform $\hat{\psi}(\boldsymbol{r}) = \sum_{\boldsymbol{p}} \hat{a}_{\boldsymbol{p}} \exp[\mathrm{i}\boldsymbol{p}\cdot\boldsymbol{r}/\hbar]\sqrt{V}$, where V is a quantization volume. This leads to the Hamiltonian in momentum space:

$$\hat{H} = \sum_{\boldsymbol{p}} \frac{p^2}{2m}\hat{a}^\dagger_{\boldsymbol{p}}\hat{a}_{\boldsymbol{p}} + \frac{1}{2V}\sum_{\boldsymbol{p}_1,\boldsymbol{p}_2,\boldsymbol{q}} \tilde{U}(\boldsymbol{q})\hat{a}^\dagger_{\boldsymbol{p}_1+\boldsymbol{q}}\hat{a}^\dagger_{\boldsymbol{p}_2-\boldsymbol{q}}\hat{a}_{\boldsymbol{p}_2}\hat{a}_{\boldsymbol{p}_1}, \tag{6.11}$$

where $\tilde{U}(\boldsymbol{q}) = g + \tilde{U}_{dd}(\boldsymbol{q})$. Assuming a condensate in $p = 0$ (i.e., an homogeneous BEC), we may approximate $\hat{a}_0 = \hat{a}_0^\dagger = \sqrt{N}$, where $N = n_0 V$ is the number of particles. Expanding up to second order $\hat{a}_{\boldsymbol{p}\neq 0}$ we find, up to a constant:

$$\hat{H} = \sum_{\boldsymbol{p}\neq 0} \frac{p^2}{2m}\hat{a}^\dagger_{\boldsymbol{p}}\hat{a}_{\boldsymbol{p}} + \frac{n_0}{2}\sum_{\boldsymbol{p}} \tilde{U}(\boldsymbol{p})\left(2\hat{a}^\dagger_{\boldsymbol{p}}\hat{a}_{\boldsymbol{p}} + \hat{a}^\dagger_{\boldsymbol{p}}\hat{a}^\dagger_{-\boldsymbol{p}} + \hat{a}_{\boldsymbol{p}}\hat{a}_{-\boldsymbol{p}}\right). \tag{6.12}$$

As one can easily see, this Hamiltonian just couples $\boldsymbol{p}$ with $-\boldsymbol{p}$. This means that we can diagonalize using a Bogoliubov transformation for each $\boldsymbol{p}$. We do not do it here in detail because this is the standard Bogoliubov-transformation procedure. It is more important to have a look to the dispersion resulting from this diagonalization:

$$\epsilon(\boldsymbol{p}) = \sqrt{\frac{p^2}{2m}\left[\frac{p^2}{2m} + 2n_0\left(g + \tilde{U}_{dd}(\boldsymbol{p})\right)\right]}. \tag{6.13}$$

Note that due to the momentum dependence of the DDI, the dispersion now has an anomalous momentum dependence. Let us see what happens for $p \to 0$. In that case we may approximate it as

$$\epsilon(\boldsymbol{p}) = pc_{s0}\sqrt{1 + \epsilon_{dd}(3\cos^2\theta_p - 1)}, \tag{6.14}$$

where $c_{s0} = \sqrt{\frac{gn_0}{m}}$ is the sound velocity in the absence of DDI. Note that for $\theta_p = \pi/2$, i.e., for momenta perpendicular to the dipolar orientation, $\epsilon(\boldsymbol{p}) = pc_{s0}\sqrt{1-\epsilon_{dd}}$. As a result, for $\epsilon_{dd} > 1$, some excitation modes are purely imaginary. Hence, the homogeneous 3D dipolar BEC is dynamically unstable against very long-wave length excitations. We will denote this instability 'phonon instability'.

6.2.3 Trapped gases: geometry-dependent stability

The phonon instability scenario resembles the case of $a < 0$, where for similar reasons, a homogeneous BEC is also unstable (that is easy to see from Eq. (6.13), taking $g < 0$ and $\tilde{U}_{dd} = 0$). However, in a trap, quantum pressure may stabilize the BEC for small atom numbers ($N < N_c$). For an isotropic harmonic trap of frequency ω, $N_c|a|/a_{ho} = 0.58$, where $a_{ho} = \sqrt{\hbar/(m\omega)}$ is the oscillator length (Ruprecht *et al.* 1995). For anisotropic traps, the dependence on the trap geometry is weak (Gammal *et al.* 2001).

The situation is very different in dipolar BECs, owing to the anisotropy of the DDI. We shall see in the following that the trap geometry crucially determines the stability of a dipole BEC. To see that, let us consider for simplicity a cylindrically symmetric trap, with a symmetry axis z coinciding with the dipole orientation. The axial and radial trapping frequencies are denoted ω_z and ω_ρ, respectively, such that $\lambda = \omega_z/\omega_\rho$. Let us consider a Gaussian variational ansatz of the form:

$$\psi(\rho, z) = \frac{\sqrt{N}}{\pi^{3/4} l_\rho l_z^{1/2}} e^{-z^2/2l_z^2} e^{-\rho^2/2l_\rho^2}. \tag{6.15}$$

Note that the cloud aspect ratio $\kappa = l_\rho/l_z$ is, in general, different from the trap aspect ratio $\lambda^{1/2}$.

We may evaluate the expression for the total energy as a function of l_ρ and l_z: $E = E_{\rm kin} + E_{\rm trap} + E_{sr} + E_{\rm dd}$, where

$$E_{\rm kin} = \frac{\hbar^2}{2m}\int \mathrm{d}^3\boldsymbol{r}|\nabla\psi|^2 = \frac{N\hbar^2}{2m}\left\{\frac{1}{l_z^2} + \frac{2}{l_\rho^2}\right\} \tag{6.16}$$

is the kinetic energy,

$$E_{\rm trap} = \int \mathrm{d}^3\boldsymbol{r}\, V_{\rm trap}(\boldsymbol{r})n(\boldsymbol{r}) = \frac{Nm}{4}\left\{2\omega_\rho^2 l_\rho^2 + \omega_z^2 l_z^2\right\} \tag{6.17}$$

is the trap energy ($n(\boldsymbol{r}) = |\psi(\boldsymbol{r})|^2$), and

$$E_{sr} = \frac{g}{2}\int \mathrm{d}^3\boldsymbol{r} n(\boldsymbol{r})^2 = \frac{gN^2}{2(2\pi)^{3/2} l_z l_\rho^2} \tag{6.18}$$

is the contact interaction energy.

Finally $E_{\rm dd}$ is the mean DDI, which is especially interesting for us:

$$E_{\rm dd} = \frac{1}{2}\int \mathrm{d}^3\boldsymbol{r}\int \mathrm{d}^3\boldsymbol{r}'\, U_{\rm dd}(\boldsymbol{r}-\boldsymbol{r}')n(\boldsymbol{r}')n(\boldsymbol{r}). \tag{6.19}$$

This expression is best evaluated in momentum space:

$$E_{\rm dd} = \frac{1}{2}\int \frac{\mathrm{d}^3\boldsymbol{k}}{(2\pi)^3}\tilde{U}_{\rm dd}(\boldsymbol{k})|\tilde{n}(\boldsymbol{k})|^2, \tag{6.20}$$

where $\tilde{n}(\boldsymbol{k}) = N\exp[-k_z^2 l_z^2/4 - k_\rho^2 l_\rho^2/4]$ is the Fourier transform of the density. Let $q_j = k_j l_j$ ($j = \rho, z$); then

$$\begin{aligned} E_{\rm dd} &= \frac{C_{\rm dd}N^2}{6(2\pi)^3 l_\rho^2 l_z}\int \mathrm{d}^3\boldsymbol{q}\mathrm{e}^{-q^2/2}\frac{2\kappa^2\cos^2\theta_q - \sin^2\theta_q}{\kappa^2\cos^2\theta_q + \sin^2\theta_q} \\ &= \frac{C_{\rm dd}N^2}{3(2\pi)^{3/2} l_\rho^2 l_z}\left\{\frac{2\kappa^2+1}{\kappa^2-1} - \frac{3\kappa^2}{(\kappa^2-1)^{3/2}}\arctan\left[\sqrt{\kappa^2-1}\right]\right\} \\ &= \frac{C_{\rm dd}N^2}{3(2\pi)^{3/2} l_\rho^2 l_z}f(\kappa). \end{aligned} \tag{6.21}$$

The function $f(\kappa)$ increases monotonically, has asymptotic values $f(0) = -1$ and $f(\infty) = 2$, and vanishes for $\kappa = 1$ (implying that for an isotropic density distribution the mean DDI averages to zero). For a cigar-shape BEC elongated along the dipole orientation ($\kappa < 1$), $E_{\rm dd} < 0$, i.e., the DDI is essentially attractive. This is intuitively easy to understand, since in that case the dipoles see each other head with tail on average. On the contrary, for an oblate trap ($\kappa > 1$) $E_{\rm dd} > 0$, i.e., the DDI is essentially repulsive. This is again intuitive, because the dipoles see each other side by side on average.

The physics of the dipolar BEC, and in particular its stability and the relation between trap and cloud aspect ratio, may be obtained by minimizing the energy E with respect to l_ρ and l_z for fixed N, ω_ρ, and ω_z. A stable BEC is characterized by the presence of a minimum value of E (at least locally) for finite values of l_ρ and l_z (a local minimum actually means a metastable solution, but we shall consider the lifetime as infinite). The disappearance of such a minimum marks the point of destabilization of the condensate.

For purely dipolar interactions (i.e., $a = 0$), there is a critical $\lambda_{\rm cr} \simeq 5.2$ (Santos *et al.* 2000; Yi and You 2001; Góral and Santos 2002; Eberlein *et al.* 2005; Koch *et al.* 2005), such that for $\lambda < \lambda_{\rm cr}$ the condensate is unstable for a sufficiently large number of particles (i.e., quite similar to the $a < 0$ case without DDI). On the contrary, if $\lambda > \lambda_{\rm cr}$ (sufficiently like a pancake-shaped trap), there is, in principle, no critical number of particles. In other words, the phonon-like instability is geometrically stabilized (however, another type of instability may occur in that case, as we shall see later). Note that, interestingly, the critical trap aspect ratio is $\lambda_{\rm cr}^{1/2} = 2.28$, i.e., a slightly pancake-shaped trap, and not, as one could naively think, a spherical trap. This is because for large N, $\kappa \to 1$ when $\lambda \to \lambda_{\rm cr}$, i.e., the trap is pancake-shaped but the BEC cloud for $\lambda_{\rm cr}$ is spherical (note that this can never happen in non-dipolar BECs).

Let us analyse what happens when $a \neq 0$. Repulsive short-range interactions ($a > 0$) may stabilize (up to some point) an unstable dipolar BEC. Note also that even if the mean DDI is repulsive, a sufficiently strong attractive contact interaction ($a < 0$) may destabilize the BEC. Hence, for non-zero contact interactions, we may expect that for a given λ, there exists a critical value $a_{\rm crit}(\lambda)$, such that for $a < a_{\rm crit}(\lambda)$ the dipolar BEC is unstable.

For a given N, the actual curve $a_{\rm crit}(\lambda)$ must be determined numerically (Bohn *et al.* 2009). This curve decreases monotonously with λ. For very prolate traps ($\omega_z \ll \omega_\rho$), $E_{\rm dd} < 0$, and one expects that a positive a is necessary to stabilize the BEC (hence $a_{\rm crit} > 0$). On the contrary, for very oblate traps ($\omega_z \gg \omega_\rho$), $E_{\rm dd} > 0$ and hence one needs a sufficiently large $a < 0$ to destabilize the BEC (i.e., $a_{\rm crit} < 0$).

In particular, in the limit $N \to \infty$, the asymptotic behaviour of the $a_{\rm crit}$ curve ($a^{\infty}_{\rm crit}(0) = a_{\rm dd} \equiv mC_{\rm dd}/12\pi\hbar^2$ and $a^{\infty}_{\rm crit}(\infty) = -2a_{\rm dd}$) can easily be understood, as only the sign of the interaction term $E_{sr} + E_{\rm dd}$ (which scales as N^2 and not as N, like $E_{\rm kin}$ and $E_{\rm trap}$) determines the stability. For an extremely pancake-shaped trap $\lambda \to \infty$, the cloud has an aspect ratio $\kappa \to \infty$, and, as $f(\kappa \to \infty) = -2$, the BEC is (meta)stable only if $a > -2a_{\rm dd}$. In the same way, one readily understands that for $\lambda \to 0$, the critical scattering length is $a_{\rm dd}$.

The influence of the trapping geometry on the stability of a ^{52}Cr BEC was investigated experimentally (Koch *et al.* 2005). In particular, that experiment determined the curve $a_{\rm crit}(\lambda)$. A typical measurement is shown in Figure 6.3. Note that, although the main qualitative features are recovered by a simple Gaussian ansatz, to calculate the exact stability threshold one needs to resort to a numerical solution of the non-local GPE, Eq. (6.10) (Bohn *et al.* 2009).

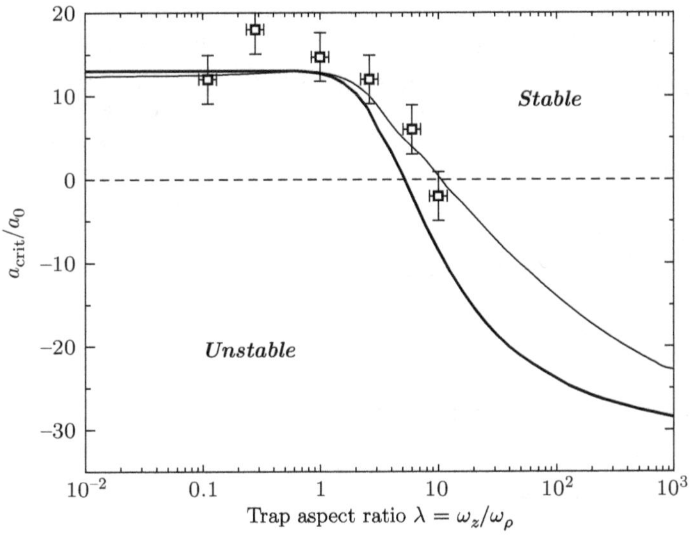

Figure 6.3 Stability diagram of a dipolar condensate in the plane (λ, a). The points with error bars correspond to the experimental data (Koch *et al.* 2005); the thick solid line to the threshold $a_{crit}(\lambda)$ obtained using the Gaussian ansatz Eq. (6.15), with $N = 20,000$; the thin solid line to the numerical solution of the GPE Eq. (6.10) (Bohn *et al.* 2009). Figure from (Lahaye *et al.* 2009).

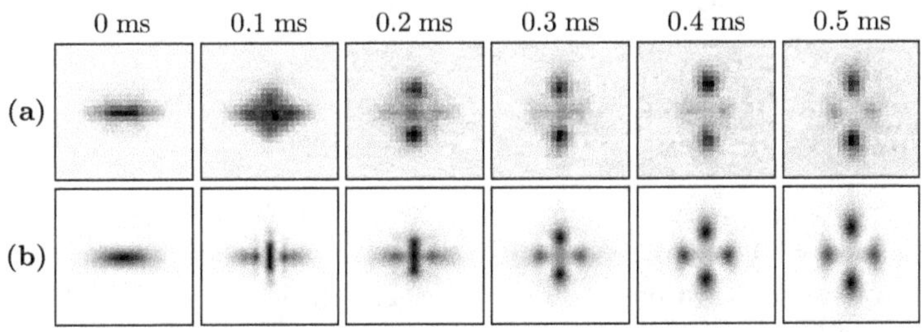

Figure 6.4 (a) Experimental images of a dipolar condensate after collapse and explosion, as a function of the time t_{hold} between the crossing of the critical scattering length for instability and the release from the trap. The time of flight is 8 ms. (b) Results of a numerical simulation of the collapse dynamics, without any adjustable parameter. The field of view is 130 μm ×130 μm. Figure from (Lahaye *et al.* 2008).

When the system becomes unstable (owing to this phonon-like instability), it collapses. This collapse (which is induced by a change in a by means of Feshbach resonances) has been recently observed experimentally (Lahaye *et al.* 2008). Interestingly, the post-collapse images present a cloverleaf pattern, which is caused by the anisotropic collapse of the system (see Figure 6.4). When the atomic density increases, owing to the attractive interaction, three-body losses predominantly occur in the high-density region. The centripetal force is then decreased, and the atoms that gathered in this narrow central region are ejected, owing to the quantum pressure arising from the uncertainty principle. The kinetic energy is supplied by the loss of the negative interaction energy. As the collapse occurs mainly in the $x-y$-direction, because of the anisotropy of the DDI (in the absence of inelastic losses, the condensate would indeed become an infinitely thin cigar-shaped cloud along z), and therefore the condensate explodes essentially radially, producing the anisotropic shape of the cloud.

Note finally, that phonon instability does not necessarily lead to collapse in 2D geometries, and may be accompanied by the formation of stable 2D solitons (Nath *et al.* 2009), which are discussed in Section 6.2.6.

6.2.4 Trapped gases: Thomas–Fermi regime

As for the case of non-dipolar BECs, for sufficiently strong interactions, we may neglect quantum pressure, and consider the Thomas–Fermi (TF) regime:

$$\mu = V_{\text{trap}}(\boldsymbol{r}) + g|\psi(\boldsymbol{r},\boldsymbol{t})|^2 + \int \mathrm{d}^3\boldsymbol{r}'\, U_{\text{dd}}(\boldsymbol{r}-\boldsymbol{r}')|\psi(\boldsymbol{r}',\boldsymbol{t})|^2. \tag{6.22}$$

Amazingly, the TF solution for the trapped BEC has the same inverted parabola shape as for the non-dipolar case (O'Dell *et al.* 2004; Eberlein *et al.* 2005). This is a quite non-trivial result, taking into account the rather complicated form of Eq. (6.22).

In particular, the ground-state density in a cylindrically symmetric trap (with frequencies ω_ρ and $\omega_z = \lambda\omega_\rho$, as before) has the form

$$n(\boldsymbol{r}) = n_0\left(1 - \frac{\rho^2}{R_\rho^2} - \frac{z^2}{R_z^2}\right) \tag{6.23}$$

for $n(\boldsymbol{r}) \geqslant 0$, where $n_0 = 15N/(8\pi R_\rho^2 R_z)$. These expressions are the same as for the non-dipolar case, but the explicit expressions for the TF radii are, of course, different. They may be obtained by substituting Eq. (6.23) into Eq. (6.22) (O'Dell *et al.* 2004; Eberlein *et al.* 2005):

$$R_\rho = \left[\frac{15gN\kappa}{4\pi m\omega_\rho^2}\left\{1 + \varepsilon_{\text{dd}}\left(\frac{3}{2}\frac{\kappa^2 f(\kappa)}{\kappa^2-1} - 1\right)\right\}\right]^{1/5}, \tag{6.24}$$

with $\kappa = R_\rho/R_z$, and $f(\kappa)$, as before.

Interestingly, one may obtain a closed transcendental equation that links the BEC aspect ratio (κ) and the trap aspect ratio ($\lambda^{1/2}$):

$$3\kappa\varepsilon_{\text{dd}}\left[\left(\frac{\lambda^2}{2}+1\right)\frac{f(\kappa)}{\kappa^2-1} - 1\right] + (\varepsilon_{\text{dd}} - 1)(\kappa^2 - \lambda^2) = 0. \tag{6.25}$$

A plot of the condensate aspect ratio as a function of ε_{dd} is shown in Figure 6.5.

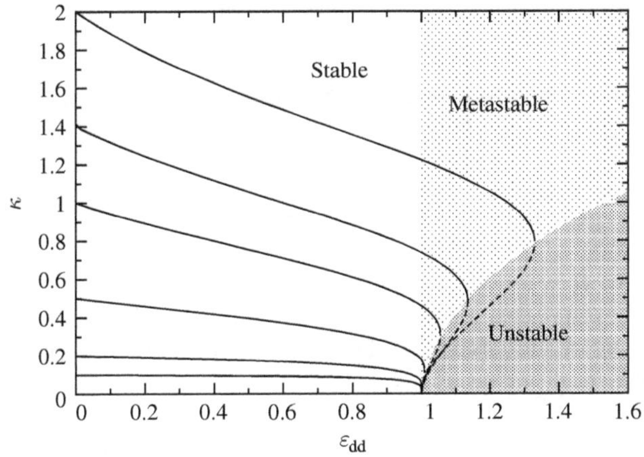

Figure 6.5 Aspect ratio κ of the condensate as a function of the dipole–dipole to s-wave coupling ratio ε_{dd}. Each line is for a different trap aspect ratio, $\lambda = \omega_z/\omega_x$, which can be read off by noting that $\kappa(\varepsilon_{dd} = 0) = \gamma$. When $0 < \kappa < 1$ the condensate is prolate; for $\kappa > 1$, it is oblate. Likewise, for $0 < \gamma < 1$, the trap is prolate, and when $\gamma > 1$, the trap is oblate. Figure courtesy of C. Eberlein.

The latter equation is quite interesting, since it may actually lead to more than one solution κ for a given lambda (see Figure 6.5). One of the solutions (let us call it the normal solution) is stable (or metastable). Interestingly, for $a = 0$, this solution only exists for $\lambda^{1/2} > \lambda_{\rm cr}^{1/2} = 2.28$, i.e., exactly the stability criterion obtained in the previous section for the Gaussian ansatz. On the contrary, the second solution, which appears for a sufficiently large DDI ($\epsilon_{dd} > 1$), is characterized by two main features. On the one hand, it corresponds to a less pancake-like solution. On the other hand, this anomalous solution is unstable. Hence, contrary to what may have been expected from our discussion of the geometric stabilization in the previous section, a TF solution is not stable for arbitrary large number of particles, being unstable against the nucleation of anomalous 'bubbles', which will eventually collapse. This means that, unlike the instability discussed in previous sections, we are not dealing here with the instability of phonon-like excitations, i.e., excitations with a wavelength comparable to that of the condensate. On the contrary, intermediate finite wavelengths become here unstable, leading to local (and not global) collapses. This instability is directly related to the roton instability, which is discussed in the next section.

6.2.5 Roton-like dispersion law

In the previous section, we have seen that a dipolar BEC may become eventually unstable even for pancake-shaped traps, i.e., the BEC may be geometrically stabilized only to some, extent. We shall now try to understand the nature of this instability. For simplicity of our discussion, we shall concentrate on a dipolar BEC harmonically confined in the dipole direction (z-axis) and uniform on the xy plane. The corresponding non-local GPE then reads

$$i\hbar\frac{\partial}{\partial t}\psi(\boldsymbol{r},t) = \left[-\frac{\hbar^2}{2m}\nabla^2 - \mu + \frac{m\omega^2 z^2}{2} + g|\psi(\boldsymbol{r},\boldsymbol{t})|^2 + \int \mathrm{d}\boldsymbol{r}'\, U_{\mathrm{dd}}(\boldsymbol{r}-\boldsymbol{r}')|\psi(\boldsymbol{r}',\boldsymbol{t})|^2\right]\psi(\boldsymbol{r},\boldsymbol{t}). \tag{6.26}$$

where ω is the trap frequency. The ground-state wavefunction is independent of the in-plane coordinates and can be written as $\psi_0(z)$. Then, integrating over the in-plane coordinates in the DDI, we obtain a 1D equation similar to the GP equation for a 1D system with short-range interactions:

$$\left[\frac{-\hbar^2}{2m}\frac{\mathrm{d}^2}{\mathrm{d}z^2} + \frac{m\omega^2 z^2}{2} + (g+g_d)|\psi_0|^2 - \mu\right]\psi_0(z) = 0, \tag{6.27}$$

where $g_d = 8\pi C_{\mathrm{dd}}/3$. In the following, we consider the case $(g+g_d) > 0$, where $\mu > 0$. For $\mu \gg \omega$, the BEC is in the TF regime with a density profile $n_0(z) = n_0(1 - z^2/L^2)$ with a central density $n_0 = \mu/(g+g_d)$ and a TF radius $L = (2\mu/m\omega^2)^{1/2}$.

We are interested in the elementary excitations on top of ψ_0, which we shall study by means of a Bogoliubov–de Gennes analysis. We look for solutions of the form:

$$\psi(\boldsymbol{r},t) = \psi_0(z) + u(z)\mathrm{e}^{\mathrm{i}\boldsymbol{q}\cdot\boldsymbol{\rho}}\mathrm{e}^{-\mathrm{i}\omega t} + v^*(z)\mathrm{e}^{-\mathrm{i}\boldsymbol{q}\cdot\boldsymbol{\rho}}\mathrm{e}^{-\mathrm{i}\omega t}, \tag{6.28}$$

where $u(z)$ and $v(z)$ are complex amplitudes of small oscillations of the condensate around the ground state. The excitations are characterized by a momentum q of the in-plane free motion. Introducing the convenient functions $f_\pm = u \pm v$, the Bogoliubov–de Gennes equations take the form:

$$\hbar\omega f_-(z) = \hat{H}_{\mathrm{kin}} f_+(z), \tag{6.29}$$

$$\hbar\omega f_+(z) = \hat{H}_{\mathrm{kin}} f_-(z) + \hat{H}_{\mathrm{int}}[f_-(z)], \tag{6.30}$$

where

$$\hat{H}_{\mathrm{kin}} = \frac{\hbar^2}{2m}\left[-\frac{\mathrm{d}^2}{\mathrm{d}z^2} + q^2 + \frac{\nabla^2\psi_0}{\psi_0}\right], \tag{6.31}$$

$$\hat{H}_{\mathrm{int}}[f_-] = 2(g+g_d)\psi_0^2(z)f_-(z) - \frac{3}{2}qg_d\psi_0(z)\int_{-\infty}^{\infty}\mathrm{d}z'\psi_0(z')\exp[-q|z-z'|]f_-(z'), \tag{6.32}$$

are the kinetic and the interaction operator, respectively. For each q we have different eigenenergies, $\hbar\omega$. The most interesting is the lowest branch $\omega_0(q)$, which provides us with the dispersion law.

The integral term of $H_{\rm int}[f_-]$ originates from the non-local character of the DDI and gives rise to the momentum dependence of an effective coupling strength. In the limit of low in-plane momenta, $qL \ll 1$, this term can be omitted. In this case, excitations of the lowest branch are essentially 2D and the effective coupling strength corresponds to repulsion. We then recover the Bogoliubov–de Gennes equations for the excitations of a trapped non-dipolar BEC with a coupling constant $(g+g_d) > 0$. In particular, at $q \to 0$, we recover phonons propagating in the xy-plane, with a sound velocity $c_{\rm s} = (2\mu/3m)^{1/2}$.

The situation is very different for $qL \gg 1$. In that case, the excitations become 3D and the effective coupling strength is reduced to $(2g - g_d)$, as one can put $z_0 = z$ in the arguments of f and ψ_0 in the integrand of Eq. (6.32). We hence recover once more the Bogoliubov–de Gennes equations for excitations of a non-dipolar condensate but now with a coupling constant $(2g - g_d)$. If the parameter $\beta = g_d/g < 2$, this coupling constant is positive and one has excitation energies that are real and positive for any momentum q and condensate density n_0. For $\beta > 2$, the coupling constant is negative, and at a sufficiently large density the condensate becomes dynamically unstable with regard to the creation of high-momentum excitations.

We hence see that something quite remarkable may happen as a result of the momentum-dependence of the DDI. For low momenta, we may have stable phonons (i.e., no phonon instability like that discussed in Section 6.2.2), but the BEC may be unstable anyway at finite momenta. Figure 6.6 shows a typical dispersion law as directly obtained from the Bogolibov–de Gennes equations. Note the significant departure when compared with the usual Bogoliubov spectrum. The usual Bogoliubov spectrum is characterized by a phonon dispersion ($\sim q$) at low q and a single-particle dispersion ($\sim q^2$) at large momenta. On the contrary, the dispersion law in a dipolar BEC may become non-monotonical. For in-plane momenta $qL \ll 1$, we have 2D phonon-like

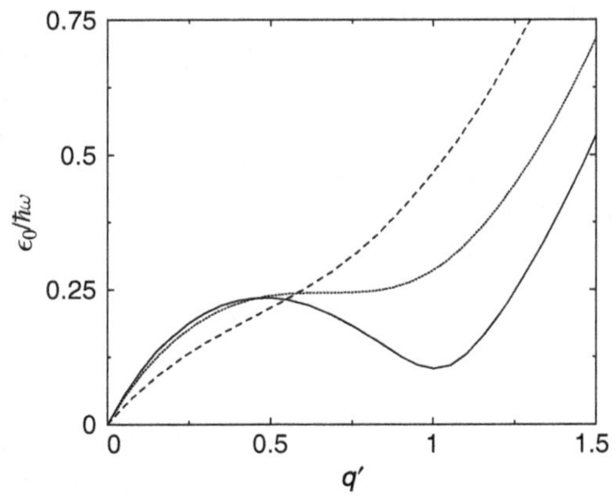

Figure 6.6 Typical dispersion law $\epsilon_0(q)$ ($q' \equiv q l_0$, with $l_0 = \sqrt{\hbar/m\omega}$) for a non-dipolar gas ($\beta = 0$) (dashed), and for a dipolar gas immediately before developing the roton (dotted) and with a developed roton-like minimum (solid). See (Santos *et al.* 2003) for more detail.

excitations; for $qL > 1$, excitations are 3D and the interparticle repulsion is reduced. This decreases the excitation energy under an increase of q. The dispersion reaches a minimum and then starts to grow as the excitations enter the single-particle regime. This minimum resembles that found in helium physics (although the physics behind it is actually rather different), and hence we shall call it a roton-like minimum. If this roton touches zero, the BEC will then become dynamically unstable.

The existence of a roton-like minimum is a remarkable characteristic feature of dipolar condensates. Dipolar BEC indeed constitutes the first example of a weakly interacting gas, which may show a roton-like minimum in the dispersion, which up to now has been observed only in the relatively more complicated physics of liquid helium. Moreover, in contrast to the helium case, by varying the density, the frequency of the confinement, and the short-range coupling, one can easily manipulate and control the spectrum, making the roton minimum deeper or shallower. One can eliminate it completely and obtain the Bogoliubov-type spectrum or, at the opposite case, reach the point of roton instability.

The roton-like minimum has not yet been observed experimentally, although on-going experiments in this direction are being performed in Stuttgart. If created, the presence of a roton-like minimum may be revealed in various ways: a reduction in the superfluid critical velocity (Santos *et al.* 2003), a 'halo' effect at finite temperature in time-of-flight images (Wang and Demler 2010), or the dramatically altered response of the system against a periodic driving in the presence of a roton-like minimum (Nath and Santos 2010).

Let us finally comment on the fate of the condensate after the roton instability sets in. The presence of dynamical instability at finite momentum seems to suggest that the system may develop a modulation with a finite wavelength provided by the inverse roton momentum. If this were so, this could open a route towards supersolidity (one of the holy grails of condensed matter physics), which is a superfluid with both diagonal and non-diagonal long-range order. Unfortunately, it seems that the ultimate answer to this problem is negative. The formation of a modulation is just a transient state, which quickly leads to the formation of local collapses (Shlyapnikov and Pedri 2006; Dutta and Meystre 2007; Komineas and Cooper 2007). However, it has been shown recently that introducing a cut-off of the dipole–dipole interaction at short distances may stabilize the supersolid pattern. This may be achieved by properly tailoring inter-molecular interactions (Wang private communication). Another possibility is provided by the dipole blockade in Rydberg gases, as recently studied in (Cinti *et al.* 2010; Henkel *et al.* 2010).

6.2.6 Solitons

The 1D Gross–Pitaevskii equation (with $a < 0$) supports the existence of solitons, i.e., localized waves that travel with neither attenuation nor change of shape, owing to the compensation between dispersion and non-linearity (Zakharov and Shabat 1972). Solitons have indeed been observed in quasi-1D condensates with $a < 0$ (Khaykovich *et al.* 2002; Strecker *et al.* 2002). The quasi-1D condition requires a tight transversal harmonic trap of frequency $\omega_\perp$, such that $\hbar\omega_\perp$ exceeds the mean-field interaction energy. This in turn demands the transversal BEC size to be smaller than the soliton width. When this condition is violated, the soliton becomes unstable against transversal modulations, and hence multidimensional solitons are not stable in non-dipolar BECs. Remarkably, the latter is not necessarily true in dipolar BECs, where, as a consequence of the non-local non-linearity, 2D bright solitary waves may become stable under appropriate conditions (Pedri and Santos 2005). In the following, we shall consider the scenario studied by Tikhonenkov *et al.* (2008), since it is closer to possible actual experiments (with chromium) than the original proposal of (Pedri and Santos 2005).

The possibility of obtaining stable solitary waves may easily be understood from a simplified discussion, where we consider no trapping in the xz-plane and a strong harmonic confinement with frequency ω_y in the y-direction. The dipoles are oriented along the z-direction, i.e., within the plane of the trap. A good insight on the stability of the solitons may be obtained from a simple Gaussian ansatz:

$$\psi(\boldsymbol{r}) = \frac{1}{l_y^{3/2}} \frac{1}{\pi^{3/4}\Lambda_x\Lambda_z} \mathrm{e}^{\frac{1}{2l_y^2}\left(\frac{x^2}{\Lambda_x^2}+\frac{z^2}{\Lambda_z^2}+y^2\right)}, \tag{6.33}$$

where $l_y = \sqrt{\hbar/m\omega_y}$ is the oscillator length along the transverse direction, and $\Lambda_{x,z}$ are variational parameters that determine the width of the width of the Gaussian (in l_y units). We may insert this ansatz into the Hamiltonian of the non-local GPE, obtaining (apart from unimportant constants):

$$\epsilon \equiv \frac{E}{N\hbar\omega_y} = \frac{1}{4(\Lambda_x^2+\Lambda_z^2)} + \frac{\tilde{g}}{4\pi\Lambda_x\Lambda_z}\left[1 + \epsilon_{\mathrm{dd}} h\left(\frac{\Lambda_x}{\Lambda_z}, \frac{1}{\Lambda_z}\right)\right], \tag{6.34}$$

where $\tilde{g} = \frac{m}{\hbar^2}\frac{Ng}{\sqrt{2\pi}l_y}$, and

$$h(\alpha,\beta) = -1 + 3\int_0^1 \mathrm{d}s \frac{3\alpha\beta s^2}{[1+\alpha^2-1)s^2]^{1/2}\,[1+\beta^2-1)s^2]^{1/2}}. \tag{6.35}$$

Let us consider first what happens for non-dipolar gases. In that case $\epsilon_{\mathrm{dd}} = 0$ and

$$\epsilon(\Lambda = \Lambda_x = \Lambda_z) = \frac{(1+\tilde{g}/2\pi)}{2\Lambda^2}, \tag{6.36}$$

hence, depending on the sign of $1 + \tilde{g}/2\pi$, the system minimizes the energy either by expanding without limits, or by contracting without limits. The localized solution is, therefore, unstable. This is once more the well-known instability of solitons in 2D.

The extra term provided by the DDI is quite interesting, since it introduces an additional dependence on $\Lambda_{x,z}$. This allows (under appropriate conditions) for a minimum in the energy, and hence for a stable self-localized solution. This minimum is characterized by its equilibrium widths, Λ_{x0} and $\Lambda_{z,0}$. Note that they are in general not equal. This asymmetry comes, of course, from the fact that the dipole is along

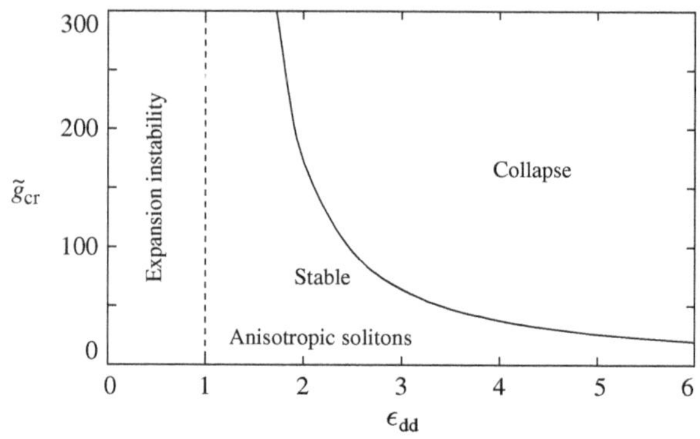

Figure 6.7 Stability diagram of an anisotropic soliton as a function of ε_{dd} and $\tilde{g}_{cr} = g_{cr}/\sqrt{2\pi}l_z$, where for $g > g_{cr}$ the soliton is unstable against collapse even for $\varepsilon_{dd} > 1$. Figure from (Nath *et al.* 2009).

the z-direction. In Figure 6.7 (Nath *et al.* 2009), we show the stability diagram as a function of $\tilde{g}$ and ϵ_{dd}. There we observe two instability regions for 2D solitons (against collapse and against unlimited expansion). For $\epsilon_{dd} > 1$, there is a critical universal value $\tilde{g}_{cr}(\beta) \equiv gN_{cr}/\sqrt{2\pi}l_z$, such that for $N > N_{cr}$ the minimum of $E(\Lambda_x, \Lambda_z)$ disappears. As a consequence, stable 2D anisotropic self-localized solutions are stable only for a number of particles per soliton below a critical number N_{cr}, which decreases for larger ϵ_{dd}. Beyond this number, the 2D soliton collapses. This result is also verified by a direct simulation of the 3D non-local Gross–Pitaevskii equation.

In this simplified discussion, we have assumed that the problem remains 2D. If the interactions increase, the problem becomes 3D, and one may show that the condensate eventually becomes unstable (Pedri and Santos 2005; Tikhonenkov *et al.* 2008).

A major difference between bright solitons in non-dipolar and dipolar BECs concerns the soliton–soliton scattering properties. Whereas 1D solitons in non-dipolar BECs scatter elastically, the scattering of dipolar solitons is inelastic, owing to the lack of integrability (Krolikowski *et al.* 2001). The solitons may transfer centre-of-mass energy into internal vibrational modes, resulting in intriguing scattering properties, including soliton fusion (Pedri and Santos 2005) (see Figure 6.8), the appearance of strong inelastic resonances (Nath *et al.* 2007), and the possibility of observing 2D-soliton spiralling, like that already observed in photo-refractive materials (Shih *et al.* 1997).

6.2.7 Dipolar effects in spinor condensates

Let us finish the discussion on dipolar BEC with some brief comments about dipolar effects in spinor BECs, i.e., in BECs formed by atoms with non-zero spin. Dipolar effects may lead to very interesting phenomena in spinor BECs (for a review see, e.g., (Lahaye *et al.* 2009)) but here we shall focus only on a particular effect, which resembles the famous Einstein–de Haas effect.

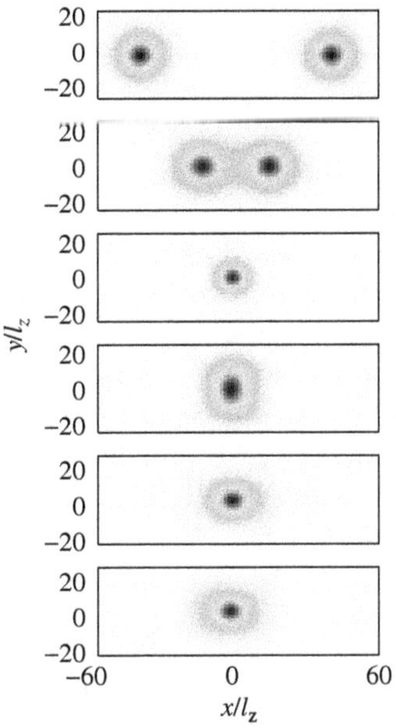

Figure 6.8 Density plot of the fusion of two dipolar 2D solitons. See (Pedri and Santos 2005) for details.

In a spinor BEC, we have various Zeeman sublevels with quantum number m. The short-range interactions may hence occur in different s-wave scattering channels with different total angular momentum, which for the case of bosons must be an even number (i.e., for spin-1 bosons we have just two scattering channels, namely $F = 0$ and $F = 2$) (Ho 1998). For each scattering channel, we have an associated s-wave scattering length a_F. The short-range interactions necessarily preserve the spin projection S_z along the quantization axis.

The DDI for a spinor BEC is of the form

$$\hat{V}_{\rm dd} = \frac{C_{\rm dd}}{2}\int \mathrm{d}\boldsymbol{r}\int \mathrm{d}\boldsymbol{r}'\frac{1}{|\boldsymbol{r}-\boldsymbol{r}'|^3}\hat{\psi}^\dagger_m(\boldsymbol{r})\hat{\psi}^\dagger_{m'}(\boldsymbol{r}')$$
$$\left[\boldsymbol{S}_{mn}\cdot\boldsymbol{S}_{m'n'} - 3(\boldsymbol{S}_{mn}\cdot\boldsymbol{e})(\boldsymbol{S}_{m'n'}\cdot\boldsymbol{e})\right]\hat{\psi}_n(\boldsymbol{r})\hat{\psi}_{n'}(\boldsymbol{r}'), \qquad (6.37)$$

where $\boldsymbol{S} = (S_x, S_y, S_z)$ are the spin-F matrices, $C_{\rm dd} = \mu_0\mu_{\rm B}^2 g_F^2/4\pi$, and $\boldsymbol{e} = (\boldsymbol{r}-\boldsymbol{r}')/|\boldsymbol{r}-\boldsymbol{r}'|$.

Interestingly, unlike the short-range interaction, the dipole–dipole interaction does not necessarily conserve the spin projection along the quantization axis, owing to the anisotropic character of the interaction. In particular, if the atoms are initially prepared in a maximally stretched state, say $m_F = -F$, short-range interactions cannot induce any spinor dynamics, owing to the previously mentioned conservation of total magnetization S_z. Dipole–dipole interactions, on the contrary, may induce a transfer into $m_F + 1$. If the system preserves cylindrical symmetry around the quantization axis, this violation of the spin projection is accompanied by a transfer of angular momentum to the centre of mass, resembling the well-known Einstein–de Haas effect (Kawaguchi *et al.* 2006; Santos and Pfau 2006). Owing to this transfer, an initially spin-polarized dipolar condensate can generate dynamic vorticity.

Unfortunately, the Einstein–de Haas effect is destroyed in the presence of even rather weak magnetic fields. Typically, magnetic fields well below 1 mG are necessary to observe the effect. Owing to the dominant role of Larmor precession, and invoking rotating-wave-approximation arguments, the physics must be constrained to manifolds of preserved magnetization (this may be overcome in tight 2D optical lattices, as suggested by recent results at Paris Nord (Pasquiou *et al.* 2011).

However, the dipole–dipole interaction may also have observable effects under conserved magnetization S_z, even for alkali spinor condensates. The mechanism for spinor dynamics in spinor BECs is the so-called spin-changing collision, i.e., a collision that conserves S_z but does not conserve the relative population of the different Zeeman components. Spin-changing collisions are characterized by an energy scale proportional to the difference between scattering lengths at different channels. This difference is typically very small, hence the energy of spin-changing collisions is very low. This is particularly so for the case of ^{87}Rb $F = 1$. As a consequence, the spinor physics may be significantly modified by the presence of other small energy scales, and in particular the DDI. Experiments at Berkeley (Vengalattore *et al.* 2008) have demonstrated the dipolar character of spin-1 ^{87}Rb spinor BECs. In particular, these experiments show the spontaneous decay of helical spin textures (externally created by magnetic field gradients) towards a spatially modulated structure of spin domains (see Figure 6.9).

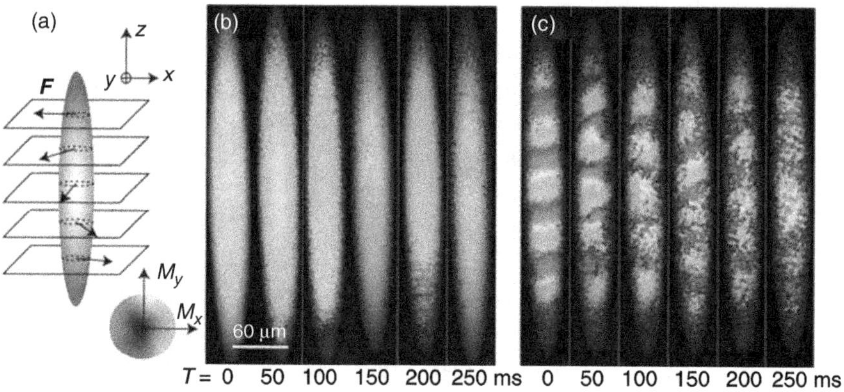

Figure 6.9 Spontaneous dissolution of helical textures in a quantum degenerate ^{87}Rb spinor Bose gas. A transient magnetic field gradient is used to prepare transversely magnetized (b) uniform or (a), (c) helical magnetization textures. The transverse magnetization column density after a variable time T of free evolution is shown in the imaged xz plane. (b) A uniform texture remains homogeneous for long evolution times, while (c) a helical texture with pitch $\lambda = 60\,\mu$m dissolves over ~200 ms, evolving into a sharply spatially modulated texture. Figure from (Vengalattore *et al.* 2008).

The formation of this modulated phase has been ascribed to magnetic dipolar interactions that energetically favour short-wavelength domains over the long-wavelength spin helix. Interestingly, the reduction of dipolar interactions (by means of radio-frequency pulses) results in a suppression of the modulation.

6.3 Dipolar gases in optical lattices

At this point, I shall discuss some of the fascinating effects introduced by the DDI in the physics of ultracold gases in optical lattices. I shall first comment on dipolar BECs in 1D optical lattices, and in particular the role of the non-local intersite interactions, a major difference between dipolar and non-dipolar gases. I shall then move away from weakly interacting systems, discussing the extended Hubbard model. This lattice model allows for various quantum phases, which are briefly discussed. Finally, I shall comment on pair-superfluidity in bilayer Bose systems, and on filament quantum gases in multilayer systems.

6.3.1 Dipolar BEC in a one-dimensional optical lattice

Let us begin our discussion of dipolar gases in optical lattices with the case of a dipolar BEC in a deep one-dimensional optical lattice of the form $V_{\text{latt}}(z) = V_0 \sin^2(\pi z/\Delta)$, where Δ is the spacing between two sites of the lattice. For simplicity, we shall assume that there is no trapping in the xy-plane.

The potential $V_{\text{latt}}(z)$ is a periodic potential, and hence the single-particle energy spectrum is characterized by the appearance of bands and gaps. We shall suppose in the following that the gap between the lowest and the second band is large enough compared with other energy scales that we may reduce our discussion to the lowest band. Since we consider a deep lattice (tight-binding regime), a good basis is provided by the Wannier functions (of the lowest band) $\phi_j(z)$, which are maximally localized at site j. For very deep lattices, $V_{\text{latt}}(z)$ may be locally approximated at a site minimum by a harmonic potential, and the corresponding Wannier lattice for the lowest band may be approximated by a Gaussian function. The Wannier functions form a complete set of functions (in the same way as the Bloch functions), and we may then project the BEC wavefunction in the Wannier basis: $\psi(\boldsymbol{r}) = \sum_j \psi_j(\boldsymbol{r}) \equiv \sum_j \Psi_j(x,y)\phi_j(z)$. Employing this projection, we may rewrite the Hamiltonian in the form

$$\begin{aligned}\hat{H} = &\sum_{j,j'} \left[\int \mathrm{d}^3 r \psi_j^*(\boldsymbol{r}) \left[\frac{-\hbar^2}{2m} \nabla^2 + V_{\text{latt}}(z) \right] \psi_{j'}(\boldsymbol{r}) \right] \\ &+ \frac{1}{2} \sum_{\substack{j,j' \\ l,l'}} \left[\iint \mathrm{d}^3 r \mathrm{d}^3 r' \psi_j^*(\boldsymbol{r}) \psi_l^*(\boldsymbol{r}') \psi_{l'}(\boldsymbol{r}') \psi_{j'}(\boldsymbol{r}) U(\boldsymbol{r} - \boldsymbol{r}') \right], \end{aligned} \tag{6.38}$$

where $U(\boldsymbol{r}) = g\delta(\boldsymbol{r}) + U_{\text{dd}}(\boldsymbol{r})$.

For sufficiently deep lattices, we may neglect in the first line all terms except for $j = j'$ and those terms where j and j' are nearest neighbours. In the same way, we may neglect in the interaction part all terms except those with $j = j'$ and $l = l'$. Owing to the Gaussian-like localization of the on-site wavefunctions $\psi_j(\boldsymbol{r})$, the neglected terms are typically exponentially smaller than the terms kept. The Hamiltonian is then of the form

$$\hat{H} = -J\sum_{\langle i,j\rangle}\Psi_i^*(\boldsymbol{\rho})\Psi_j(\boldsymbol{\rho})$$
$$+\sum_j\int d^2\rho\Psi_j^*(\boldsymbol{\rho})\left[\frac{-\hbar^2}{2m}\nabla_{xy}^2+g\delta_{i,j}\delta(\boldsymbol{\rho}-\boldsymbol{\rho}')+\int d^2\rho' U_{ij}(\boldsymbol{\rho}-\boldsymbol{\rho}')|\Psi_j(\boldsymbol{\rho}')|^2\right]\Psi_j(\boldsymbol{\rho}), \tag{6.39}$$

where

$$J \equiv -\int \mathrm{d}^3 r\phi_i^*(z)\left[\frac{-\hbar^2}{2m}\partial_z^2 + V_{\mathrm{latt}}(z)\right]\phi_j(z), \tag{6.40}$$

$$U_{ij}(\boldsymbol{\rho}-\boldsymbol{\rho}') \equiv \int \mathrm{d}z\mathrm{d}z' U_{\mathrm{dd}}(\boldsymbol{r}-\boldsymbol{r}')|\phi_i(z)|^2|\phi_j(z')|^2 \tag{6.41}$$

describe, respectively, hopping between nearest neighbours, and the DDI between the site i and the site j. Note that the on-site interactions ($i = j$) result from the interplay between short-range and dipole–dipole interactions. On the contrary, the intersite interaction stems directly from the DDI.

Indeed, in non-dipolar gases, intersite interactions are exponentially suppressed, since the range of the interactions is typically much smaller than the intersite distance. On the contrary, *strong intersite interactions are a characteristic novel feature introduced by the DDI in the physics of dipolar lattice gases.* In particular, in the case of vanishing hopping ($J = 0$), i.e., in the case of very deep lattices, we would obtain in the case of non-dipolar BECs a set of independent 2D condensates at each layer. Indeed, this has been employed in several experiments (Hadzibabic *et al.* 2006) for achieving 2D gases. On the contrary, the non-local intersite interaction due to the DDI ensures that the BECs at different layers are not independent, in spite of the absence of hopping.

This has important consequences, for example for the excitations and the stability of dipolar condensates. Let us analyse this.

Dipolar BEC in a single two-dimensional layer Let us consider first the case of a single layer, i.e., of a quasi-2D dipolar condensate. As mentioned, we can approximate the lattice potential by a harmonic oscillator $V_{\mathrm{latt}}(\vec{r}) \simeq m\omega_z^2\, z^2/2$. The wavefunction along z may be approximated by a Gaussian $\phi(z) = \exp(-z^2/2l_z^2)/\pi^{1/4}l_z^{1/2}$ ($l_z = \sqrt{\hbar/m\omega_z}$). The effective oscillator length l_z is related to the lattice constant Δ as $l_z \approx \Delta(V_0/E_r)^{-1/4}/\pi$, where $E_r = \hbar^2\pi^2/2m\Delta^2$ is the so-called recoil energy. The ground state of the homogeneous 2D BEC is of the form $\Psi_\perp(\vec{\rho}, t) = \exp(-\mathrm{i}(\mu/\hbar + \omega_z)t)\sqrt{n_0}$, where n_0 is the 2D density, and μ is the 2D chemical potential. Introducing this form into the non-local GPE, one obtains $\mu = (g + g_d)n_0/\sqrt{2\pi}l_z$. Note that the 2D condition is satisfied for $\mu \ll \hbar\omega_z$.

We may now evaluate the excitation spectrum following the same Bogoliubov–de Gennes procedure employed in previous sections. We insert a plane-wave ansatz

$$\Psi(\vec{r},t) = \Phi_0(z)\exp(-\mathrm{i}(\mu/\hbar+\omega_z)t)(\sqrt{n_0} + u_q\exp(\mathrm{i}\vec{q}\cdot\vec{\rho} - \mathrm{i}\epsilon t/\hbar) - v_q^*\exp(-\mathrm{i}\vec{q}\cdot\vec{\rho} + \mathrm{i}\epsilon t/\hbar)) \tag{6.42}$$

into the GPE, and linearize in u_q, v_q, obtaining the Bogoliubov spectrum:

$$\epsilon(q) = \left\{\frac{\hbar^2 q^2}{2m}\left[\frac{\hbar^2 q^2}{2m} + 2A\right]\right\}^{1/2}, \tag{6.43}$$

where $A = \mu - \tilde{g}_d F(ql_z/\sqrt{2})$, with $\tilde{g}_d = g_d n_0/\sqrt{2\pi}l_z$, and $F(x) = \frac{3\sqrt{\pi}}{2}|x|\mathrm{erfc}(x)\mathrm{e}^{x^2}$. Note that without DDI ($\beta = 0$) we recover the usual Bogoliubov spectrum for a non-dipolar 2D BEC. In the following, we consider $a < 0$ (this is because we would like to have the possibility of a roton-like minimum in the quasi-2D Bogoliubov spectrum, although, of course, intersite effects are also remarkable even if $a > 0$). If $a < 0$ and $\beta = 0$, $\epsilon(q)^2 < 0$ for $q \to 0$, recovering the phonon instability in homogeneous BEC with $a < 0$. If the dipole is sufficiently large, such that $g + g_d > 0$, then the DDI prevents the instability at $q \to 0$. However, owing to the q-dependence of the DDI (given by the monotonously increasing character of the function F), the dispersion $\epsilon(q)$ may show for intermediate g_d values a roton-like minimum at a finite value of ql_z (this is similar to Section 6.2.5). For sufficiently low DDI, $\epsilon(q)^2 < 0$ at the roton-like minimum, leading to the roton-like instability. For $|\beta| > \beta_{\mathrm{cr}}$ (with β_{cr} dependent on the ratio $gn_0/l_z\hbar\omega_z$) roton instability is prevented, and the 2D homogeneous BEC is stable.

Two layers Let us consider now the case of two layers with no hopping. The system is then described by two coupled GPEs of the form:

$$\mathrm{i}\hbar\frac{\partial}{\partial t}\Psi_i(\boldsymbol{\rho},t) = \left[-\frac{\hbar^2}{2m}\nabla^2 + g|\Psi_i(\boldsymbol{\rho},t)|^2 + \sum_j\int \mathrm{d}^3r'\, U_{ij}(\boldsymbol{\rho}-\boldsymbol{\rho}')|\Psi_j(\boldsymbol{\rho}',t)|^2\right]\Psi(\vec{r},t), \tag{6.44}$$

where $i, j = 1, 2$. Note that, crucially, the DDI now couples the ith layer to the jth one. As with the single-site discussion, we consider a strong z-confinement at each site, and hence we may employ a quasi-2D ansatz $\Psi_i(\vec{r}) = \Psi_{\perp,i}(\vec{\rho},t)\phi(z-z_i)$, where $\phi(z)$ has the form discussed, and z_i is the position of the ith lattice node. The ground state of the condensates at the two layers is given by $\Psi_j = \sqrt{n_0}$. Introducing this ansatz into the NLSE, Eq. (6.44), we obtain the 2D chemical potential $\tilde{\mu} = \mu + \lambda(\Delta)$, with μ the chemical potential of an individual well and $\lambda(\Delta) = (g_d n_0/\sqrt{2\pi}l_z)\mathrm{e}^{-\Delta^2/2l_z^2}$. Note that the intersite interaction is a Gaussian function of the intersite spacing and not of the form $1/\Delta^3$. The Gaussian dependence appears because the intersite interactions are actually between two planes with an extension much larger (in our homogeneous approximation, infinitely larger) than the intersite distance.

As before, we are interested in the elementary excitation of these systems. For $\Delta \to \infty$, the Bogoliubov modes at each site are independent and described by the single-site expression, Eq. (6.43). For finite Δ the intersite coupling leads to a hybridization of the modes at both sites with significant consequences, as we discuss next. As for the single-site discussion, we insert a plane-wave ansatz $\Psi_i(\vec{r},t) = (\sqrt{n_0} + u_{qi}\exp(\mathrm{i}\vec{q}\cdot\vec{\rho} - \mathrm{i}\epsilon t/\hbar) - v_{qi}^*\exp(-\mathrm{i}\vec{q}\cdot\vec{\rho} + \mathrm{i}\epsilon t/\hbar))\Phi_0(z - z_i)\exp(-\mathrm{i}(\tilde{\mu}/\hbar + \omega_z)t)$ into the NLSE, Eq. (6.44), and linearize in u_{qi}, v_{qi}. In this way, we obtain four coupled Bogoliubov–de Gennes equations for $\{u_{1,2}, v_{1,2}\}$, which may be diagonalized to obtain the Bogoliubov modes:

$$\epsilon_{\pm}(q) = \{E_q\left[E_q + 2A \pm 2C(\Delta)\right]\}^{1/2}, \tag{6.45}$$

where

$$C(\Delta) = \lambda(\Delta) - \tilde{g}_d \tilde{F}\left(\frac{ql_z}{\sqrt{2}}, \frac{\Delta}{\sqrt{2}l_z}\right), \tag{6.46}$$

with $\tilde{F}(x,y) = \frac{3\sqrt{\pi}xe^{x^2}}{4}\sum_{\alpha=\pm1} \mathrm{e}^{-2\alpha xy}\mathrm{erfc}(x - \alpha y)$.

Note that for $\Delta \to \infty$, $C(\Delta) = 0$ and we recover two degenerate independent modes. For finite Δ the modes at the two wells hybridize, and two different branches appear for each q, one stiffer than the modes for $\Delta \to \infty$, and the other softer. The latter is particularly interesting, since the soft mode is more prone to rotonization (Figure 6.10). Interestingly, under proper conditions, two parallel non-overlapping BECs may become roton-unstable even if they are stable separately. As a consequence, a larger β_{cr} is necessary to stabilize the two-well system.

Multilayer case The hybridization (and consequent destabilization) in two-well potentials becomes even more pronounced for the case of dipolar BECs at $N_{\mathrm{s}} > 2$ sites of a 1D optical lattice, since a site i couples with all its neighbours j (naturally, with

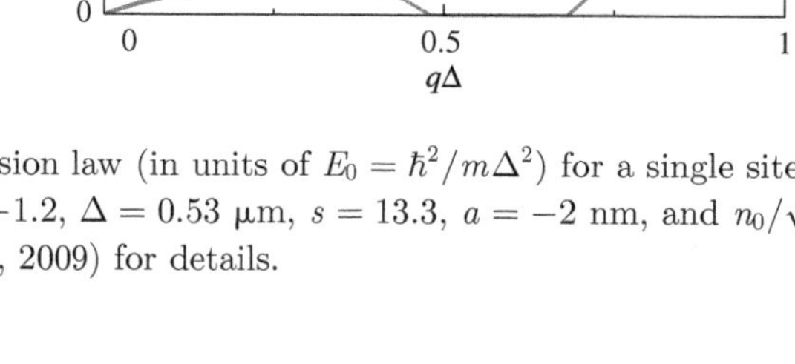

Figure 6.10 Dispersion law (in units of $E_0 = \hbar^2/m\Delta^2$) for a single site (dashed) and for two wells (solid) for $\beta = -1.2$, $\Delta = 0.53$ μm, $s = 13.3$, $a = -2$ nm, and $n_0/\sqrt{2\pi}l_z = 10^{14}/\mathrm{cm}^3$. See (Klawunn and Santos, 2009) for details.

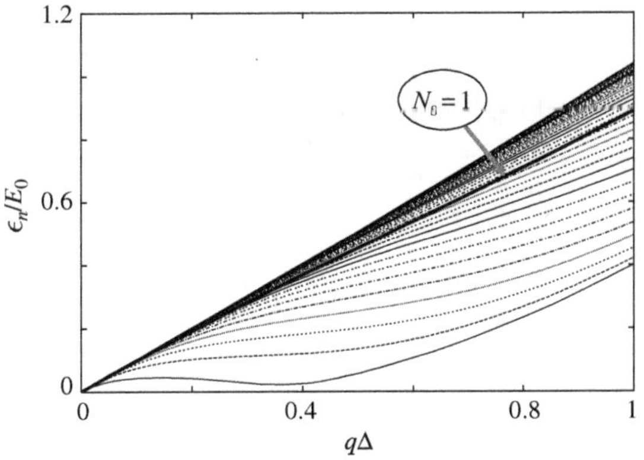

Figure 6.11 Band-like dispersion (in units of $E_0 = \hbar^2/m\Delta^2$) for $N_s = 40$, $\beta = -2.44$, $\Delta = 0.53$ μm, $s = 13.3$, $a = -2$ nm, and $n_0/\sqrt{2\pi}l_z = 10^{14}/\text{cm}^3$. The dispersion law for $N_s = 1$ is indicated. See (Klawunn and Santos, 2009) for details.

decreasing strength for growing $|i - j|$). For simplicity of our analysis, we consider the case in which all lattice sites present the same 2D density n_0. In that case, one may easily generalize the two-site analysis to the multisite case, to produce a set of coupled Bogoliubov–de Gennes equations:

$$\epsilon^2 f_{qi} = E_q(E_q + 2A)f_{qi} + 2E_q \sum_{j \neq i} C(\Delta|i - j|)f_{qj}, \tag{6.47}$$

where $f_{qi} = u_{qi} + v_{qi}$ After diagonalizing the matrix of coefficients at the r.h.s. of Eq. (6.47), one may obtain numerically the corresponding band-like set of N_s elementary excitations (Figure 6.11). Note that the band-like spectrum has an upper phonon-like boundary, which for large N_s has an approximate sound velocity $c_s \simeq \sqrt{(A + \sum_n C(\Delta|n|))/m}$. The lower mode of the N_s manifold becomes significantly softer than the individual modes for independent sites. As a consequence, the roton instability extends to larger β_{cr} when N_s increases, until saturating for a sufficiently large N_s (owing to the decreasing DDI for increasing distance between sites).

Figure 6.12 summarizes the behaviour of the stability as a function of β (we recall that $g < 0$). As mentioned, if $g + g_d < 0$ ($|\beta| < 1$) the system is unstable against phonon instability. For $1 < |\beta| < |\beta_{\text{cr}}(N_s)|$, the system is unstable against roton instability. $|\beta_{\text{cr}}|$ increases with N_s, saturating for sufficiently large N_s. For $|\beta| > |\beta_{\text{cr}}(N_s)|$, the quasi-2D BECs are stable.

Summarizing this section, we have seen that interlayer interactions may have rather dramatic effects in dipolar BECs even in the absence of hopping. On the one hand, the system presents band-like excitations collectively shared by the BECs at different layers. On the other hand, a BEC that would be stable in a single layer may become unstable if other layers are present.

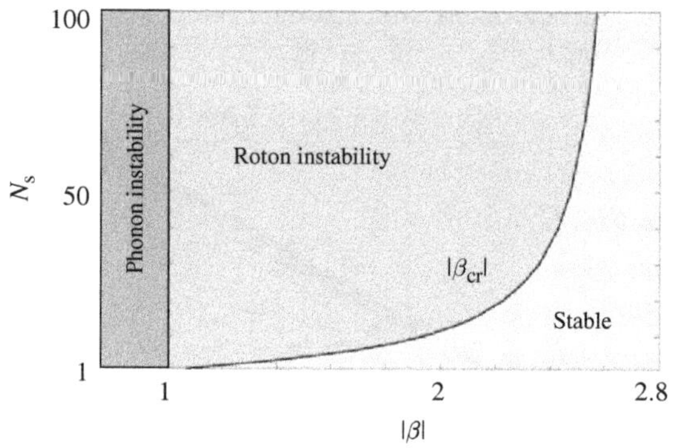

Figure 6.12 Stable and unstable regimes for N_s dipolar 2D BECs. Same parameters as in Figure 6.11. See (Klawunn and Santos, 2009) for details.

6.3.2 Dipole–dipole interactions in strongly correlated lattice gases: extended Hubbard model

In the previous section, we have seen that the intersite interactions may have quite dramatic effects on the physics of BECs in deep lattices. The DDI may also change quite significantly the physics of strongly correlated gases in optical lattices. In this section, I shall briefly comment on that.

Let us consider a collection of dipolar bosons loaded in a 2D square lattice $V_{\text{latt}}(x, y)$ (fermions will also have interesting physics but we shall focus here only on bosons). In the transverse direction, we assume a harmonic confinement with frequency ω_z, which is sufficiently tight that the z wavefunction is the ground state of the harmonic oscillator $\varphi_0(z)$. We shall assume that the dipoles are oriented perpendicularly to the lattice.

The system is described by the Hamiltonian (note that now we keep the operator character of the fields since we are interested in strongly correlated systems):

$$\hat{H} = \int \mathrm{d}^3 r \hat{\psi}^\dagger(\boldsymbol{r}) \left[\frac{-\hbar^2}{2m} \nabla^2 + V_{\text{latt}}(x, y) + \frac{1}{2} m \omega_z z^2 + \frac{1}{2} \int \mathrm{d}^3 r' \, U(\boldsymbol{r} - \boldsymbol{r}') \hat{\psi}^\dagger(\boldsymbol{r}') \hat{\psi}(\boldsymbol{r}') \right] \hat{\psi}(\boldsymbol{r}), \tag{6.48}$$

where $U(\boldsymbol{r}) = g\delta(\boldsymbol{r}) + U_{\text{dd}}(\boldsymbol{r})$, and $\hat{\psi}(\boldsymbol{r})$ is the field operator that annihilates a boson in $\boldsymbol{r}$. As in the discussion of the previous section, since V_{latt} is a periodic potential, the single-particle energy spectrum is characterized by the appearance of bands and gaps. As previously, we consider again that we can reduce to the lowest band, and project in the basis of Wannier functions $\Phi_j(\boldsymbol{r})$, which are maximally localized at site j (for a 2D square lattice $j \equiv (j_x, j_y)$). We may then express the field operator as $\hat{\psi}(\boldsymbol{r}) = \sum_j \phi_j(\boldsymbol{r}) \hat{a}_j$, where $\phi_j = \Phi_j(x, y)\varphi_0(z)$ and $\hat{a}_j$ annihilates a particle at site j. Employing this projection we may rewrite the Hamiltonian in the form

$$\hat{H} = \sum_{j,j'} \left[\int \mathrm{d}^3 r \phi_j^*(\boldsymbol{r}) \left[\frac{-\hbar^2}{2m} \nabla^2 + V_{\mathrm{latt}}(\boldsymbol{r}) \right] \phi_{j'}(\boldsymbol{r}) \right] \hat{a}_j^\dagger \hat{a}_{j'}$$
$$+ \frac{1}{2} \sum_{\substack{j,j' \\ l,l'}} \left[\iint \mathrm{d}^3 r \mathrm{d}^3 r \phi_j'^*(\boldsymbol{r}) \phi_l^*(\boldsymbol{r}') \phi_{l'}(\boldsymbol{r}') \phi_{j'}(\boldsymbol{r}) U(\boldsymbol{r} - \boldsymbol{r}') \right] \hat{a}_j^\dagger \hat{a}_l^\dagger \hat{a}_{l'} \hat{a}_{j'}. \tag{6.49}$$

We proceed now as in the previous section. For sufficiently deep lattices, we neglect in the first line all terms except for $j = j'$ (this term merely leads to a global energy shift of the sites and will be neglected) and those terms where j and j' are nearest neighbours. In the same way, we may neglect in the interaction part all terms except those with $j = j'$ and $l = l'$. We may then reduce the Hamiltonian to the so-called extended Bose–Hubbard model:

$$\hat{H} = -t \sum_{\langle j,j' \rangle} \hat{a}_j^\dagger \hat{a}_{j'} + \frac{U_0}{2} \sum_{j,j'} \hat{n}_j (\hat{n}_j - 1) + \sum_{\substack{\delta \\ |\delta|>0}} \frac{U_\delta}{2} \sum_j \hat{n}_j \hat{n}_{j+\delta}, \tag{6.50}$$

where

$$t \equiv - \int \mathrm{d}^3 r \phi_j^*(\boldsymbol{r}) \left[\frac{-\hbar^2}{2m} \nabla^2 + V_{\mathrm{latt}}(\boldsymbol{r}) \right] \phi_{j'}(\boldsymbol{r}), \tag{6.51}$$

$$U_0 \equiv g \int \mathrm{d}^3 r |\phi_j(\boldsymbol{r})|^4 + \iint \mathrm{d}^3 r \mathrm{d}^3 r' |\phi_j(\boldsymbol{r}')|^2 |\phi_j(\boldsymbol{r})|^2 U(\boldsymbol{r} - \boldsymbol{r}'), \tag{6.52}$$

$$U_\delta \equiv \iint \mathrm{d}^3 r \mathrm{d}^3 r' |\phi_j(\boldsymbol{r}')|^2 |\phi_{j+\delta}(\boldsymbol{r})|^2 U(\boldsymbol{r} - \boldsymbol{r}') \tag{6.53}$$

describe, respectively, hopping between nearest neighbours, on-site interactions, and intersite interactions. As in Section 6.3.1, note again that, crucially, the on-site interactions $U_0 \equiv U_0^{(sc)} + U_0^{(\mathrm{dd})}$ result from the interplay between short-range ($U_0^{(sc)}$) and dipole–dipole interactions ($U_0^{(\mathrm{dd})}$), but the intersite interaction (U_δ) stems directly from the DDI. Hence, the possibly quite strong interactions between sites are a novel effect introduced by the DDI, which may radically change the physics of strongly correlated gases in optical lattices, as we shall see.

Note that dipolar gases in 2D lattices allow for different forms of control of the coupling constants of the extended Hubbard Hamiltonian:

- As for any other lattice gases, we may change the hopping t by changing the depth of the lattice.
- Again, as for any lattice gas, we may change the short-range interactions by employing Feshbach resonances. Note that this will affect the on-site interactions (U_0).
- Note that because of the anisotropy of the DDI, the spatial dependence of the on-site wavefunction may significantly modify $U_0^{(\mathrm{dd})}$. In particular, depending on the

relation between the transversal oscillator length and the on-site extension in the xy plane, $U_0^{(\mathrm{dd})}$ may be positive or negative. Hence, remarkably, by controlling the geometry of the trapping we may control $U_0^{(\mathrm{dd})}$, allowing for the change between different interaction regimes. For example, we may arrive at a situation in which $U_0 = 0$ although $U_\delta \neq 0$, i.e., we may get an extended Hubbard model without on-site interactions.

- Note also that the U_δ coupling constants decay with δ (for zero-dimensional sites as $1/\delta^3$). As a consequence, it depends on the dipole strength and the lattice constant whether U_δ is relevant (compared with other energy scales in the system, in particular the on-site interaction U_0). Under some conditions, one may consider just the nearest neighbour (U_1) or the next-to-nearest neighbour (U_2), but for a strong dipole (as that of very polar molecules) even further neighbours contribute significantly.
- Although we do not use it here, we may also modify the angle between the vector normal to the lattice and the dipole direction. In that case, the intersite interactions would be anisotropic on the 2D lattice.

Therefore, in addition to standard control possibilities, dipolar gases offer novel possibilities for the control of quantum gases in optical lattices.

6.3.3 Quantum phases of dipolar bosons in optical lattices

For the case $U_\delta = 0$ (i.e., the non-dipolar case) we recover the usual Bose–Hubbard model. It is quite well known that in that case, and depending on the chemical potential, the tunnelling, and the on-site interactions, the system may present two distinct quantum phases: a superfluid phase, and a Mott-insulator phase (with a commensurate filling per site). In dipolar gases, the extended Hubbard model leads to a much richer physics of possible quantum phases, which we can only briefly consider here. Let us briefly discuss some of the most remarkable consequences.

Supersolid Since the dipole is perpendicular to the lattice plane, $U_\delta > 0$, i.e., we have repulsive intersite interactions. As a result, it seems quite intuitive that the bosons will 'dislike' being at neighbouring sites, and hence it is not surprising that for strong enough interactions the density of particles may present a modulation (which is not the trivial modulation induced by the lattice). Actually, for a sufficiently large U_δ (and a commensurate filling $\bar{n} = 1/2$), the system may enter into a chequerboard phase (Góral *et al.* 2002); i.e., the bosons are placed as in the dark squares of a chequerboard. This phase is obviously an insulating phase (and actually one may call it a Mott insulator with half filling).

However, a much more intriguing phase may occur (Góral *et al.* 2002), in which the system remains superfluid but the density presents a modulation. This phase is called a supersolid, and constitutes one of the holy grails of condensed-matter physics. In principle, this phase may be unstable against phase separation into an insulating crystal and a superfluid. It has recently been shown by means of quantum Monte Carlo calculations that a supersolid may be stabilised against phase separation in an extended Bose–Hubbard model with just nearest-neighbour interactions, as long as the filling of the lattice $\bar{n} > 1/2$ and $U_1 > U_0/z$ (where z is the coordination number of the lattice; $z = 4$ for a square lattice) (Sengupta *et al.* 2005). Note that, as mentioned before, we may easily control the ratio U_1/U_0 by Feshbach resonances or by modifying the transversal confinement, and hence the supersolid phase may be stabilized. Not only is a chequerboard supersolid possible, but also under proper circumstances (e.g., including next-to-nearest neighbour interactions) one may achieve supersolids with other patterns (striped and quarter-filled crystals) (Chen *et al.* 2008).

Haldane insulator Another interesting phase induced by the DDI has been discussed for the case of bosonic dipoles in a 1D optical lattice by Dalla Torre *et al.* (2006). Let us consider the case of a filling factor $\bar{n} = 1$ per site. Let us assume that we may have sites with zero particles but consequently also sites with $n = 2$. Let us assume that only these three possibilities are possible; $n = 0, 1, 2$. One may cut at $n = 2$ if U_0 is sufficiently large, however one may check that the presence of sites with $n = 3$ does not spoil the physics discussed next. We may then introduce a pseudo-spin $S_z = n - \bar{n}$, which may acquire values $-1, 0, 1$. Hence, interestingly, the system is, to a large extent, equivalent to a spin-1 gas system with extended interactions, realizing a so-called Haldane spin-1 chain. It is known that such system may present a rather intriguing phase characterized by string correlations, in which $S_z = \pm 1$ appear in alternating order along the chain separated by strings of $S_z = 0$ of arbitrary length. In other words, we have a site with zero (two) particles, followed by whatever number of sites with one particle, and then we have a site with two (zero) particles. So something like:

$$\ldots \mathbf{1\,0\,1} \ldots \mathbf{1\,2\,1} \ldots \mathbf{1\,0\,1} \ldots \mathbf{1\,2\,1} \ldots$$

This phase is characterized by

$$\langle \delta n_i e^{i\pi \sum_{k=i}^{j} \delta n_k} \delta n_j \rangle \rightarrow \text{const.} \neq 0. \tag{6.54}$$

In (Dalla Torre *et al.* 2006) it has been shown that this rather subtle phase may be probed by parametrically modulating the lattice, and determining the absorption rate associated with the linear response (a delta-like peak appears in the absorption spectrum).

Metastable states In addition to the ground-state phases discussed previously (and some other more exotic phases, which we shall not discuss) the DDI lead to another interesting phenomenon, namely the proliferation of metastable states, characterized by a non-uniform particle distribution in the lattice (Menotti *et al.* 2007). These states correspond to local energy minima with a very long lifetime. The existence of these metastable states may be understood from the competing length scales in the problem, i.e., the combination of different non-local interactions at different neighbours. As a result, disordered configurations of atoms lead to a disordered self-induced interaction potential. This self-induced disorder largely resembles extent the so-called structural glasses in condensed-matter physics (Schmalian and Wolynes 2000).

Because of the presence of those very many metastable states, in an experiment it may be hard to reach the ground state or a given metastable configuration. One can, however, use superlattices to prepare the atoms in configurations of preferential symmetry. Note that the configurations obtained in such a way will also remain stable once the superlattice is removed, thanks to the DDI.

6.3.4 Pair superfluidity in bilayer systems

Up to now we have considered the case of a 2D lattice gas of polar molecules. Some very interesting physics may appear if we have more than one such 2D layer. As mentioned in Section 6.3.1 for the case of BECs, this is true even if there is no hopping between layers. We shall illustrate this by first looking at the idea of pair superfluidity in bilayer systems. In the next section, we shall consider how DDI may lead to filament quantum gases in multilayers.

In the following, we consider dipolar bosons placed at two neighbouring, but disconnected, 1D traps (wires). Along the 1D systems we assume an additional lattice equal for both 1D traps. We hence have a ladder-like configuration. To illustrate clearly the effects of interlayer interactions, we consider a configuration for which only the (attractive) dipole–dipole interaction between sites at the same rung plays a significant role. This is physically possible for particular 1D dipole configurations (see (Argüelles and Santos 2007) for details), but here we just assume it for simplicity of the discussion. Of course, in general the (repulsive) DDI between sites belonging to the same wire cannot be neglected, and interesting physics will result from there, as we shall briefly mention later.

Under the previous conditions, the system is described by a Bose–Hubbard Hamiltonian similar to that of Eq. (6.50), but now with two wires

$$\hat{H} = -J \sum_{\alpha=1,2} \sum_{<i,j>} \left\{ \hat{b}_i^{(\alpha)\dagger} \hat{b}_j^{(\alpha)} + H.c. \right\} - \mu \sum_{\alpha=1,2} \hat{n}_i^{(\alpha)}$$
$$+ \frac{U_0}{2} \sum_{\alpha=1,2} \sum_i \hat{n}_i^{(\alpha)} (\hat{n}_i^{(\alpha)} - 1) - |U'| \sum_i \hat{n}_i^{(1)} \hat{n}_i^{(2)}, \tag{6.55}$$

where $\hat{b}_i^{(\alpha)}$, $\hat{b}_i^{(\alpha)\dagger}$, and $\hat{n}_i^{(\alpha)}$ are, respectively, the annihilation, creation, and number operators for the site i at the wire α. J describes the hopping between neighbouring sites i and j in each wire, U_0 describes the on-site interactions (as already mentioned, this is a combination of short-range and dipolar contributions), and we consider the same chemical potential μ in both wires. Atoms in sites at the same rung interact attractively by the DDI, which is characterized by a coupling $-|U'|$.

The ground-state phases are characterized by the correlation functions $G_1(\Delta) = \langle \hat{b}_{0,\alpha}^\dagger \hat{b}_{\Delta,\alpha} \rangle$ and $G_2(\Delta) = \langle \hat{b}_{0,1}^\dagger \hat{b}_{0,2}^\dagger \hat{b}_{\Delta,1} \hat{b}_{\Delta,2} \rangle$. Note that G_1 is the single-particle density matrix along an individual wire, whereas G_2 is related to pairs placed in the same rung of the ladder but at opposite wires. The phase diagram of the system presents three distinct phases (Figure 6.13):

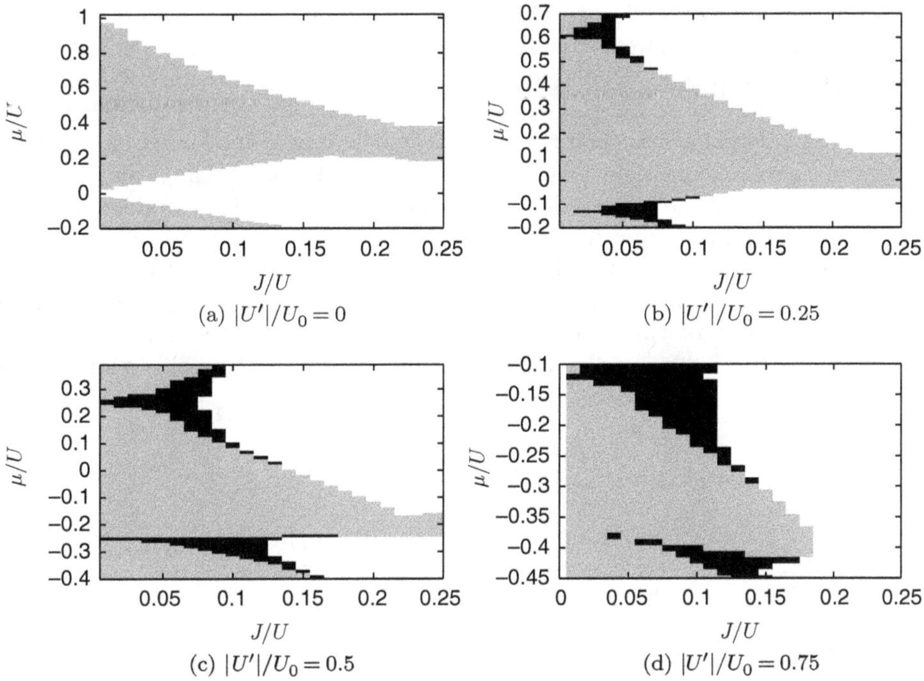

Figure 6.13 Phase diagram for $|U'|/U_0 = 0$ (a), 0.25 (b), 0.5 (c), and 0.75 (d), where white represents 2SF, grey MI, and black PSF. Figure from (Argüelles and Santos 2007).

Mott-insulator (MI) (shaded regions) This is characterized by a commensurate occupation at the sites of both wires. Figure 6.13 shows DMRG results for the surroundings of the lowest MI lobe (with occupation $\langle n_i^\alpha \rangle = 1$) for $|U'|/U_0 = 0$ (a), 1/4 (b), 1/2 (c), and 3/4 (d). Note that to avoid collapse in a single site, $|U'| < U_0$. For the case of $U' = 0$, the usual Mott lobes (shaded regions) are recovered. This is an insulating region, in which both correlations G_1 and G_2 decay exponentially.

Pair-superfluid (PSF) (black regions) In this phase, G_1 still decays exponentially but G_2 decays polynomially. We hence have a superfluid phase, but the superfluidity is given by the formation of pairs.

Two-superfluid phase (2SF) (white regions) In this phase, both G_1 and G_2 decay polynomially.

The MI and 2SF phases correspond to the usual MI and SF phases known for the Bose–Hubbard Hamiltonian. The PSF phase in bilayer systems is, however, rather peculiar for dipolar gases, resembling the situation known for two-component Bose gases (Kuklov *et al.* 2004).

Actually, the presence of the PSF phase drastically modifies the MI lobes (Figure 6.13). Let us briefly consider why this is so. The boundaries of the MI lobes are provided by the energy gap between the MI state and the lowest excited state conserving the particle number. In the usual Bose–Hubbard Hamiltonian (one wire, local interactions) (Fisher *et al.* 1989), this lowest excitation is provided by particle–hole excitations. The MI boundaries can then be calculated by a strong-coupling expansion (SCE) (Freericks and Monien 1996), estimating the energy of a state with an extra particle and a state with an extra hole. This is indeed the case of $U' = 0$, where the lowest excitations are given by uncorrelated particle–hole excitations in both wires. The situation changes for $|U'| > 0$, since for sufficiently low tunnelling, there is a direct transition between MI and PSF phases, i.e., superfluid phases of composites (or composite holes). In that case, the first excitation of the MI lobe is given by the correlated creation of pairs of particles (or holes) at opposite sites of the two wires, explaining the qualitative change in the shape of the lobe boundaries. In particular, a second-order SCE in $J/|U'|$ provides the following dependence for sufficiently low tunnelling for the lowest boundary of the MI lobe with n_0 particles per site:

$$\frac{\mu}{U_0} = n_0 - 1 + \frac{|U'|}{U_0}\left(\frac{1}{2} - n_0\right) - 4\left(\frac{J}{U_0}\right)^2\left[n_0(n_0+1) - \frac{(n_0^2-1)/2}{2-|U'|/U_0} - \frac{n_0^2 U_0}{|U'|}\right]. \tag{6.56}$$

From Eq. (6.56) it becomes clear that for any $U' > 0$ the gap boundaries are quadratic (and not linear) in J for sufficiently low J (this is so because the motion goes in pairs). Interestingly, the lowest boundary of the first MI region ($n_0 = 1$) inverts its slope at $J = 0$ for $|U|' > U_0/2$, in agreement with our numerical results. One may also observe that an inversion of the slope of the lowest boundary is expected also for $n_0 = 2$ at $|U'|/U_0 \simeq 0.85$, but is not expected for $n_0 > 2$.

The re-entrant shape of the MI lobes leads to a non-trivial behaviour of the MI plateaus (wedding-cake structure) in experiments with an axial harmonic confinement. In particular, the MI plateaus may (for low hopping) become insensitive to the hopping, or even counterintuitively grow for larger tunnelling. Actually, this result may be extended to 2D lattices at unconnected layers where the first MI lobe follows at low J the relation of Eq. (6.56) but substituting $2(J/U_0)^2$ by $z(J/U_0)^2$, where z is the coordination number.

To finish this section, let me point out that if the repulsion along the wires (or within the layers in 2D) is considered, the bending of the Mott lobes is preserved, and a pair-supersolid phase may be obtained (Trefzger *et al.* 2009). In addition, the presence of interlayer hopping may lead to interesting effects (see, e.g., (Wang 2007)). I would also like to note that a number of recent works have analysed the related case of polar Fermi molecules in bilayer systems when the electric dipoles are oriented perpendicular to the layer planes, and the interlayer hopping is negligible. This situation is particularly interesting for current experiments in JILA, since chemical recombination may be strongly reduced in 2D configurations (Ni *et al.* 2010), as would be the case for the different layers. The interlayer interaction is attractive and may allow for interlayer superfluidity, and the equivalent of a BCS–BEC crossover (Baranov *et al.* 2010; Klawunn *et al.* 2010b; Pikovski *et al.* 2010; Potter *et al.* 2010).

6.3.5 Quantum filament gases in multilayer systems

In the previous section, we saw that the interlayer interactions may lead to the formation of interlayer superfluidity given by pairs of particles placed at opposite layers. Interestingly, it is clear that similar reasoning suggests that filaments may be formed in multilayer systems. These filaments may present some very interesting physics, which is only partially understood. For simplicity of the discussion, we shall look here at perhaps the simplest scenario, which already contains many of the ingredients that could make filament quantum gases so interesting. We consider polar Fermi molecules in a three-layer system (actually, for this discussion we shall consider three quasi-1D wires), without interwire hopping, and with harmonic trapping along the wire. We shall consider the attractive interaction between dipoles at different wires, but neglect interactions between dipoles along the wires. This is, of course, a rather rough approximation, although in 1D it may be the case under proper conditions (Klawunn *et al.* 2010a). In general, intralayer interactions will play an important role, but at least part of the physics discussed here will be qualitatively maintained. Polar bosonic molecules in multilayers (under similar approximations as those discussed here) were considered by Wang *et al.* (2006).

The attraction between polar molecules placed on top of each other may be strong enough to bind two or more polar molecules into self-assembled chains (Figure 6.14). Whereas for bosonic molecules, these chains are in any case bosons (Wang *et al.* 2006), for fermionic molecules the fermionic or bosonic character of the filaments depends on the odd or even number of molecules in a given chain. In particular, the three-well configuration allows for fermionic trimers (and of course monomers), and two different kinds of bosonic dimer, namely those between two molecules at nearest neighbours (type I dimer), and those between two molecules at the highest and lowest sites (type II dimers) (Figure 6.14). Note that dimers I are actually pseudo-spin-1/2 bosons, since dimers in sites 1 and 2 are not equivalent to dimers in sites 2 and 3.

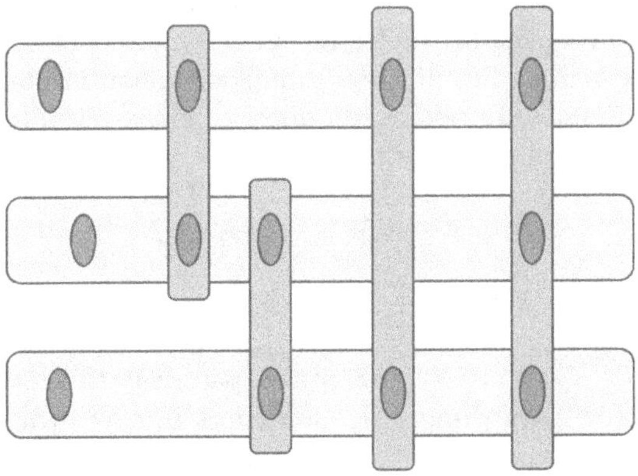

Figure 6.14 Polar fermionic molecules in a three-well potential may remain unpaired, form fermionic trimers, or bosonic dimers between nearest neighbours or next-nearest neighbours.

In addition, it is important to note that transverse filament excitations contribute to the gas entropy, being relevant at finite temperature T. These modes are string-like perturbations of the straight filaments. In contrast with the case of bosonic molecules (Wang *et al.* 2006), transverse modes are important for fermionic molecules even at very low T, since they may significantly reduce the trimer Fermi energy.

In the following, we consider the filament statistics, assuming an ideal filament gas. This rough approximation largely simplifies the analysis of the problem, while allowing for the discussion of key qualitative features of these systems, in particular the competition between different Bose and Fermi composites. The fermionic or bosonic character of the chains is reflected by the average occupations for trimers, dimers I, dimers II, and monomers:

$$N_{\mathrm{T}}(n,\nu_{\mathrm{T}}) = \left[\mathrm{e}^{\beta[-E_{\mathrm{T}}+\xi_{\nu_{\mathrm{T}}}+\epsilon_n-(2\mu_1+\mu_2)]}+1\right]^{-1}, \tag{6.57}$$

$$N_{\mathrm{D,I}}(n,\nu_{\mathrm{D,I}}) = \left[\mathrm{e}^{\beta[-E_{\mathrm{D,I}}+\xi_{\nu_{\mathrm{D,I}}}+\epsilon_n-(\mu_1+\mu_2)]}-1\right]^{-1}, \tag{6.58}$$

$$N_{\mathrm{D,II}}(n,\nu_{\mathrm{D,II}}) = \left[\mathrm{e}^{\beta[-E_{\mathrm{D,II}}+\xi_{\nu_{\mathrm{D,II}}}+\epsilon_n-2\mu_1]}-1\right]^{-1}, \tag{6.59}$$

$$N_{\mathrm{S},j}(n) = \left[\mathrm{e}^{\beta[\epsilon_n-\mu_j]}+1\right]^{-1}, \tag{6.60}$$

where $N_{\mathrm{S},j}$ denotes the average occupation of individual molecules at wire j, $-E_{\mathrm{T}}$, $-E_{\mathrm{D,I}}$, and $-E_{\mathrm{D,II}}$ are the binding energies for, respectively, trimers, dimers I, and dimers II, $\xi_{\nu_{\mathrm{T;D,I;D,II}}}$ are the transverse filaments modes of the different composites, $\epsilon_n = \hbar\omega(n+1/2)$ are the harmonic oscillator levels of the trap along the wires, and $\beta = 1/k_{\mathrm{B}}T$ the inverse temperature. In the previous expressions we have assumed symmetric configurations, such that the number of dimers I in sites 1–2 is the same as the number of dimers I in sites 2–3, and equal to $N_{\mathrm{D,I}}(n,\nu_{\mathrm{D,I}})$. Note that $\mu_1 = \mu_3$ is the chemical potential for molecules at the highest and lowest sites, whereas μ_2 denotes the chemical potential for molecules in the middle site. These different chemical potentials are necessary to fulfil the normalization conditions, in which we assume N molecules per lattice site. Imposing symmetry between the highest and the lowest sites, these conditions acquire the form:

$$N = N_{\mathrm{T}} + N_{\mathrm{D,I}} + N_{\mathrm{D,II}} + N_{\mathrm{S},1}, \tag{6.61}$$

$$N = N_{\mathrm{T}} + 2N_{\mathrm{D,I}} + N_{\mathrm{S},2}, \tag{6.62}$$

where N_{T}, $N_{\mathrm{D,I}}$, $N_{\mathrm{D,II}}$, $N_{\mathrm{S},1}$ and $N_{\mathrm{S},2}$ denote respectively the total number of trimers, dimers I in sites 1–2 (or 2–3), dimers II, monomers in site 1 (or 3), and monomers in site 2. From Eqs (6.61) and (6.62), we obtain $\mu_1(N,T)$ and $\mu_2(N,T)$, and from Eqs (6.57–6.60), the occupation numbers.

Owing to the attractive DDI between molecules in the filament, the most bound chain is the trimer. The difference in binding between dimers and trimers means that for sufficiently small N and at low enough T the DCL becomes a degenerate Fermi gas of trimers. The trimers fill up oscillator levels (and also transverse trimer modes) up to the corresponding Fermi energy $E_{\rm F}(N)$, which equals $N\hbar\omega$ for rigid filaments but is actually smaller, owing to the transverse trimer modes. However, if the number of molecules per site is sufficiently large, the growth in Fermi energy overcomes the binding energy difference. This transition may easily be estimated by comparing the average energy per molecule for the case of two trimers and that for the case of two dimers I and one dimer II. This leads to a condition for the critical number of molecules per site $N_{\rm c}(U_0, \omega)$, $E_{\rm F}(N_{\rm c}) = 2E_{\rm T} - 3(E_{\rm D,I} + E_{\rm D,II})/2$ (which we have confirmed numerically). Note that $N_{\rm c}$ grows with growing interlayer attraction and decreasing trap frequency ω. For $N < N_{\rm c}$, the quantum gas is a degenerate trimer gas, whereas for $N > N_{\rm c}$ the trimer gas coexists with a mixture of pseudo-spin-1/2 bosons (dimers I) and spin-less bosons (dimers II).

The peculiar properties of the DCL translate into the spatial molecular distribution integrated over the three sites. For $N < N_{\rm c}$ and $N < \xi_{1_{\rm T}}/\omega$, only trimers in their internal ground state are formed, and hence the gas behaves as a spin-less Fermi gas of particles of mass $3m$, presenting a Thomas–Fermi density profile $(1 - (x/R)^2)^{1/2}$ with $R/l_{\rm HO} = \sqrt{2N/3}$. For $\xi_{1_{\rm T}}/\omega < N < N_{\rm c}$, the DCL is still a trimer gas, but transversal trimer modes may be populated. In that case, the density profile departs from the Thomas–Fermi profile (Figure 6.15, top), owing to the appearance of internally excited trimers in low harmonic oscillator levels. For $N > N_{\rm c}$, the density profile changes dramatically. Note that since we consider 1D gases, the dimer BEC is, strictly speaking, precluded. However, because of the finite size of the system, the dimers quasi-condense (at low enough T), occupying the few lowest levels of the harmonic oscillator. Hence when N surpasses $N_{\rm c}$ a Bose cloud nucleates at the trap centre. As a result the distribution of the polar molecules shows a Gaussian-like peak at the trap centre (Figure 6.15, bottom).

For $N \gg N_{\rm c}$ and $U_0 > U_0^*(\text{D,II})$ the quantum gas at low T is a basically pure Bose gas of dimers I and II (except for a small trimer fraction). Since both dimers have double mass, the difference between them cannot be discerned from the analysis of the integrated density profile of the molecules. However, the different binding energy and excited dimer modes for both types of dimers may be studied spectroscopically to reveal the dual nature of the mixture.

The filament gas also presents intriguing finite temperature physics, owing to the role of filament modes and the different binding energy of dimers and trimers. This is particularly clear from a finite T analysis with $N < N_{\rm c}$. While at very low T the DCL is purely a trimer Fermi gas, at finite T it becomes more favourable to populate dimers than to populate higher excited trimer states. As a consequence, the system presents a striking thermal enhancement of the bosonic modes. Interestingly, in contrast with the standard situation, this leads to a maximal central peak density for a given finite T. For even larger T, the central density decreases again, owing to the occupation of dimers at higher oscillator modes, and the breaking of the filaments into individual molecules.

Let us finally point out that for the case of bosonic polar particles, the filaments are obviously always bosonic. As a result, at sufficiently low temperature there is a

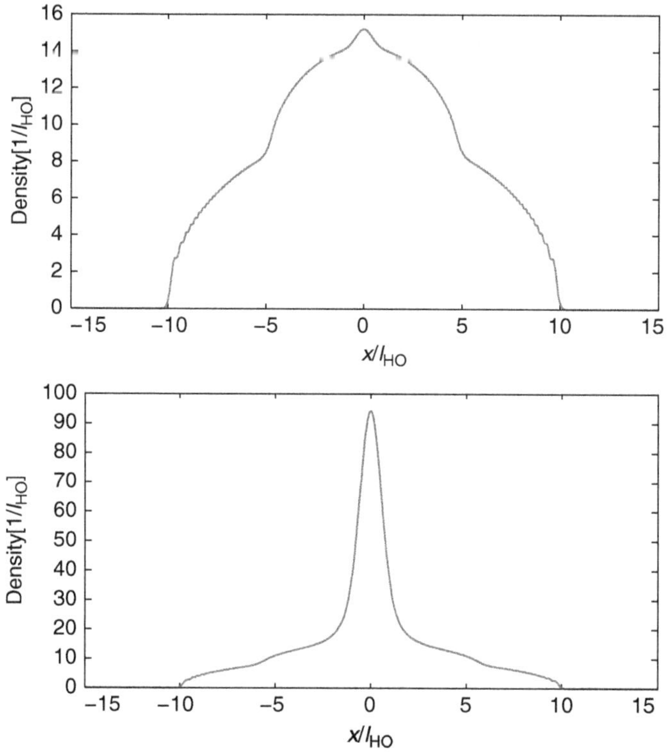

Figure 6.15 Integrated density profiles of the molecules, for $\xi_{1T}/\omega < N < N_c$ (top), and $N > N_c$ (bottom). We consider $U_0 = 2$, $\omega/2\pi = 1$ Hz, $m = 100$ amu, which lead to $N_c = 230$. Figure from (Klawunn *et al.* 2010a).

Bose–Einstein condensation of the largest filaments (which, we recall, are the ones with the tightest binding) (Wang *et al.* 2006).

Although this simple theory contains very suggestive ingredients, including the possibility of filaments BEC, the competition between fermionic and bosonic filaments, and the role of the transversal modes, it is clear that interfilament interactions will lead to interesting novel properties, which are still not well understood.

6.4 Conclusions

Since the properties of ultracold gases are to a large extent determined by interparticle interactions, the presence of dipole–dipole interactions, in addition to the usual van-der-Waals-like interactions, introduces qualitative changes in the physics of dipolar gases. In this chapter, we have reviewed some of this interesting physics, which concerns a large variety of areas, ranging from non-linear atom optics to condensed-matter physics. In this sense, although recent experiments on atomic chromium condensates have already unveiled part of the expected richness of dipolar gases, it is expected that a new generation of experiments on polar molecules (and possibly also Rydberg gases) may soon reach quantum degeneracy, opening the path for strongly polar quantum gases. As mentioned in this chapter, these systems are expected to provide radically new scenarios for ultracold gases. One should hence expect exciting developments on dipolar gases in the future.

References

Argüelles, A and Santos, L (2007). *Phys. Rev. A*, **75**, 053613.

Baranov, MA (2008). *Phys. Rep.*, **464**, 71.

Baranov, MA, Micheli, A, Ronen, S, and Zoller, P (2010). arXiv:1012.5589.

Beaufils, Q, Chicireanu, R, Zanon, T, *et al.* (2008). *Phys. Rev. A*, **77**, 061601.

Bohn, J, Wilson, R, and Ronen, S (2009). *Laser Phys.*, **19**, 547.

Bortolotti, DCE, Ronen, S, Bohn, JL, and Blume, D (2006). *Phys. Rev. Lett.*, **97**, 160402.

Chen, Y-C, Melko, RG, Wessel, S, and Kao, Y-J (2008). *Phys. Rev. B*, **77**, 014524.

Cinti, F, Jain, P, Boninsegni, M, Micheli, A, Zoller, P, and Pupillo, G (2010). *Phys. Rev. Lett.*, **105**, 135301.

Conti, C, Peccianti, M, and Assanto, G (2003). *Phys. Rev. Lett.*, **91**, 073901.

Dalla Torre, EG, Berg, E, and Altman, E (2006). *Phys. Rev. Lett.*, **97**, 260401.

Deb, B and You, L (2001). *Phys. Rev. A*, **64**, 022717.

Deiglmayr, J, Grochola, A, Repp, M, *et al.* (2008). *Phys. Rev. Lett.*, **101**, 133004.

Dutta, O and Meystre, P (2007). *Phys. Rev. A*, **75**, 053604.

Eberlein, C, Giovanazzi, S, and O'Dell, DHJ (2005). *Phys. Rev. A*, **71**, 033618.

Fattori, M, Roati, G, Deissler, B, *et al.* (2008). *Phys. Rev. Lett.*, **101**, 190405.

Fisher, MPA, Weichman, PB, Grinstein, G, and Fisher, DS (1989). *Phys. Rev. B*, **40**, 546.

Freericks, JK and Monien, H (1996). *Phys. Rev. B*, **53**, 2691.

Gammal, A, Frederico, T, and Tomio, L (2001). *Phys. Rev. A*, **64**, 055602.

Góral, K and Santos, L (2002). *Phys. Rev. A*, **66**, 023613.

Góral, K, Santos, L, and Lewenstein, M (2002). *Phys. Rev. Lett.*, **88**, 170406.

Griesmaier, A, Werner, J, Hensler, S, Stuhler, J, and Pfau, T (2005). *Phys. Rev. Lett.*, **94**, 160401.

Griesmaier, A, Stuhler, J, Koch, T, Fattori, M, Pfau, T, and Giovanazzi, S (2006). *Phys. Rev. Lett.*, **97**, 250402.

Hadzibabic, Z, Kruger, P, Cheneau, M, Battelier, B, and Dalibard, J (2006). *Nature*, **441**, 1118.

Henkel, N, Nath, R, and Pohl, T (2010). *Phys. Rev. Lett.*, **104**, 195302.

Ho, T-L (1998). *Phys. Rev. Lett.*, **81**, 742.

Kawaguchi, Y, Saito, H, and Ueda, M (2006). *Phys. Rev. Lett.*, **96**, 080405.

Khaykovich, L, Schreck, F, Ferrari, G, *et al.* (2002). *Science*, **296**, 1290.

Klawunn, M and Santos, L (2009). *Phys. Rev. A*, **80**, 013611.

Klawunn, M, Duhme, J, and Santos, L (2010a). *Phys. Rev. A*, **81**, 013604.

Klawunn, M, Pikovski, A, and Santos, L (2010b). *Phys. Rev. A*, **82**, 044701.
Koch, T, Lahaye, T, Metz, J, Frohlich, B, Griesmaier, A, and Pfau, T (2005). *Nat. Phys.*, **4**, 218.
Komineas, S and Cooper, NR (2007). *Phys. Rev. A*, **75**, 023623.
Kotochigova, S, Julienne, PS, and Tiesinga, E (2003). *Phys. Rev. A*, **68**, 022501.
Krolikowski, W, Bang, O, Rasmussen, JJ, and Wyller, J (2001). *Phys. Rev. E*, **64**, 016612.
Kuklov, A, Prokof'ev, N, and Svistunov, B (2004). *Phys. Rev. Lett.*, **92**, 050402.
Lahaye, T, Metz, J, Fröhlich, B, *et al.* (2008). *Phys. Rev. Lett.*, **101**, 080401.
Lahaye, T, Menotti, C, Santos, L, Lewenstein, M, and Pfau, T (2009). *Rep. Prog. Phys.*, **72**, 126401.
Landau, LD and Lifshitz, EM (1977). *Quantum Mechanics*(2nd edn). Butterworth-Heinemann, New York.
Litvak, AG, Mironov, VA, Fraiman, GM, and Yunakoskii, AD (1975). *Sov. J. Plasma Phys.*, **1**, 60.
Marinescu, M and You, L (1998). *Phys. Rev. Lett.*, **81**, 4596.
Menotti, C, Trefzger, C, and Lewenstein, M (2007). *Phys. Rev. Lett.*, **98**, 235301.
Micheli, A, Pupillo, G, Büchler, HP, and Zoller, P (2007). *Phys. Rev. A*, **76**, 043604.
Nath, R and Santos, L (2010). *Phys. Rev. A*, **81**, 033626.
Nath, R, Pedri, P, and Santos, L (2007). *Phys. Rev. A*, **76**, 013606.
Nath, R, Pedri, P, and Santos, L (2009). *Phys. Rev. Lett.*, **102**, 050401.
Ni, K-K, Ospelkaus, S, de Miranda, MHG, *et al.* (2008). *Science*, **322**, 231.
Ni, K-K, Ospelkaus, S, Wang, D, *et al.* (2010). *Nature*, **464**, 1324.
O'Dell, DHJ, Giovanazzi, S, and Eberlein, C (2004). *Phys. Rev. Lett.*, **92**, 250401.
Pasquiou, B, Bismut, G, Maréchal, E, *et al.* (2011). *Phys. Rev. Lett.*, **106**, 015301.
Pedri, P and Santos, L (2005). *Phys. Rev. Lett.*, **95**, 200404.
Pikovski, A, Klawunn, M, Shlyapnikov, GV, and Santos, L (2010). *Phys. Rev. Lett.*, **105**, 215302.
Pollack, SE, Dries, D, Junker, M, Chen, YP, Corcovilos, TA, and Hulet, RG (2009). *Phys. Rev. Lett.*, **102**, 090402.
Potter, AC, Berg, E, Wang, D-W, Halperin, BI, and Demler, E (2010). *Phys. Rev. Lett.*, **105**, 220406.
Ruprecht, PA, Holland, MJ, Burnett, K, and Edwards, M (1995). *Phys. Rev. A*, **51**, 4704.
Santos, L and Pfau, T (2006). *Phys. Rev. Lett.*, **96**, 190404.
Santos, L, Shlyapnikov, GV, Zoller, P, and Lewenstein, M (2000). *Phys. Rev. Lett.*, **85**, 1791.
Santos, L, Shlyapnikov, GV, and Lewenstein, M (2003). *Phys. Rev. Lett.*, **90**, 250403.
Schmalian, J and Wolynes, PG (2000). *Phys. Rev. Lett.*, **85**, 836.
Schmidt, PO, Hensler, S, Werner, J, *et al.* (2003). *Phys. Rev. Lett.*, **91**, 193201.
Sengupta, P, Pryadko, LP, Alet, F, Troyer, M, and Schmid, G (2005). *Phys. Rev. Lett.*, **94**, 207202.
Shih, MF, Segev, M, and Salamo, G (1997). *Phys. Rev. Lett.*, **78**, 2551.
Shlyapnikov, GV and Pedri, P (2006). Conference on Many-Body Phenomena in Dipolar Systems (Dresden).

Strecker, KE, Partridge, GB, Truscott, AG, and Hulet, RG (2002). *Nature*, **417**, 150.
Stuhler, J, Griesmaier, A, Koch, T, *et al.* (2005). *Phys. Rev. Lett.*, **95**, 150406.
Tikhonenkov, I, Malomed, BA, and Vardi, A (2008). *Phys. Rev. Lett.*, **100**, 090406.
Trefzger, C, Menotti, C, and Lewenstein, M (2009). *Phys. Rev. Lett.*, **103**, 035304.
Vengalattore, M, Leslie, SR, Guzman, J, and Stamper-Kurn, DM (2008). *Phys. Rev. Lett.*, **100**, 170403.
Wang, DW (2007). *Phys. Rev. Lett.*, **98**, 060403.
Wang, DW and Demler, E (2010). arXiv:0812.1838.
Wang, D-W, Lukin, MD, and Demler, E (2006). *Phys. Rev. Lett.*, **97**, 180413.
Yi, S and You, L (2001). *Phys. Rev. A*, **63**, 053607.
Zakharov, VE and Shabat, AB (1972). *JETP*, **61**, 118.

7
Ultracold polar molecules

Deborah S. Jin and Jun Ye

JILA, NIST and University of Colorado,
Department of Physics, University of Colorado
Boulder, CO 80309-0440, USA

Chapter Contents

7.1 Motivation and challenges

Ultracold polar molecules represent an exciting and challenging research direction for experiments. There are many different motivations for this work, ranging from potential applications in quantum information and precision measurements to studies of ultracold chemistry and exotic quantum matter. With this last motivation, we seek to build on the tremendous success of ultracold atom gases in realizing interesting many-body physics in a clean, controllable model system. Extending this work to polar molecules can provide access to new types of many-body systems because of the strong, long-range, and anisotropic dipole–dipole interaction.

The production of ultracold polar molecules at temperatures and densities required for the quantum gas regime has proven to be very difficult [1]. Innovative techniques for the direct cooling of molecules have been demonstrated, reaching temperatures of about 10 mK and densities of about 10^6 molecules per cubic centimetre. An alternative approach, using photoassociation to create molecules from laser-cooled atoms, has reached similar phase-space densities. However, the net result is still a gas over 12 orders of magnitude away from quantum degeneracy.

Recently, a new approach starting from ultracold atoms, which are laser cooled and then evaporatively cooled to the quantum regime, has resulted in ultracold molecules near quantum degeneracy [2; 3; 4; 5]. In what follows, we give an overview of the experiments on KRb ground-state polar molecules done at JILA. These experiments were performed by a team of wonderful students and postdoctoral fellows (see Figure 7.1). More information about the experiments can be found in [2; 3; 6; 7; 8; 9; 10].

Figure 7.1 JILA ultracold polar molecule experiment team. In the front row, from left to right, are Kang-Kuen Ni, Marcio de Miranda, Dajun Wang, and Josh Zirbel. In the back are Silke Ospelkaus and Brian Neyenhuis. The current experimental team includes Amodsen Chotia and Steven Moses, who are not pictured.

7.2 Making ultracold polar molecules

We produce ground-state polar $^{40}K^{87}Rb$ molecules in a two-step process. First, weakly bound Feshbach molecules are created from an ultracold atom gas mixture simply by ramping an applied magnetic field over a heteronuclear Fano–Feshbach resonance. These large, weakly bound molecules are not themselves useful as ultracold polar molecules because they have essentially no dipole moment. Moreover, they are lost from the trap (a far-detuned optical dipole trap) in a few milliseconds, owing to collisional relaxation to lower energy vibrational states [6].

To reach the ground rotational and vibrational state, we use a two-photon coherent optical transfer that spans an energy difference between the initial and final molecular states of about 6000 K, in units of temperature. This binding energy difference is about 10 orders of magnitude larger than the translational temperature of the gas. This points to the need for the transfer to be perfectly coherent, and very efficient, to preserve the low temperature (a few hundred nK) and reasonably high density (about $10^{12}/\text{cm}^3$) of the initial gas.

We use stimulated Raman adiabatic passage (STIRAP) [11] to accomplish this coherent state transfer. One challenge in doing this was that of finding a good electronically excited state to use as the intermediate state for the two-photon process. This intermediate state must have reasonably large wavefunction overlaps with the initial and target states. Here, having some spectroscopic data [12] and the help of theory collaborators (Svetlana Kotochigova and Paul Julienne, in particular) was critical. A second challenge for the experiment was that of having the required phase coherence between two continuous-wave lasers that are used for Raman transfer. The two lasers are separated by a frequency difference of 125 THz. For this, we locked the two Raman lasers to different teeth of a stabilized femtosecond laser frequency comb. In the end, the STIRAP transfer could be made as high as 90% efficient from the Feshbach molecules to the ro-vibrational ground state, with no heating of the molecular gas observed.

The end result is that about 40 000 KRb molecules are held in an optical dipole trap with a temperature of a few hundred nK and a peak density of about $10^{12}/\text{cm}^3$. This represents an increase of about 12 orders of magnitude in the phase-space density compared with previous results for a cold gas of polar molecules. Since our molecules are made from fermionic ^{40}K and bosonic ^{87}Rb, they are fermionic and our best results for the temperature and density of the gas correspond to a temperature of the molecules that is only about 1.4 times the Fermi temperature.

7.3 Characterizing the ultracold polar molecules

In our set-up, detection of the ground-state molecules is most easily accomplished by reversing the coherent state transfer and using STIRAP to bring the molecules back to the Feshbach molecule state. These weakly bound molecules can then be imaged using absorption imaging and light resonant with the atomic transition. However, we also showed that direct detection of the ground-state molecules could be accomplished using absorption imaging on a single molecular transition [10].

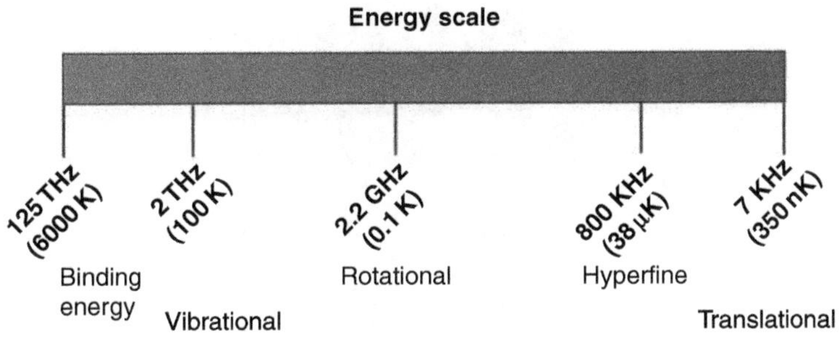

Figure 7.2 Energy scales covered in the ultracold KRb experiment. Starting with free atoms, we produce molecules at a temperature of a few hundred nK in their lowest energy state in all internal degrees of freedom, including electronic, vibrational, rotational, and nuclear spins. These internal molecular degrees of freedom cover an enormous range of energies.

We have made a number of measurements characterizing the KRb molecules. Using the two-photon transition, we spectroscopically measured the permanent dipole moment of KRb to be 0.57 Debye [3]. We also measured the rotational energy splitting using two-photon spectroscopy of the second rotationally excited level ($N=2$) instead of the ground state ($N=0$).

With a combination of spectral resolution and angular momentum selection rules, we showed that the two-photon STIRAP process could produce a gas where all the molecules are in just one of the 36 possible molecular hyperfine states. In addition, we demonstrated that molecules could be transferred from this state into the lowest energy hyperfine state (in a magnetic field) using microwaves to drive rotational transitions up to $N=1$ and then back down to the rotational ground state $N=0$, but now in a different hyperfine state [7]. With this control over the molecular hyperfine state, we reach the absolute lowest energy internal state of the molecules, which requires control over an enormous range of energies that correspond to electronic, vibrational, rotational, nuclear, and translational degrees of freedom (see Figure 7.2).

7.4 Ultracold chemistry, dipolar interactions, and reduced dimensionality

Most of our recent work has focused on ultracold chemical reactions. One might guess that ultracold KRb molecules, in their absolute lowest energy internal state, would remain essentially forever in the trap. However, we observe a lifetime in the trap of the order of 1 s, which is much shorter than the vacuum-limited lifetime. This loss is due to the following chemical reaction [8]:

$$\mathrm{KRb} + \mathrm{KRb} \rightarrow \mathrm{K_2} + \mathrm{Rb_2}.$$

The quantum statistics of our fermionic molecules play an important role in how this chemical reaction occurs. When the molecules are all prepared in a single internal

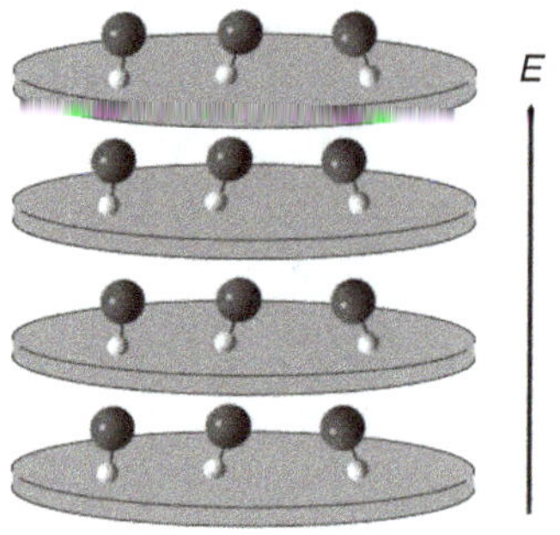

Figure 7.3 Schematic depiction of polar molecules in a 1D optical lattice. A 1D optical lattice can be used to confine polar molecules in a stack of isolated 2D layers. Tight confinement along the direction of the dipoles can help reduce the rate of head-to-tail collisions that lead to bimolecular chemical reactions.

state, the bimolecular reaction occurs through p-wave collisions, and we observe a linear dependence on temperature, as expected for an inelastic p-wave process at low collision energy. When the molecules are instead prepared in a 50/50 mixture of two hyperfine states, we observe a much higher trap loss rate (as much as 100 times higher at our lowest temperature) and no temperature dependence. This is consistent with the expected s-wave collisions for ultracold fermions in different internal states.

When we apply an electric field (up to 5 kV/cm) to the polar molecules, we induce a dipole moment in the lab frame. This turns on the dipole–dipole interactions, and we measure a large increase in the rate of chemical reactions [9; 13]. Roughly speaking, this increase in the rate can be understood in terms of the attractive part of the dipole–dipole interaction reducing the centrifugal barrier for p-wave collisions. Of course, more precisely, the dipole–dipole interaction is spatially anisotropic and therefore mixes different partial wave collision channels.

The spatial anisotropy of the dipole–dipole interaction also provides a way of potentially suppressing the large chemical reaction rate for the polar KRb molecules in an electric field. Recognizing that the attractive part of the dipole–dipole interaction occurs primarily for molecules that collide in a head-to-tail orientation, we are currently working on preventing this type of collision. Specifically, we have loaded the ground-state KRb molecules into a one-dimensional optical lattice. This traps the molecules in a stack of isolated two-dimensional (2D) pancakes (see Figure 7.3). By applying an electric field perpendicular to this stack, we orient the molecules along the direction of tight confinement, which helps to suppress head-to-tail collisions [14].

Acknowledgements

The experimental work on ultracold polar molecules at JILA has been supported financially by NIST, NSF, AFOSR-MURI, DOE, and DARPA.

References

1. Carr, LD, DeMille, D, Krems, RV, and Ye, J Cold and ultracold molecules: science, technology and applications. *New J. Phys.* **11**, 055049 (2009).
2. Ospelkaus, S, Pe'er, A, Ni, K-K, *et al.* Efficient state transfer in an ultracold dense gas of heteronuclear molecules. *Nat. Phys.* **4**, 622 (2008).
3. Ni, K-K, Ospelkaus, S, de Miranda, MHG, *et al.* A high-phase-space-density gas of polar molecules. *Science* **322**, 231 (2008).
4. Danzl, JG, Haller, E, Gustavsson, M, *et al.* Quantum gas of deeply bound ground state molecules, *Science* **321**, 1062 (2008).
5. Danzl, JG, Mark, MJ, Haller, E, *et al.* An ultracold high-density sample of ro-vibronic ground-state molecules in an optical lattice. *Nature Phys.* **6**, 265 (2010).
6. Zirbel, JJ, Ni, K-K, Ospelkaus, S, *et al.* Collisional stability of fermionic Feshbach molecules *Phys. Rev. Lett.* **100**, 143201 (2008).
7. Ospelkaus, S, Ni, K-K, Quéméner, G, *et al.* Controlling the hyperfine state of rovibronic ground state polar molecules. *Phys. Rev. Lett.* **104**, 030402 (2010).
8. Ospelkaus, S, Ni, K-K, Wang, D, *et al.* Quantum-state controlled chemical reactions of ultracold KRb molecules. *Science* **327**, 853 (2010).
9. Ni, K-K, Ospelkaus, S, Wang, D, *et al.* Dipolar collisions of polar molecules in the quantum regime. *Nature* **464**, 1324 (2010).
10. Wang, D, Neyenhuis, B, de Miranda, MHG, *et al.* Direct absorption imaging of ultracoldpolar molecules. *Phys. Rev. A* **81**, 061404(R) (2010).
11. Bergmann, K, Theuer, H, and Shore, BW Coherent population transfer among quantum states of atoms and molecules. *Rev. Mod. Phys.* **70**, 1003 (1998).
12. Amiot, C The KRb $(2)^3\Sigma^+$ electronic state. *J. Mol. Spec.* **203**, 126 (2000).
13. Quéméner, G and Bohn, JL Strong dependence of ultracold chemical rates on electric dipole moments. *Phys. Rev. A* **81**, 022702 (2010).
14. de Miranda, MHG, Chotia, A, Neyenhuis, B, *et al.* Controlling the quantum stereodynamics of ultracold bimolecular reactions. *arXiv:*1010.3731 (2010).

8

Ultracold Fermi gases as quantum simulators of condensed matter physics

Frédéric Chevy

Laboratoire Kastler Brossel, Département de Physique de l'ENS,
24 rue Lhomond, 75231 Paris cedex 05, France

Chapter Contents

8.1 Introduction

From the microscopic world of quantum mechanics to astrophysics, the greatest challenge faced by modern physics is the understanding of interacting many-body systems. In quantum mechanics, quantum correlations enlarge the phase space of the system so much that even the tremendous progress in computing capabilities is insufficient to tackle some of the most important current open questions reliably, among which the understanding of the microscopic origin of high-critical-temperature superconductors is probably the most famous.

Faced with this computational barrier, Feynman proposed in the early 1980s changing scope radically (Feynman, 1982) and suggested replacing algorithmic schemes used in digital (super)-computers with *analogue* simulation, where one solves a given mathematical problem by finding a physical system obeying the very equations under study. By experimenting on this system, one may therefore obtain the result of the original problem. Before the development of modern electronics, the scheme was actually widely used to solve numerical problems: for instance, to solve differential equations, one would design an electronic circuit whose time evolution would follow the same equations.

Recent advances in ultracold gases have suggested that these systems may constitute good simulators for quantum many-body systems (Jaksch *et al.* 1998; Bloch *et al.* 2008). In particular, the cooling of atomic fermionic gases to quantum degeneracy was achieved in 1999 and opened the path to simulations of electrons in solids (DeMarco and Jin 1999). This pioneering experiment bridged the gap between atomic and condensed matter physics and initiated a new field of research, which mutually enriched these two major fields of modern physics.

I present here a brief account of some recent experiments on attractive Fermi gases, as well as an introduction to experimental techniques used to probe these systems. I will focus on the problem of superfluidity in spin 1/2 fermionic systems, and I will show how experiments with ultracold atoms gave invaluable insights to this problem. The issues addressed here constitute only a small fraction of the recent works on ultracold gases, and for interested readers, more details can be found in Varenna's summer school proceedings (Inguscio *et al.* 2006).

8.2 The non-interacting Fermi gas

8.2.1 Evaporative cooling of a fermionic gas

Historically, the first successful technique used to achieve Bose–Einstein condensation in a sample of ultracold atoms was the evaporative cooling of a cloud of spin polarized bosons trapped in a magnetic potential (Anderson *et al.* 1995; Davis *et al.* 1995; Cornell and Wieman 2002; Ketterle 2002). In the case of fermions, this technique is made unfruitful by the Pauli exclusion principle, which, prevents cold collisions in an even-wave channel for fermionic particles. Indeed, the scattering cross-section in ℓ-wave processes, scales as $E^{2\ell}$, where E is the kinetic energy in the centre of mass frame, meaning that only s-wave ($\ell = 0$) processes survive at zero temperature.

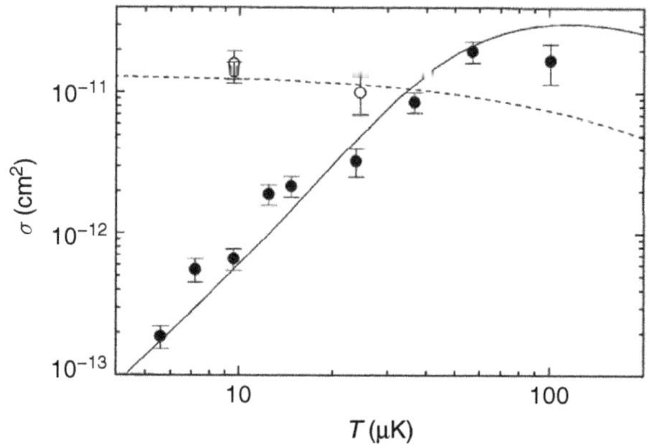

Figure 8.1 Evolution of the elastic scattering cross-section σ for a gas of fermions (^{40}K) polarized in a single spin state (full circles) or in a mixture of two spin states (empty circles) (data from (de Marco *et al.* 1999)). In the first case, s-wave collisions are permitted, and the cross-section does not vary significantly at low temperature. In the second case, the Pauli exclusion principle forbids s-wave processes, and p-wave scattering leads to a quadratic decrease of σ with temperature.

For the lowest partial wave, $\ell = 1$, permitted to fermions polarized in a single spin state, we thus expect a quenching of collisions at low temperature, as demonstrated experimentally by the groupe of Jin at JILA (DeMarco *et al.* 1999) and displayed in Figure 8.1. To circumvent this obstacle, several strategies have been successfully implemented:

1. As observed in Figure 8.1, s-wave collisions between fermions with different spins are permitted at low temperature. Evaporative cooling is thus possible, in principle, for this kind of mixture. However, for most alkali, the trappable spin states are not ground states of the hyperfine manifold and, as soon as two different spin states are trapped simultaneously, inelastic collisions associated with hyperfine changing are possible. Since the hyperfine splitting is of the order of several hundred MHz, i.e., several mK, the energy released in this process is enough to expel the atoms from the trap. In practice, this works only with ^{40}K, which possesses an 'inverted' hyperfine structure that protects it from this kind of loss. Using the same strategy, mixtures of potassium with a second fermionic species can be evaporatively cooled in a magnetic trap, as demonstrated by recent experiments on ^{6}Li-^{40}K fermionic mixtures at Innsbruck and Amsterdam (Wille *et al.* 2008). An alternative scheme is to trap the atoms at the focus of a far-detuned laser beam using dipolar forces. When the detuning is large enough, the trapping potential is the same for all Zeeman sublevels. It therefore allows for the trapping of an arbitrary number of Zeeman sublevels of the hyperfine ground-state manifold, as demonstrated first for Bose–Einstein condensates (Barrett *et al.* 2001), and later with fermions.

2. Another route consists of trapping both fermionic and bosonic species, for instance alkali mixtures of ^{6}Li and ^{7}Li (Schreck *et al.* 2001; Truscott *et al.* 2001), ^{6}Li and ^{33}Na (Hadzibabic *et al.* 2003), ^{40}K and ^{87}Rb (Roati *et al.* 2002), ^{6}Li and ^{87}Rb (note, however, that in this later case the relatively weak inter-species scattering length makes sympathetic cooling less efficient than for other configurations—see (Silber *et al.* 2005)), or even the triple species mixture ^{6}Li, ^{40}K and ^{87}Rb (Taglieber *et al.* 2008). In this so-called *sympathetic cooling* scheme, the bosons are evaporatively cooled using the standard procedure, and act as a refrigerator for the fermions. As a consequence, this strategy requires a large excess of bosons to cool the fermionic atoms efficiently; this may limit their number.

Let us consider a gas of particles of mass m and density n. A convenient way of quantifying the cooling efficiency is to compare the final temperature T with the Fermi temperature T_{F}, such that $E_{\mathrm{F}} = k_{\mathrm{B}} T_{\mathrm{F}}$, with $E_{\mathrm{F}} = \hbar^2 k_{\mathrm{F}}^2/2m$ and $k_{\mathrm{F}} = (6\pi^2 n)^{1/3}$: quantum degeneracy corresponds to $T/T_{\mathrm{F}} \lesssim 1$. Using the technique described previously, it is now possible to achieve $T/T_{\mathrm{F}} \sim 10^{-2}$, where the system is deeply in the quantum regime. This value is both large, since it corresponds to the degeneracy level of electrons in a solid at room temperature, and small, since it allows one to probe quantum statistics phenomena and also to observe fermionic superfluidity, as we will see later.

8.2.2 Density profile of the non-interacting Fermi gas

As mentioned in the previous section, an ultracold cloud of polarized fermions constitutes a textbook experimental realization of an ideal Fermi gas. In this case, the density profile of the cloud can be easily described in the so-called local density approximation (LDA), where one assumes the existence of a mesoscopic scale, much smaller than the size of the cloud but much larger than the interparticle distance, where the system can be considered homogeneous. As a consequence, it is possible to define a local chemical potential $\mu(\boldsymbol{r})$ given by

$$\mu(\boldsymbol{r}) = \mu_{\mathrm{hom}}(n(\boldsymbol{r}), T) + V(\boldsymbol{r}), \tag{8.1}$$

where $\mu_{\mathrm{hom}}(n, T)$ is the chemical potential of a gas of density n and temperature T, and V is the trapping potential. At equilibrium, the second principle of thermodynamics imposes a uniform chemical potential $\mu(\boldsymbol{r}) = \mu_0$. In this case, the density profile of the cloud containing N atoms is simply given by the implicit equation

$$\mu_{\mathrm{hom}}(n(\boldsymbol{r}), T) = \mu_0 - V(\boldsymbol{r}), \tag{8.2}$$

where μ_0 is determined by the normalization condition $\int \mathrm{d}^3\boldsymbol{r}\, n(\boldsymbol{r}) = N$. In the case of an ideal Fermi gas, the homogeneous equation of state is given by

$$n\lambda_{\mathrm{th}}^3 = -\mathrm{Li}_{3/2}(-\mathrm{e}^{\mu_{\mathrm{hom}}/k_{\mathrm{B}} T}), \tag{8.3}$$

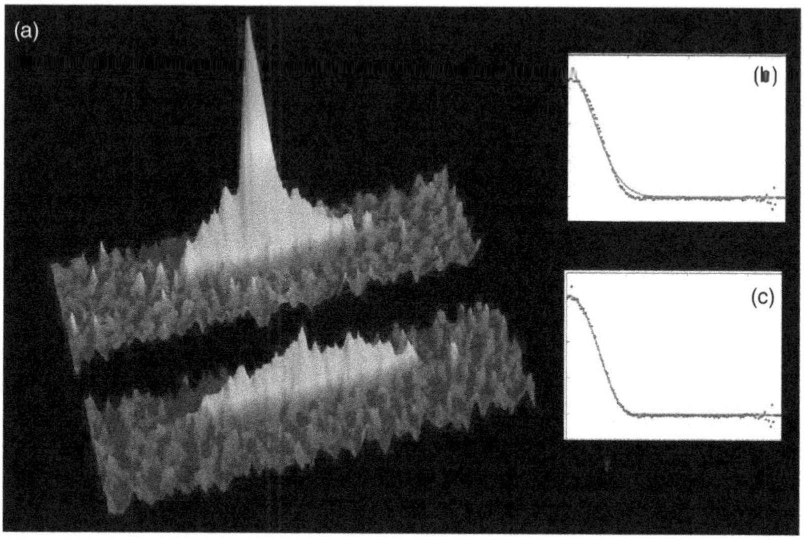

Figure 8.2 (a) Atomic density profiles in a harmonic trap. A mixture of bosonic (^{7}Li), top, and fermionic (^{6}Li), bottom, lithium atoms were simultaneously sympathetically cooled in a magnetic trap. The narrow peak at the centre of the bosonic cloud is the signature of Bose–Einstein condensation. The density profile of the fermionic cloud is fitted using a Gaussian distribution (b) and a Fermi–Dirac distribution (c), yielding a temperature $T/T_F \sim 0.05$.

where $\lambda_{\text{th}} = \sqrt{2\pi m k_B T/h^2}$ is the thermal de Broglie wavelength and $\text{Li}_{3/2}(z) = \sum_{n>0} z^n/n^{3/2}$ is the polylogarithm function of order 3/2. Comparison with experiment is presented in Figure 8.2, which shows that the experimental profile can be very accurately fitted using a Fermi–Dirac distribution, Eq. (8.2), by contrast with the Gaussian classical density profile. Since T is one of the fitting parameters, this measurement gives access to the temperature of the sample.

Although very powerful at high temperature, this thermometry procedure is limited to temperatures above $T/T_F \sim 0.05$. Indeed, in the classical regime corresponding to high temperature, the temperature of the cloud is actually obtained from the size of the cloud that scales as $\sqrt{T}$, according to the equipartition theorem in a harmonic trap, while at low temperature, the size of the cloud actually saturates to a radius R_F (the Thomas–Fermi radius) given by the balance between the Fermi pressure and the trapping potential.

More precisely, at low temperature ($T/T_F \ll 1$) Eq. (8.3) takes a simpler form. Indeed, in this case one can take for μ_{hom} the equation of state of a zero temperature Fermi gas $\mu_{\text{hom}} = \hbar^2 k_F^2/2m$, with the Fermi wavevector $k_F = (6\pi^2 n)^{1/3}$. The density profile in the trap is then

$$n(\boldsymbol{r}) = \frac{1}{6\pi^2}\left(\frac{2m}{\hbar^2}\left(\mu_0 - V(\boldsymbol{r})\right)\right)^{3/2}. \tag{8.4}$$

The value of μ_0 is then obtained by the normalization condition $\int d^3\boldsymbol{r}n(\boldsymbol{r}) = N$, where N is the number of fermions which yields for a harmonic trap of frequencies $\omega_{x,y,z}\mu_0 = \hbar\bar{\omega}\,(6N)^{1/3}$, with $\bar{\omega}^3 = \omega_x\omega_y\omega_z$.

8.3 Fermionic superfluidity and the BEC–BCS crossover

8.3.1 Introduction

The previous section demonstrated that a vapour of ultracold polarized fermions provided an experimental realization of an ideal Fermi gas but the physics becomes even more interesting when one considers the influence of interactions between particles. For instance, it has been known since the famous work of Bardeen, Cooper, and Schrieffer, (BCS, (Bardeen *et al.* 1957)) on the microscopic origin of superconductivity in metals that any attractive interaction would destabilize the Fermi surface of an ensemble of electrons at low temperature, leading to a gapped excitation spectrum and superconductivity. This BCS pairing is based on electron pairing in momentum space and, these pairs being composed of two fermions, it is tempting to describe them as composite bosons that form a Bose–Einstein condensate (BEC) at low temperature. Despite its simplicity, this analogy is actually misleading, since BCS and BEC theories differ in many aspects: most notably, Bose–Einstein condensation can happen in an ideal gas, and the associated critical temperature depends only on the particle density, and not the interaction strength, by contrast with the BCS scenario where attractive interactions are required, and for which the critical temperature vanishes for an ideal gas. In what follows, I will show that the relationship between BEC and BCS theories is actually subtler, since they rather constitute two limiting cases of a single theory, the so-called BEC–BCS crossover, which was developed in the early 1980s by Leggett, Noziéres, and Schmitt-Rink (Leggett 1980; Nozieres and Schmitt-Rink 1985).

8.3.2 Qualitative description of the crossover theory

Let us consider an ensemble of fermionic atoms interacting through a generic interatomic potential, shown in Figure 8.3. When the depth V_0 is big enough, the potential well possesses bound states thanks to which the atoms can pair up to form dimers. Since the fermionic atoms carry a half-integer spin, the molecules are integer spin bosons. When the density of dimers is small enough, and in particular when the mean distance between two of them is much bigger than their size, we can describe the system as an ideal gas of point-like boson, which will Bose-condense at low temperature: quite surprisingly, Bose–Einstein statistics are thus more suited for describing the physics of a strongly attractive Fermi gas! With hindsight, however, this scenario is very generic, since in practice we only observe Bose–Einstein condensates of atoms (helium, alkali...), which are composite particles made by the assembly of more fundamental fermionic particles (protons, neutrons, and electrons). In some sense, ^{4}He condensation can be described by the formation of 'dimers' of electrons and He^+ ions.

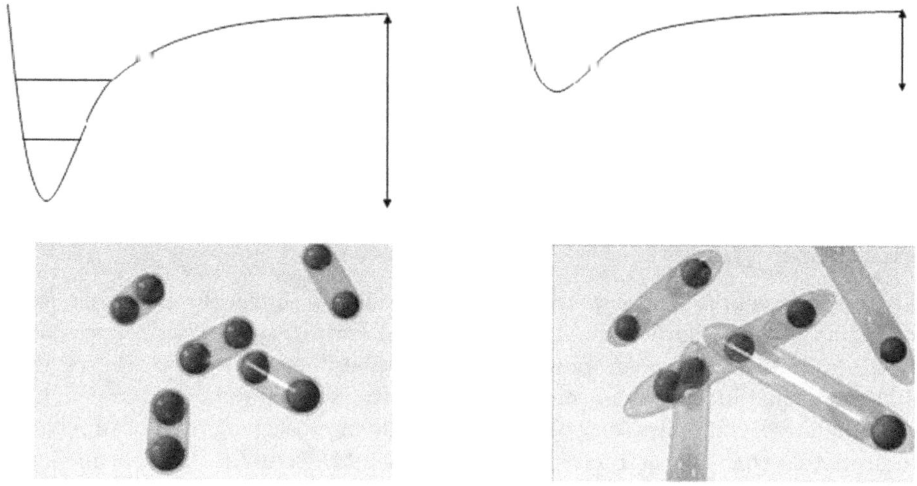

Figure 8.3 The BEC–BCS crossover scenario. Left: BEC side of the resonance (scattering length $a > 0$). A two-body state of binding energy $-\hbar^2/ma^2$ exists and the ground state of the many-body system is a Bose–Einstein condensate of dimers. Right: BCS side of the resonance ($a < 0$). For $a < 0$, there is no low-lying bound state and the Cooper pairs are stabilized by the presence of the Fermi sea. In contrast with the BEC case, the origin lies in many-body, and not two-body, effects.

Below a certain depth, a shallow 3D-potential does not possess any bound state, and the physical picture derived from the previous argument is no longer applicable. Naively, one would even conclude that in this regime, no superfluidity should be observed, since bosonic dimers cannot be formed. This argument is obviously wrong since in the case of electrons, superconductivity originates from the extremely weak attractive interaction mediated by phonons, which does not give rise to an electronic bound state. Fifty years ago, Bardeen, Cooper, and Schrieffer (Bardeen *et al.* 1957) demonstrated that, in analogy with the strongly attractive case, superconductivity does indeed arise from fermionic pairing but that Pauli blocking created by the presence of the whole Fermi sea is necessary to stabilize these pairs, in contrast with the strongly attractive regime, where only two-body interactions are sufficient to create pairs.

In the two regimes of weak and strong interactions, the attractive Fermi gas is intuitively simple, since only two-body correlations in dimers or Cooper pairs are necessary to capture the main features of the superfluid. By contrast, the intermediate regime has remained a challenge to quantum many-body theoreticians for a very long time, since in this regime weakly bound dimers have a size becoming of the order of their mutual distance, and can no longer be described as point-like bosons.

Noziére, Schmitt-Rink, Leggett, and Eagles proposed that the BEC (strongly attractive) and BCS (weakly attractive) regimes could be smoothly connected and were simply two limiting cases of a more general theory called the BEC–BCS crossover

(Leggett 1980; Nozieres and Schmitt-Rink 1985). This model is based on an extension of the BCS theory, which includes the fluctuations of the centre of mass motion of the pairs to describe their Bose–Einstein condensation properly. However, to test it experimentally it is necessary to be able to change the strength of the interactions from weak to strong, which is impossible in condensed matter system and explains why it had for a long time been impossible to discriminate between this mechanism and alternative theories predicting, for instance, a demixing instability in the intermediate regime (Houbiers *et al.* 1997).

In the language of atomic physics, it is more convenient to express the BEC–BCS crossover in terms of the scattering length a. Indeed using general scattering theory it is possible to demonstrate that the disappearance of a bound state is a resonant process associated with a divergence of the scattering length. The BEC–BCS crossover physics can, therefore, be explored in the vicinity of a Feshbach resonance. The region $a > 0$ corresponds to the BEC side of the crossover, with a two-body bound state of energy $E = -\hbar^2/ma^2$, while the BCS side is associated with $a < 0$.

8.3.3 Feshbach resonances with ultracold fermions

In condensed matter physics, it is possible to change the strength of interactions over only a small range and the BEC–BCS scenario remained unverified experimentally for a very long time. This deadlock was broken in 1998 by the observation of Feshbach resonances in cold atoms that allow one to change the scattering length characterizing s-wave interactions by applying an external magnetic field. The detailed physical mechanisms underlying this phenomenon are provided in (Vignolo, 2011), and we just recall here that a divergence of the scattering length is observed when a bound state of one of the molecular potentials becomes resonant with the energy of two incoming atoms. Feshbach resonances were predicted in (Tiesinga *et al.* 1992; Tiesinga *et al.* 1993; Moerdijk *et al.* 1995; Vogels *et al.* 1997) and observed for the first time on ultracold bosonic atoms (Courteille *et al.* 1998; Roberts *et al.* 1998), for which a huge inelastic loss rate was observed at the resonance. These preliminary observations did not bode well for an exploration of the BEC–BCS crossover physics with ultracold fermions (Loftus *et al.* 2002), as apparently confirmed by the observation of huge losses of lithium atoms close to the resonance at about 720 G for ^{6}Li atoms, see Figure 8.4 (Bourdel *et al.* 2003). One surprising feature was that, contrary to bosons, the maximum loss took place at a magnetic field significatively below the calculated position of the Feshbach resonance (Figure 8.4). In this region, the scattering length is positive, and the potential possesses a low-lying bound state: it was then understood that the atoms were not actually lost, but were binding together to form dimers that could not be observed by absorption imaging because the interaction between the two atoms of the pair shift the absorption lines and bring the molecules out of resonance with the imaging laser beam. The formation of the molecules was demonstrated experimentally, again by tuning the interactions. Indeed, by ramping the magnetic field slowly to a value where the scattering length is negative, the molecules are adiabatically dissociated and the atoms become observable. Using this ramping technique, it was possible

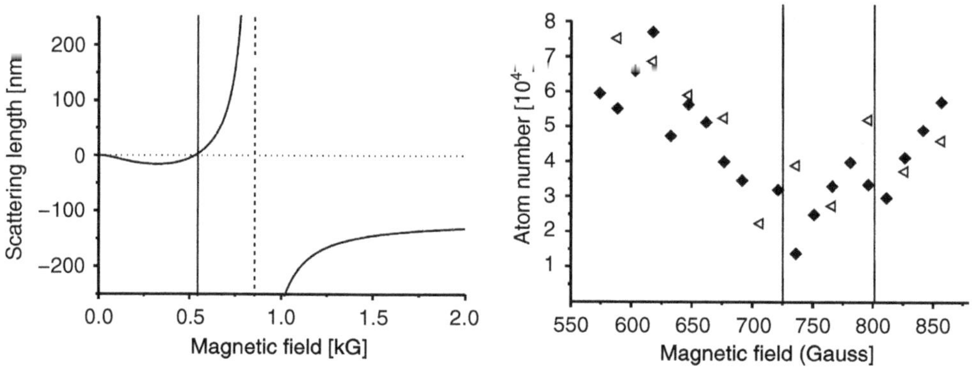

Figure 8.4 (a) Feshbach resonance for ^{6}Li. (b) Loss rate in ^{6}Li. The vertical lines indicate the position of the predicted and observed Feshbach resonances (Bourdel *et al.* 2004).

to demonstrate that unlike the bosonic case, molecules formed by pairing of two fermions have a long lifetime (Cubizolles *et al.* 2003; Jochim *et al.* 2003a; Regal *et al.* 2004). The reason for this stability is a consequence of the Pauli exclusion principle: first, Fermi pressure prevents collapse in the regime of strong attractive interactions, where gaseous Bose–Einstein condensates are unstable. The second reason is more subtle. Indeed, in principle, ultracold gases are metastable systems, since at low temperature, the ground state of the system is a solid. The lifetime of a trapped gas is thus ultimately limited by the condensation of the vapour through three-body losses. Petrov *et al.* (2004) demonstrated that in the case of a two-spin mixture, a three-body process would need to have at least two particles with identical spins within a few angstrom of each other, which is prevented by the Pauli exclusion principle. A more precise calculation shows that the three-body loss rate scales as $a^{-2.55}$ (Petrov *et al.* 2004), a behaviour confirmed by experiments (Bourdel *et al.* 2003; Regal *et al.* 2004) that show that inelastic losses vanish at the Feshbach resonance, where by contrast the elastic cross section increases. These observations paved the way to the Bose–Einstein condensation of molecules that was observed soon after (Greiner *et al.* 2003; Jochim *et al.* 2003b; Zwierlein *et al.* 2003; Bourdel *et al.* 2004) and confirmed the existence of a smooth crossover between the BEC and BCS sides of the resonance.

8.4 Probing the fermionic superfluid

8.4.1 Fermi gas at unitarity

As mentioned in the previous section, one of the greatest challenges in the theory of fermionic superfluids is the understanding of the unitary regime ($a = \infty$) where strong quantum correlations prevent the use of perturbative techniques. However, although the microscopic details of the system are complex, its macroscopic behaviour at low temperatures is surprisingly simple, and is actually, quite paradoxically, very similar to that of an ideal Fermi gas. Indeed, let us consider, for instance, the equation of state relating the chemical potential μ to the density n of atoms, which using LDA yields the density profile in a trap (Eq. (8.2)). Since at zero temperature the particle mass, the scattering length, and $\hbar$ are the only parameters describing the system, we have generally $\mu = f(n, a, m, \hbar)$. Using dimensional analysis, one finds that this equation can be expressed in terms of a single dimensionless parameter $1/k_F a$,[1] where $k_F = (3\pi^2 n)^{1/3}$ is the Fermi wavevector of an ideal gas of spin 1/2 fermions of density n. More precisely,

$$\mu = E_F g\left(\frac{1}{k_F a}\right), \tag{8.5}$$

where $E_F = \hbar^2 k_F^2/2m$ is the Fermi energy and g is a dimensionless function that must be determined either experimentally or theoretically. Let us now consider the special case $a = \infty$. In this case, Eq. (8.5) takes the simple form $\mu = \xi E_F$, with $\xi = g(0)$ a pure number. In other words, except for the scaling factor ξ, the equation of state of the unitary Fermi case is identical to that of an ideal gas $\mu = E_F$. In particular, the density profile can be calculated in the local density approximation using Eq. (8.5), which yields

$$n(\boldsymbol{r}) = \frac{1}{3\pi^2}\left(\frac{2m}{\xi\hbar^2}\left(\mu_0 - V(\boldsymbol{r})\right)\right)^{3/2}. \tag{8.6}$$

The measurement of the density profile of a strongly interacting Fermi gas therefore gives access experimentally to the value of the parameter ξ. Various versions of this scheme were implemented. These measurements converge towards the value $\xi = 0.42(1)$ (Bourdel *et al.* 2004; Bartenstein *et al.* 2004; Kinast *et al.* 2005; Partridge *et al.* 2006; Stewart *et al.* 2006), in agreement with the most recent Monte-Carlo simulations of the problem (Astrakharchik *et al.* 2004; Carlson *et al.* 2003; Carlson and Reddy 2005).

The extension of this result to the whole crossover and finite temperature is more involved. First, usual thermometric techniques based on time-of-flight expansion of ideal gases are no longer possible in the strongly interacting regime and, second, the inhomogeneous density profile associated with the integration over the line of sight makes it difficult to determine the local properties of the system. Several attempts were made to overcome these difficulties: use the low-lying excitation spectrum of the gas as an indirect probe of the equation of state (Altmeyer *et al.* 2007); replace measurement of temperature by that of entropy by ramping the magnetic field to a value where the scattering length cancels out, (Stewart *et al.* 2006; Luo *et al.* 2007); use ideal gas thermometry in a weakly spin-imbalanced Fermi gas (Zwierlein *et al.* 2006b), for which the wings are known to behave as an ideal gas; or reconstruct the local three-dimensional density profile using inverse Abel transforms (Shin 2008).

Recently, it was pointed out that, paradoxically, the trap could actually be considered as a valuable tool for measuring the fundamental equation of state $P(\mu_i, T)$

[1] Of course $k_F a$ is also a legitimate dimensionless parameter. However, $1/k_F a$ stresses the continuity of the physics of the system across the Feshbach resonance, despite the divergence of the scattering length.

in a single shot, relating the pressure P of the system to the chemical potentials and temperature of a *homogeneous* system (Ho and Zhou 2009). Indeed, let us consider the double integrated density of spin species i defined by $\bar{n}_i(z) = \int dxdyn(x, y, z)$. Using the Gibbs–Duhem relation $n_i = \partial_{\mu_i} P$, we can write that

$$\sum_i \bar{n}_i(z) = \int dxdy \sum_i \frac{\partial P}{\partial \mu_i}.$$

In the local density approximation, the chemical potential varies as $\mu_i = \mu_i^0 - V(\boldsymbol{r})$, where V is the trapping potential. Owing to the dependence on position of the chemical potentials, one can turn the spatial integral into an integral over μ_i reading

$$\sum_i \bar{n}_i(z) = \int \sum_i \frac{2\pi d\mu_i}{m\omega_r^2} \frac{\partial P}{\partial \mu_i},$$

where $\omega_r^2 = \omega_x \omega_y$ is the transverse trapping frequency.

$$P(\mu_i(z)) = \frac{m\omega_r^2}{2\pi} \sum_i \bar{n}_i(z). \tag{8.7}$$

Interestingly, this scheme is very general, and can in principle be used in any kind of situation (finite temperature, mixtures of atomic species, Bose or Fermi gases...), as long as the trapping is harmonic and the gas can be described using local density approximation. This scheme was successfully implemented in an experiment on ^{6}Li (Nascimbene *et al.* 2010), where the temperature was measured by adding a small quantity of bosonic ^{7}Li in the trap. At unitarity, the equation of state of the gas can be expressed dimensionally as

$$P = P_0(\mu, T)h(z),$$

where P_0 is the pressure of an ideal Fermi gas and h is some dimensionless function of the fugacity $z = \exp(\mu/k_B T)$. The resulting dimensionless equation of state is displayed in Figure 8.5, where it is compared with some of the latest theoretical calculations, using advanced finite-temperature schemes. Interestingly, the equation of state can be fitted by a T^2 law, $h = a + bT^2$, reminiscent of an ideal Fermi gas at low temperature. More precisely, in this picture, the equation of state can be used to extract the effective mass of the quasi-particles of the corresponding Landau Fermi liquid, yielding $m^* = 1.13m$.

8.4.2 Single-particle excitation spectrum and superfluidity

Although powerful, the scheme described in the previous section is not really suitable for the direct study of pairing and superfluidity in Fermi gases. Several specific probes were therefore designed for this particular purpose and have clarified the superfluid nature of these systems.

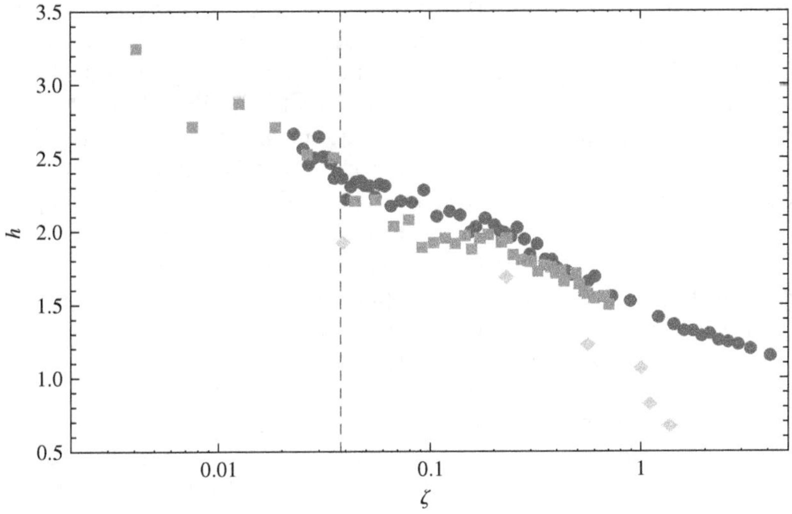

Figure 8.5 Experimental determination of the equation of state of a finite temperature unitary Fermi gas. Disks: experimental results; squares: dynamical mean field Monte Carlo; Diamonds: diagrammatic Monte Carlo. The dashed line indicates the normal–superfluid transition.

Let us first concentrate on fermionic pairing, which is the source of superfluidity in fermionic systems. The main investigation tool here is radio-frequency spectroscopy, where one flips atomic spins using an oscillating magnetic field and monitors the resonance shift due to interactions with the other atoms. Pioneering results using this method were obtained at JILA and MIT on the formation and characterization of uncondensed molecules on the BEC side of the resonance and were later extended to the case of fermionic superfluids (Chin *et al.* 2004). These first experiments gave strong evidence for the existence of a pairing gap at low temperature, but quantitative analysis was plagued by several artefacts, in particular, inhomogeneity of the cloud and interaction effects in the initial and final states. On the one hand, important theoretical effort was made in the understanding of these side effects (Punk and Zwerger 2007; Basu and Mueller 2008; Massignan *et al.* 2008; Perali *et al.* 2008; Veillette *et al.* 2008). On the other hand, by choosing a target state interacting weakly with the superfluid and developing a tomographic spectroscopy scheme, a spectroscopic measurement devoid of these artefacts could be obtained and permitted a clear and precise determination of the excitation gap of the system (Schirotzek *et al.* 2008). As for the chemical potential, the gap at unitarity is dimensionally proportional to Fermi energy and these experiments yielded $\Delta = 0.44E_{\mathrm{F}}$, in agreement with Monte-Carlo calculations (Carlson *et al.* 2003).

Another interesting application of this method is momentum-resolved spectroscopy, which provides a generalization of cold atom systems of the ARPES (angle resolved photoemission spectroscopy) that was developed for condensed matter systems and, in particular, helped to determine the abnormal structure of the Fermi surface in the normal phase of high-critical-temperature superconductors (Damascelli *et al.* 2003). As demonstrated in (Dao *et al.* 2008) and implemented experimentally in (Stewart *et al.* 2008), it allows one to measure the spectral function of the system directly; this is one of the key quantities calculated in quantum many-body theories and most notably allows one to characterize the structure of the Fermi surface of the system.

It should, nevertheless, be stressed that observing an excitation gap is different from actually confirming the existence of the superfluid order. For instance, in the BEC side of the phase diagram, the gap can be interpreted as the binding energy of the dimers that are already stable at a temperature higher T^* than the Bose–Einstein condensation threshold T_{c}. In the region $T^* > T > T_{\mathrm{c}}$, it is thus possible to observe an excitation gap, without any superfluidity.

Historically, the first definition of superfluidity is the absence of viscous (for liquid) or ohmic (for metals and alloys) dissipation. This property is intimately connected to the excitation spectrum of the cloud, as first discussed by Landau. Indeed, if the energy transfer between a moving obstacle and the superfluid is associated with the formation of elementary excitations of the superfluid, then a simple energy–momentum balance shows that no excitation can be produced if the velocity of the obstacle is smaller than the critical value $v_{\mathrm{c}} = \min_k \omega_k/k$, where ω_k is the dispersion relation of the excitations of the superfluid. Although intuitively simple, this argument is valid only for small objects, creating weak perturbations in the superfluid. For larger objects, the energy transfer processes through the formation of macroscopic excitations of the superfluid, such as quantum analogues of von Karman vortex alleys (Frisch *et al.* 1992). This first aspect of superfluidity can be probed in cold-atom systems by moving an obstacle inside the cloud and measuring the subsequent heating. This scheme was first applied to bosons to probe superfluidity in Bose–Einstein condensates and was later extended to the case of fermions (see Figure 8.6). In the work described in (Miller *et al.* 2007), the cloud is excited by a moving optical lattice and, as expected by Landau's scenario or its extensions, we observe that below a critical velocity v_{c} no atom loss associated with heating is observed (Figure 8.6).

In the case of fermions, the interpretation of the dependence of critical velocity with $k_{\mathrm{F}}a$ is slightly more involved than for bosons, owing to the existence of two excitation branches: collective modes associated with sound waves on the one hand and pair-breaking single particle excitations on the other hand. Just like atomic BEC, superfluidity is limited by phonon excitations in the molecular condensate sector, while in the BCS region, the vanishingly small gap limits the velocity of the superfluid flow. This competition between the two branches is responsible for the maximum critical velocity close to unitarity, which was predicted in (Sensarma *et al.* 2006; Combescot *et al.* 2006; Spuntarelli *et al.* 2007) and clearly observed experimentally (see Figure 8.6).

Another definition of superfluidity is the existence of a (complex) order parameter characterizing the 'macroscopic' pair wavefunction—in the BCS theory this is the superfluid gap. Based on general arguments, it can be demonstrated that phase singularities of this order can be interpreted as vortices with quantized circulation that can be observed after stirring the cloud. At high rotation frequencies, theses vortices form

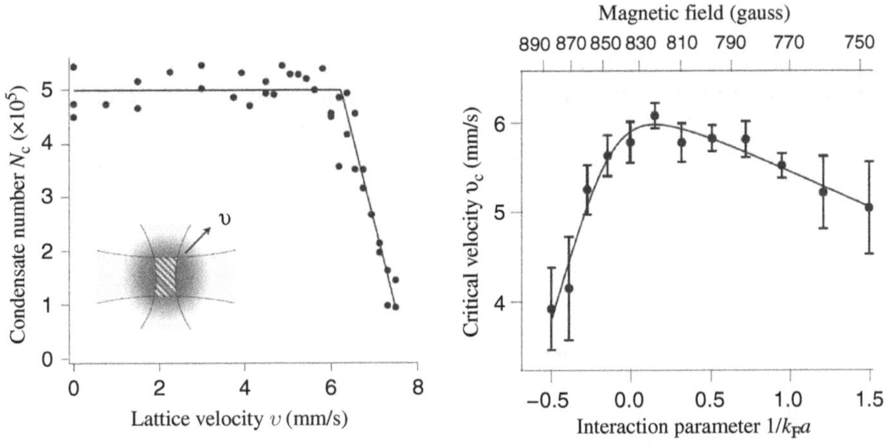

Figure 8.6 Observation of the onset of dissipation in a fermionic superfluid. Left: evolution of the number of condensed atoms as a function of velocity. Below a critical velocity v_c, the condensed fraction stays constant; it decreases above v_c. Right: Evolution of v_c with interaction parameter $1/k_F a$. (Miller *et al.* 2007).

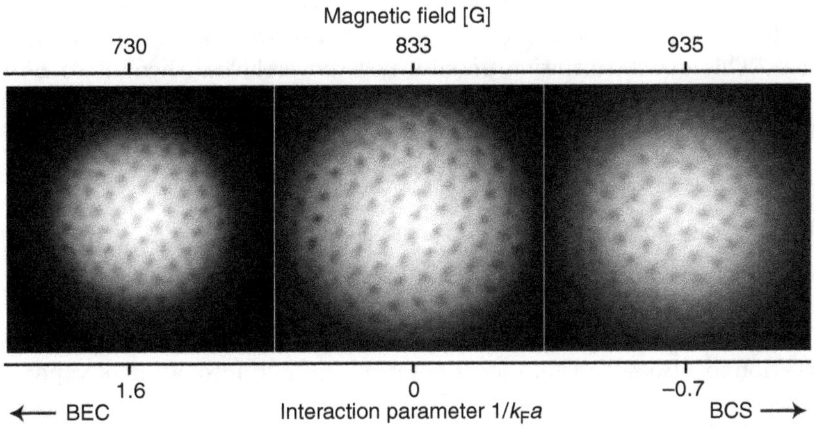

Figure 8.7 Generation of vortex lattices in rotated Fermi gases of ^{6}Li along the BEC–BCS crossover. The dips in the density profiles indicate the position of the vortex cores. Picture from (Zwierlein *et al.* 2005).

triangular arrays, known as Abrikosov lattices in the case of Type II superconductors. In a dramatic series of experiments the group of W. Ketterle at MIT managed to generate vortex lattices in the whole crossover, demonstrating without any ambiguity the existence of quantum order in this system (See Figure 8.7).

8.5 Conclusion

The results presented in this chapter demonstrate that the latest developments in the experimental study of ultra-Fermi gases have brought us a very precise knowledge of the physical properties of strongly correlated Fermi gases, which would probably not have been achieved using theoretical tools only. I have restricted my discussion to the case of the BEC–BCS crossover that was the first issue tackled with these new systems, but many other interesting problems are and will be addressed in the future. Even now, studies performed at Rice, MIT and later at ENS (Partridge *et al.* 2006; Zwierlein *et al.* 2006a; Nascimbene *et al.* 2009) have shed new light on the fate of a fermionic superfluid when spin populations are imbalanced, a question that had remained unsettled for nearly forty years.

The next endeavour will be the exploration of the properties of ultracold Fermi gases in optical lattices, where the atoms would simulate the famous Hubbard lattice Hamiltonian that constitutes one of the most favoured models in condensed matter physics. Recent progress in this direction has been reported recently with the observation of the fermionic Mott transition independently at Mainz and Zurich (Jördens *et al.* 2008; Schneider *et al.* 2008). However, the next challenges will require cooling down cold atom samples to temperatures never yet achieved, but will introduce us to a realm of new physical phenomena. Among these, are the possibility of studying quantum magnetism, and in particular of solving open questions on the nature of antiferromagnetism in some non-trivial lattice geometries. The main goal, however, would be to determine whether the *repulsive* Hubbard model can sustain a superfluid phase. This counterintuitive possibility is nevertheless thought to be the main microscopic explanation for the high-critical-temperature superconductivity in cuprate compounds.

References

Altmeyer, A, Riedl, S, Kohstall, C, *et al.* (2007). Precision measurements of collective oscillations in the BEC–BCS crossover. *Phys. Rev. Lett.*, **98**, 040401.

Anderson, MH, Ensher, JR, Matthews, MR, Wieman, CE, and Cornell, EA (1995). Observation of Bose–Einstein condensation in a dilute atomic vapor. *Science*, **269**, 198.

Astrakharchik, GE, Boronat, J, Casulleras, J, and Giorgini, S (2004). Equation of state of a Fermi gas in the BEC–BCS crossover: a quantum Monte Carlo study. *Phys. Rev. Lett.*, **93**, 200404.

Bardeen, J, Cooper, LN, and Schrieffer, JR (1957). Theory of superconductivity. *Physical Review*, **108**, 1175.

Barrett, MD, Sauer, JA, and Chapman, MS (2001). All-optical formation of an atomic Bose–Einstein condensate. *Phys. Rev. Lett.*, **87**, 010404.

Bartenstein, M, Altmeyer, A, Riedl, S, *et al.* (2004, Mar). Crossover from a molecular Bose–Einstein condensate to a degenerate Fermi gas. *Phys. Rev. Lett.*, **92**, 120401.

Basu, S and Mueller, EJ (2008). Final-state effects in the radio frequency spectrum of strongly interacting fermions. *Phys. Rev. Lett.*, **101**, 60405.

Bloch, I, Dalibard, J, and Zwerger, W (2008). Many-body physics with ultracold gases. *Rev. Mod. Phys.*, **80**, 885.

Bourdel, T, Cubizolles, J, Khaykovich, L, *et al.* (2003). Measurement of the interaction energy near a Feshbach resonance in a ^{6}Li fermi gas. *Phys. Rev. Lett.*, **91**, 020402.

Bourdel, T, Khaykovich, L, Cubizolles, J, *et al.* (2004). Experimental study of the BEC–BCS crossover region in lithium 6. *Phys. Rev. Lett.*, **93**, 050401.

Carlson, J and Reddy, S (2005). Asymmetric two-component fermion systems in strong coupling. *Phys. Rev. Lett.*, **95**, 060401.

Carlson, J, Chang, S-Y, Pandharipande, VR, and Schmidt, KE (2003). Superfluid Fermi gases with large scattering length. *Phys. Rev. Lett.*, **91**, 050401.

Chin, C, Bartenstein, M, Altmeyer, A, *et al.* (2004). Observation of the pairing gap in a strongly interacting Fermi gas. *Science*, **305**, 1128.

Combescot, R, Kagan, MY, and Stringari, S (2006). Collective mode of homogeneous superfluid Fermi gases in the BEC–BCS crossover. *Phys. Rev. A*, **74**, 042717.

Cornell, EA and Wieman, CE (2002). Nobel Lecture: Bose–Einstein condensation in a dilute gas, the first 70 years and some recent experiments. *Rev. Mod. Phys.*, **74**, 875.

Courteille, P, Freeland, RS, Heinzen, DJ, van Abeelen, FA, and Verhaar, BJ (1998). Observation of a Feshbach resonance in cold atom scattering. *Phys. Rev. Lett.*, **81**, 69.

Cubizolles, J, Bourdel, T, Kokkelmans, SJJMF, Shlyapnikov, GV, and Salomon, C (2003, Dec). Production of long-lived ultracold Li_2 molecules from a Fermi gas. *Phys. Rev. Lett.*, **91**, 240401.

Damascelli, A, Hussain, Z, and Shen, Z-X (2003). Angle-resolved photoemission studies of the cuprate superconductors. *Rev. Mod. Phys.*, **75**, 473.

Dao, TL, Georges, A, Dalibard, J, Salomon, C, and Carusotto, I (2007). Measuring the one-particle excitations of ultracold fermionic atoms by stimulated Raman Spectroscopy. *Phys. Rev. Lett.*, **98**, 240402.

Davis, KB, Mewes, MO, Andrews, MR, *et al.* (1995). Bose–Einstein condensation in a gas of sodium atoms. *Phys. Rev. Lett.*, **75**, 3969.

DeMarco, B, Bohn, JL, Burke, JP, Holland, M, and Jin, DS (1999). Measurement of *p*-wave threshold law using evaporatively cooled fermionic atoms. *Phys. Rev. Lett.*, **82**, 4208.

DeMarco, B and Jin, DS (1999). Onset of Fermi degeneracy in a trapped atomic gas. *Science*, **285**, 1703.

Feynman, RP (1982). Simulating physics with computers. *Int. J. Theor. Phys.*, **21**, 467.

Frisch, T, Pomeau, Y, and Rica, S (1992). Transition to dissipation in a model of superflow. *Phys. Rev. Lett.*, **69**, 1644.

Greiner, M, Regal, CA, and Jin, DS (2003). Emergence of a molecular Bose–Einstein condensate from a Fermi gas. *Nature*, **426**, 537.

Hadzibabic, Z, Gupta, S, Stan, CA, *et al.* (2003). Fiftyfold improvement in the number of quantum degenerate fermionic atoms. *Phys. Rev. Lett.*, **91**, 160401.

Ho, T and Zhou, Q (2009). Obtaining phase diagram and thermodynamic quantities of bulk systems from the densities of trapped gases. *arXiv:0901:0018*.

Houbiers, M, Ferwerda, R, Stoof, HTC, McAlexander, WI, Sackett, CA, and Hulet, RG (1997). Superfluid state of atomic ^{6}Li in a magnetic trap. *Phys. Rev. A*, **56**, 4864.

Inguscio, M, Ketterle, W, and Salomon, C (2006). Ultracold Fermi gases. *Proceedings of the International School of Physics Enrico Fermi, Course CLXIV, Varenna.*

Jaksch, D, Bruder, C, Cirac, JI, Gardiner, CW, and Zoller, P (1998). Cold bosonic atoms in optical lattices. *Phys. Rev. Lett.*, **81**, 3108.

Jochim, S, Bartenstein, M, Altmeyer, A, *et al.* (2003a). Pure gas of optically trapped molecules created from fermionic atoms. *Phys. Rev. Lett.*, **91**, 240402.

Jochim, S, Bartenstein, M, Altmeyer, A, *et al.* (2003b). Bose–Einstein condensation of molecules. *Science*, **302**, 2101.

Jördens, R, Strohmaier, N, Günter, K, Moritz, H, and Esslinger, T (2008). A Mott insulator of fermionic atoms in an optical lattice. *Nature*, **455**, 204.

Ketterle, W (2002). Nobel Lecture: when atoms behave as waves: Bose–Einstein condensation and the atom laser. *Rev. Mod. Phys.*, **74**, 1131.

Kinast, J, Turlapov, A, Thomas, JE, Chen, Q, Stajic, J, and Levin, K (2005). Heat capacity of a strongly interacting Fermi gas. *Science*, **307**, 1296.

Leggett, AJ (1980). Cooper pairing in spin-polarized Fermi systems. *J. Phys. Coll. (Paris)*, **41**C7.

Loftus, T, Regal, CA, Ticknor, C, Bohn, JL, and Jin, DS (2002). Resonant control of elastic collisions in an optically trapped Fermi gas of atoms. *Phys. Rev. Lett.*, **88**, 173201.

Luo, L, Clancy, B, Joseph, J, Kinast, J, and Thomas, JE (2007). Measurement of the entropy and critical temperature of a strongly interacting Fermi gas. *Phys. Rev. Lett.*, **98**, 80402.

Massignan, P, Bruun, GM, and Stoof, HTC (2008). Twin peaks in RF spectra of Fermi gases at unitarity. *Phys. Rev. A*, **77**, 31601.

Miller, DE, Chin, JK, Stan, CA, *et al.* (2007). Critical velocity for superfluid flow across the BEC–BCS crossover. *Phys. Rev. Lett.*, **99**, 070402.

Moerdijk, AJ, Verhaar, BJ, and Axelsson, A (1995). Resonances in ultracold collisions of ^{6}Li, ^{7}Li, and ^{23}Na. *Phys. Rev. A*, **51**, 4852.

Nascimbene, S, Navon, N, Jiang, K, *et al.* (2009). Collective oscillations of an imbalanced Fermi gas: axial compression modes and polaron effective mass. *Phys. Rev. Lett.*, **103**, 170402.

Nascimbene, S, Navon, N, Jiang, K, Chevy, F, and Salomon, C (2010). Exploring the thermodynamics of a universal Fermi gas. *Nature*, **463**, 1057.

Nozieres, P and Schmitt-Rink, S (1985). Bose condensation in an attractive fermion gas: from weak to strong coupling superconductivity. *J. Low Temp. Phys.*, **59**, 195.

Partridge, GB, Li, W, Kamar, RI, Liao, Y, and Hulet, RG (2006). Pairing and phase separation in a polarized Fermi gas. *Science*, **311**, 503.

Perali, A, Pieri, P, and Strinati, GC (2008). Competition between final-state and pairing-gap effects in the radio-frequency spectra of ultracold Fermi atoms. *Phys. Rev. Lett.*, **100**, 10402.

Petrov, DS, Salomon, C, and Shlyapnikov, GV (2004). Weakly bound dimers of fermionic atoms. *Phys. Rev. Lett.*, **93**, 90404.

Punk, M and Zwerger, W (2007). Theory of RF-spectroscopy of strongly interacting fermions. *Phys. Rev. Lett.*, **99**, 170404.

Regal, CA, Greiner, M, and Jin, DS (2004). Lifetime of molecule-atom mixtures near a Feshbach resonance in ^{40}K. *Phys. Rev. Lett.*, **92**, 083201.

Roati, G, Riboli, F, Modugno, G, and Inguscio, M (2002). Fermi–Bose quantum degenerate ^{40}K$-^{87}$Rb mixture with attractive interaction. *Phys. Rev. Lett.*, **89**, 150403.

Roberts, JL, Claussen, NR, Burke, JP, Greene, CH, Cornell, EA, and Wieman, CE (1998). Resonant magnetic field control of elastic scattering in cold ^{85}Rb. *Phys. Rev. Lett.*, **81**, 5109.

Schirotzek, A, Shin, Y, Schunck, CH, and Ketterle, W (2008). Determination of the superfluid gap in atomic Fermi gases by quasiparticle spectroscopy. *Phys. Rev. Lett.*, **101**, 140403.

Schneider, U, Hackermuller, L, Will, S, *et al.* (2008). Metallic and insulating phases of repulsively interacting fermions in a 3D optical lattice. *Science*, **322**, 1520.

Schreck, F, Ferrari, G, Corwin, KL, *et al.* (2001). Sympathetic cooling of bosonic and fermionic lithium gases towards quantum degeneracy. *Phys. Rev. A*, **64**, 011402.

Sensarma, R, Randeria, M, and Ho, T-L (2006). Vortices in superfluid fermi gases through the BEC to BCS crossover. *Phys. Rev. Lett.*, **96**, 090403.

Shin, Y (2008). Determination of the equation of state of a polarized fermi gas at unitarity. *Phys. Rev. A*, **77**, 041603.

Silber, C, Günther, S, Marzok, C, Deh, B, Courteille, PW, and Zimmermann, C (2005). Quantum-degenerate mixture of fermionic lithium and bosonic rubidium gases. *Phys. Rev. Lett.*, **95**, 170408.

Spuntarelli, A, Pieri, P, and Strinati, GC (2007). Josephson effect throughout the BCS–BEC crossover. *Phys. Rev. Lett.*, **99**, 040401.

Stewart, JT, Gaebler, JP, Regal, CA, and Jin, DS (2006). Potential energy of a ^{40}K Fermi gas in the BCS–BEC crossover. *Phys. Rev. Lett.*, **97**, 220406.

Stewart, JT, Gaebler, JP, and Jin, DS (2008). Using photoemission spectroscopy to probe a strongly interacting Fermi gas. *Nature*, **454**, 744.

Taglieber, M, Voigt, A-C, Aoki, T, Hänsch, TW, and Dieckmann, K (2008). Quantum degenerate two-species fermi-fermi mixture coexisting with a Bose–Einstein condensate. *Phys. Rev. Lett.*, **100**, 010401.

Tiesinga, E, Moerdijk, AJ, Verhaar, BJ, and Stoof, HTC (1992). Conditions for Bose–Einstein condensation in magnetically trapped atomic cesium. *Phys. Rev. A*, **46**, R1167.

Tiesinga, E, Verhaar, BJ, and Stoof, HTC (1993). Threshold and resonance phenomena in ultracold ground-state collisions. *Phys. Rev. A*, **47**, 4114.

Truscott, AG, Strecker, KE, McAlexander, WI, Partridge, GB, and Hulet, RG (2001). Observation of Fermi pressure in a gas of trapped atoms. *Science*, **291**, 2570.

Veillette, M, Moon, EG, Lamacraft, A, Radzihovsky, L, Sachdev, S, and Sheehy, DE (2008). Radio-frequency spectroscopy of a strongly imbalanced Feshbach-resonant Fermi gas. *Phys. Rev. A*, **78**, 33614.

Vignolo, P (2011). Degenerate Fermi gases. In *Ultracold Gases and Quantum Information*. Lecture Notes of the Les Houches Summer School in Singapore,

Vol. **91**, July, 2009. Ed. C Miniatura, L-C Kwek, M Ducloy *et al.* Oxford University Press, Oxford, Chapter 2.

Vogels, JM, Tsai, CC, Freeland, RS, Kokkelmans, SJJMF, Verhaar, BJ, and Heinzen, DJ (1997). Prediction of Feshbach resonances in collisions of ultracold rubidium atoms. *Phys. Rev. A*, **56**, R1067.

Wille, E, Spiegelhalder, FM, Kerner, G, *et al.* (2008). Exploring an ultracold Fermi–Fermi mixture: interspecies Feshbach resonances and scattering properties of ^{6}Li and ^{40}K. *Phys. Rev. Lett.*, **100**, 053201.

Zwierlein, MW, Abo-Shaeer, JR, Schirotzek, A, Schunck, CH, and Ketterle, W (2005). Vortices and superfluidity in a strongly interacting Fermi gas. *Nature*, **435**, 1047.

Zwierlein, MW, Stan, CA, Schunck, CH, *et al.* (2003). Observation of Bose–Einstein condensation of molecules. *Phys. Rev. Lett.*, **91**, 250401.

Zwierlein, MW, Schirotzek, A, Schunck, CH, and Ketterle, W (2006a). Fermionic superfluidity with imbalanced spin populations. *Science*, **311**, 492.

Zwierlein, MW, Schunck, CH, Schirotzek, A, and Ketterle, W (2006b). Direct observation of the superfluid phase transition in ultracold Fermi gases. *Nature*, **442**, 54.

9
Competing instabilities in quench experiments with ultracold Fermi gases near a Feshbach resonance

David PEKKER

California Institute of Technology, Pasadena, CA 91125, USA

Eugene DEMLER

Harvard University, Cambridge, MA 02138, USA

Chapter Contents

9.1 Overview

The tunability of effective two-body interactions near Feshbach resonances is a powerful, experimental tool in systems of ultracold atoms. It has been used to explore a variety of intriguing phenomena in recent experiments. However, not all of the many-body properties of such systems can be understood in terms of an effective model with the contact interaction given by the scattering length of the two particles in vacuum. For example, when a two-component Fermi mixture is quenched to the BEC side of the Feshbach resonance, a positive scattering length suggests that interactions are repulsive and thus collective dynamics should be dominated by the Stoner instability towards a spin-polarized ferromagnetic state. On the other hand, the existence of low energy two particle bound states suggests a competing instability driven by molecule formation. Competition between spontaneous magnetization and pair formation is determined by the the interplay of two-particle and many-body phenomena. In this chapter, we summarize our recent theoretical results, which analysed this competition from the point of view of unstable collective modes. We also comment on the relevance of this theoretical analysis to recent experiments reported in (Jo *et al.* 2009).

9.2 Introduction

It is often effective to characterize many-body systems in terms of the emergent collective modes that describe their low-energy excitations. Examples of collective modes include sound waves in interacting gases, spin (magnetization) waves in magnetic systems, and phase modes in superconducting systems. The strength of the collective modes approach is that for many cases, such as Fermi liquids, ferromagnets, superconductors, etc., these collective modes are long lived and weakly interacting. Moreover, collective modes can be useful for understanding not only the properties of systems near equilibrium but also the dynamics of systems away from equilibrium.

The purpose of this chapter is to discuss how we can utilize the analysis of unstable collective modes to study the dynamics of fermionic systems quenched from the weakly interacting state to the strongly interacting regime in the vicinity of the Feshbach resonance. Conceptually, this approach is similar to the spinodal decomposition in statistical physics (Chaikin and Lubensky 2000) and analysis of the domain formation in unstable bosonic systems, such as immiscible two-component mixtures (Stamper-Kurn and Ketterle 2001), quenched ferromagnetic spinor Bose condensates (Sadler *et al.* 2006; Lamacraft 2007), and the dynamics of spiral states in ferromagnets (Cherng 2008; Conduit and Altman 2010). The main new feature of the fermionic problem that we consider is the need to provide careful regularization of the strong two-body interactions arising from Feshbach resonances. As we show, the interplay of strong two-body interactions and many-body effects, such as Pauli blocking by the Fermi sea, lead to some very intriguing aspects of dynamics.

The non-equilibrium dynamics of quantum systems has been studied for a long time. However, until recently, the experimental ability to control quantum systems has been rather limited. The progress in parametric control of quantum Hamiltonians of ultracold atom systems has brought renewed interest to the field of non-equilibrium dynamics. One such recent experiment, performed by the MIT group, studied ramps of a two-component Fermi gas from the weakly repulsive to the strongly repulsive regime (Jo *et al.* 2009). The ramp was accomplished by sweeping the magnetic field to a value close to the Feshbach resonance, but on the BEC side. Strongly repulsive fermions are known to be susceptible to the Stoner instability to ferromagnetism (Stoner 1933). In fact, this very instability is believed to be responsible for itinerant ferromagnetism in metals. Intriguingly, the MIT experiment observed several surprising phenomena, which were interpreted as signatures of ferromagnetism, yet did not find any ferromagnetic domains. However, strongly repulsive ultracold fermions come with a price—the repulsion is a byproduct of a shallow bound state. Hence, there is an alternative explanation for the MIT observations—susceptibility to pair formation (Pekker *et al.* 2011). In this chapter our goal is to understand many-body fermionic systems during and immediately after a quench into a strongly interacting regime near a Feshbach resonance. The emphasis of our discussion will be on understanding the interplay of two-particle and many-body aspects of dynamics.

Our study can be placed in a general perspective of quenches that take a quantum system from a disordered phase to an ordered phase described by some order parameter (e.g., magnetization or pairing). We show that the dynamics is initially driven by local instabilities towards ordering. A useful way to understand these instabilities is in terms of collective modes. In equilibrium, all collective modes of a system are, by definition, stable. Thus the dynamics following a small perturbation of a collective coordinate is either periodic or damped but not growing. In other words, in equilibrium, the imaginary part of the collective mode frequency is either zero or negative (we adopt the convention that $O(t) \sim O_\omega e^{-i\omega t}$). On the other hand, consider the dynamics of a system following a parametric ramp or a quench across a phase transition. To come to equilibrium on the ordered side, the system must develop a finite-order parameter following the quench. Generically, following the quench, some of the collective modes of the system become unstable, that is, they acquire a positive imaginary frequency. This local collective mode instability corresponds to the growth of the order parameter, and leads to its 'nucleation'. For the case of the MIT system, the quench leads to two distinct types of incompatible instability: pair formation and ferromagnetism. The competition between the two ultimately determines the fate of the quantum system.

Although a complete description of dynamic processes following a ramp across a phase transition remains to be found, we will present some aspects of the problem. The problem of understanding the dynamics of large quantum systems is that there are no generic methods of attack. Therefore, one may want to start by considering small systems that are susceptible to methods like exact diagonalization. In small quantum systems, there are no true phase transitions. However, one can consider parametric tuning of the Hamiltonian that leads to a level crossing associated with the phase transition of the thermodynamic system. For a finite system, such level crossings are typically associated with the changing of an approximate symmetry of the ground state and therefore they are avoided. The dynamics of the quantum system are then described by the Landau–Zener process (Landau 1932; Zener 1932): depending on the ramp rate, the system may either remain in the ground state or jump to the excited state. If we try to scale up these arguments to the case of a large many-body system we immediately run into a problem. During a parametric ramp, the exponentially large number of eigenstates of the many-body system undergo 'spectral-flow': some of these eigenstates move up and some down. The result of the spectral-flow is a large number of Landau–Zener processes, which generically take an exponentially large Hilbert space to describe. Therefore, a method that would keep only the 'important' excitations is very useful.

The approach of studying the collective modes in order to describe the dynamics following a parametric ramp or quench has a long history. Kibble (1976) and Zurek (1985) were the first to point out that a short time after a thermal quench to an ordered state (e.g., superconducting state), distant parts of the systems could not have exchanged information and therefore the order parameter at large distances cannot be correlated. Hence, shortly after a thermal quench it is possible to find locations in space around which the order parameter winds, thus the dynamics following a quench result in the formation of topological defects (e.g., vortices for the superconducting case). Further dynamics involves the motion and recombination of the topological defects produced at short timescales. The specific problem of the motion and recombination of topological defects has been studied extensively for the case of thermodynamic phase transitions in various systems, including liquid crystals. Liquid crystals, in particular, have attracted attention owing to both their technological application in displays and the relative accessibility of their dynamics from the experimental perspective as the collective modes and topological defects associated with the various ordered states can be observed using polarized light (Chuang *et al.* 1991). The dynamics of these systems following a quench conform to the notion of initial growth followed by power-law slow defect recombination (Bray 2002).

To summarize, the dynamics can be split into two timescales. There is an initial short timescale, during which the various unstable collective modes grow exponentially. As these modes are largely non-interacting, they can be treated as independent modes in momentum space. As the amplitudes of the unstable collective modes grow, they start to interact with each other. When the interactions become strong, the resulting order parameter pattern produced by the initial growth 'freezes in'. Consequently, the dynamics is controlled by the motion of topological defects (as well as non-topological excitations).

In this chapter, we shall apply the method of collective mode instabilities to the concrete example of quenches of the non-interacting Fermi gas to the strongly interacting regime. The motivations for this are:

1. The existence of ferromagnetism in an itinerant fermionic system is a long-standing open question in condensed matter physics. While ferromagnetic transition can be predicted based on a simple mean-field analysis (Stoner 1933), several objections to this argument can be raised. Kanamori was the first to point out that screening was essential to understanding the Stoner transition, and may even prevent it (Kanamori 1963). For example, the Stoner criterion suggests ferromagnetic instability even in cases when rigorous theorems forbid such a transition, such as one-dimensional systems (Lieb and Mattis 1962). Recent theoretical investigations have found that the transition survives in 3D but becomes of first-order type at low temperatures (Houbiers *et al.* 1997; Duine and MacDonald 2005; Zhang and Sarma 2005; Tanaka and Tasaki 2007; Conduit and Simons 2009a;

Conduit *et al.* 2009; Zhang *et al.* 2010). Therefore, if results of the MIT experiments can indeed be interpreted as Stoner-type instability, then they resolve a fundamental matter of principle question in physics. However, before we accept this interpretation, it is important that we rule out competing scenarios, such as the non-equilibrium dynamics of pairing. Other arguments against the possibility of observing Stoner instability in ultracold Fermi gases near Feshbach resonance have been given in (Zhai 2009; Barth and Zwerger 2011; Zhang and Ho 2011).

2. The true ground state in such system is a condensate of molecules. It was proposed that a Stoner ferromagnetic state can be created dynamically if the rate at which magnetic correlations develop is considerably faster than the rate of molecule formation (Jo *et al.* 2009; Pilati *et al.* 2010; Chang *et al.* 2011). Knowing whether such a hierarchy of instabilities really takes place is crucial for interpreting experimental results.
3. For this reason, studying the dynamics following a quench is relevant to the current experiments.
4. Studying the Stoner and pairing instabilities is a good way to discuss Feshbach resonance in many-body systems. Throughout, we shall discuss the pairing and the Stoner instabilities from the perspective of a geometrical Feshbach resonance (in which the quantum numbers of fermions in the molecular state are identical to those of free fermions).
5. Our analysis has important general implications for the idea of quantum simulations with ultracold atoms. In many cases one is interested in using strong repulsive interactions between atoms to create analogues of condensed matter systems. Our analysis provides a warning that to get any meaningful results, one needs to make sure that molecule formation does not overwhelm dynamics determined by the repulsive interactions.

The chapter is organized as follows. In Section 9.3 we begin by discussing the relation between linear response and collective modes. We also show how to compute the pairing and the ferromagnetic responses using the equation of motion formalism. Next, in Section 9.4, we show how to describe a Feshbach resonance using a pseudo-potential model. In Section 9.5, we apply the pseudo-potential model to compute the many-body T-matrix and thus obtain the pairing collective mode. We comment on how to incorporate the many body T-matrix into the ferromagnetic susceptibility in Section 9.6. Finally, in Section 9.7 we summarize the results for the pairing versus Stoner competition in the context of the MIT experiments, and we make concluding remarks in Section 9.8.

9.3 Linear response and collective modes

9.3.1 From poles of the response functions to collective modes

A useful approach to thinking about collective modes is via their link to linear response susceptibilities. A linear response susceptibility $\chi_A(q,\omega)$ is defined as the link response of some property of the system $\langle A(q,\omega)\rangle$ under the influence of an external perturbation of strength $h_A(q,\omega)$ that is thermodynamically conjugate to it,

$$\langle A(q,\omega)\rangle = \chi_A(q,\omega)h_A(q,\omega). \tag{9.1}$$

Here, by thermodynamically conjugate, we mean that the external perturbation $h_A(q,\omega)$ contributes a term to the Hamiltonian

$$\mathcal{H}_{\text{external}} = h_A(q,\omega)e^{i\omega t}\hat{A}(-q,\omega) + h.c. \tag{9.2}$$

Generally, to obtain a finite value of $\langle A(q,\omega)\rangle$ one needs to have a non-zero value of the external field $h_A(q,\omega)$. An important exception is when the response function $\chi_A(q,\omega)$ has a pole and is therefore infinite. Writing the integral equation for the response of the system

$$\langle A(q,t)\rangle = \int dt'\, \chi_A(t-t',\omega)h_A(q,t'), \tag{9.3}$$

we see that poles of $\chi_A(q,\omega)$ correspond to long-lived modes of the system, which are identified as the collective modes (Abrikosov *et al.* 1975; Pines and Nozières 1989). This argument is conventionally used for systems with time-independent parameters. In the next section, we will discuss how it can be extended to the case when the interaction strength is changing in time.

9.3.2 Dynamics of the pairing amplitude using the time-dependent Hartree approximation

One approach to obtaining a response function, and thus the associated collective mode spectrum, is to use the equation of motion formalism for the operator $\hat{A}(q,t)$. We shall now follow this procedure in detail for the case of the pairing mode of a two-component Fermi gas near a Feshbach resonance. In doing so, we shall demonstrate the RPA approximation that is often used to bring the equation of motion $\dot{A} = i[H, A]$ into a closed form.

Consider an interacting gas composed of two species of fermion. Suppose that the gas may be described by the Hamiltonian

$$H = \sum_{k,\sigma} \epsilon_{k,\sigma} c^\dagger_{k,\sigma} c_{k,\sigma} + \sum_q \hat{\rho}_{q,\uparrow}\, V_q(t)\, \hat{\rho}_{q,\downarrow}, \tag{9.4}$$

where, $c^\dagger_{k,\sigma}$ and $c_{k,\sigma}$ are the creation and annihilation operators for a fermion of species σ, momentum k and non-interacting energy $\epsilon_{k,\sigma} = k^2/2m_\sigma - \mu$. $V_q(t)$ describes the time dependent interatomic potential undergoing parametric tuning (we shall assume

s-wave scattering) and $\hat{\rho}(q,\sigma) = \sum_k c^\dagger_{k+q,\sigma} c_{k,\sigma}$ is the density operator. To probe the pairing susceptibility, we add the thermodynamically conjugate external perturbation

$$H_{\text{ext}} = \mathcal{P}^{\text{ext}}_q(t) \sum_k c^\dagger_{q/2+k,\uparrow}(t) c^\dagger_{q/2-k,\downarrow}(t) + h.c., \tag{9.5}$$

and measure the pairing amplitude $\mathcal{P}_q(t) = \sum_k \langle c^\dagger_{q/2+k,\uparrow}(t) c^\dagger_{q/2-k,\downarrow}(t)\rangle$.

At this point, it is useful to introduce the operator $B(k,q,t) = c^\dagger_{q/2+k,\uparrow}(t)$ $c^\dagger_{q/2-k,\downarrow}(t)$, which is related to $\mathcal{P}_q(t)$ by summing over k and taking the expectation value. The equation of motion for the operator B is

$$\frac{\mathrm{d}}{\mathrm{d}t} B(k,q,t) = \mathrm{i}[H, B(k,q,t)]. \tag{9.6}$$

Evaluating the various commutators, we obtain

$$\mathrm{i}\frac{\mathrm{d}}{\mathrm{d}t} B(k,q,t) = (\epsilon_{q/2+k,\uparrow} + \epsilon_{q/2-k,\downarrow}) B(k,q,t) + [1 - \hat{n}_\uparrow(q/2+k,t) - \hat{n}_\downarrow(q/2-k,t)] \times \left[\mathcal{P}^{\text{ext}}_q(t) + \int đp\; V_{k-p}(t) B(p,q,t)\right], \tag{9.7}$$

where $\hat{n}_\sigma(q,t) = c^\dagger_{q,\sigma}(t) c_{q,\sigma}(t)$ is the number operator, and $đp$ stands for $\mathrm{d}^3p/(2\pi)^3$. By taking the expectation value of Eq. (9.7), we find that the equation of motion for the two-fermion expectation value $\langle B(k,q,t)\rangle$ will be coupled to expectation values containing four-fermion operators. This coupling to higher-order expectation values is a general feature of equations of motion for interacting theories. To obtain a closed-form equation, we must cut off the equation of motion at some point. A typical approach is called the random phase approximation (RPA), which states that the pairing amplitude must oscillate at the drive frequency, and therefore one can decouple the pairing amplitude in four-fermion terms. That is, within the RPA, we replace the four-fermion expectation values with the product of the two-fermion ones: $\langle \hat{n}_\uparrow(q/2+k,t) B(p,q,t)\rangle \to \langle \hat{n}_\uparrow(q/2+k,t)\rangle\langle B(p,q,t)\rangle$. Further, as the expectation value $\langle \hat{n}_\uparrow(q/2+k,t)\rangle$ is presumed to be stationary by the RPA, we can replace it by its value in the initial state $n^{\text{F}}_\uparrow(q/2+k)$. The RPA assumption is consistent with the assumption that only the unstable collective mode has interesting dynamics, and therefore we can ignore the dynamics of other expectation values. Taking the expectation value of the equation of motion and making the RPA approximation, we obtain

$$\mathrm{i}\frac{\mathrm{d}}{\mathrm{d}t}\langle B(k,q,t)\rangle = (\epsilon_{q/2+k,\uparrow} + \epsilon_{q/2-k,\downarrow})\langle B(k,q,t)\rangle + [1 - n^{\text{F}}_\uparrow(q/2+k) - n^{\text{F}}_\downarrow(q/2-k)] \left[\mathcal{P}^{\text{ext}}_q(t) + \int đp\; V_{k-p}(t)\langle B(p,q,t)\rangle\right]. \tag{9.8}$$

Taking the expectation value of the equation of motion, making the RPA approximation, and Fourier transforming the result, we obtain the much simplified equation

$$\omega\langle B(k,q,\omega)\rangle = (\epsilon_{q/2+k,\uparrow}+\epsilon_{q/2-k,\downarrow})\langle B(k,q,\omega)\rangle + \left[1-n^{\mathrm{F}}_{\uparrow}(q/2+k)-n^{\mathrm{F}}_{\downarrow}(q/2-k)\right]\left[\mathcal{P}^{\mathrm{ext}}_{q}(\omega)+\int \text{đ}p\; V_{k-p}\langle B(p,q,\omega)\rangle\right]. \tag{9.9}$$

Reorganizing the terms in Eq. (9.9), we bring it to the form

$$\langle B(k,q,\omega)\rangle\left[\mathcal{P}^{\mathrm{ext}}_{q}(\omega)+\int \text{đ}p\; V_{k-p}\langle B(p,q,\omega)\rangle\right] = \frac{1-n^{\mathrm{F}}_{\uparrow}(q/2+k)-n^{\mathrm{F}}_{\downarrow}(q/2-k)}{\omega-\epsilon_{q/2+k,\uparrow}-\epsilon_{q/2-k,\downarrow}}. \tag{9.10}$$

Integrating both sides over k, we obtain the integral form of the differential Eq. (9.8)

$$\mathcal{P}(q,t)=\int \mathrm{d}t'\chi^{(0)}_{\mathrm{pair}}(q,t-t')\left[V(t')\mathcal{P}(q,t')+\mathcal{P}^{\mathrm{ext}}_{q}(t')\right], \tag{9.11}$$

where $\chi^{(0)}_{\mathrm{pair}}(q,t)$ is the Fourier transform of the the bare susceptibility, and

$$\chi^{(0)}_{\mathrm{pair}}(q,\omega)=\int \text{đ}k\,\frac{1-n^{\mathrm{F}}_{\uparrow}(q/2+k)-n^{\mathrm{F}}_{\downarrow}(q/2-k)}{\omega-\epsilon_{q/2+k,\uparrow}-\epsilon_{q/2-k,\downarrow}}, \tag{9.12}$$

where, within the RPA approximation, $\chi^{(0)}_{\mathrm{pair}}(q,t)$ is evaluated with fermions in the initial state.

A particularly simple case, is the one in which the interaction strength changes in a stepwise fashion $V(t)=V\theta(t)$. In this case, small fluctuations induced by the external field after the quench will be governed by the poles of the familiar RPA-like susceptibility

$$\chi^{\mathrm{RPA}}_{\mathrm{pair}}(\omega,q)=\frac{\chi^{(0)}_{\mathrm{pair}}(\omega,q)}{1-V\chi^{(0)}_{\mathrm{pair}}(\omega,q)}. \tag{9.13}$$

Here, the difference between the usual RPA susceptibility and Eq. (9.13) is that in the former the bare susceptibility given by Eq. (9.12) is evaluated in the equilibrium fermionic state, while in the latter it is evaluated in the initial fermionic state before the quench. If the final interaction strength falls in the pairing regime, then $\chi^{\mathrm{RPA}}_{\mathrm{pair}}(\omega,q)$ will have a line of poles $\omega_q=\Omega_q+\mathrm{i}\Delta_q$ with a positive imaginary part, corresponding to the exponential growth of small fluctuations. When the time reaches $t\sim 1/\Delta_{\max}$, where $\Delta_{\max}$ corresponds to the fastest growing mode, the amplitude of the fastest growing mode will become large. At this point in time, the various unstable modes begin to interact strongly with each other and therefore our instability analysis begins to fail as a result of the initial exponential growth of the modes becomes saturated and the topological defects freeze in.

To get a simple physical picture of these dynamics, it is useful to consider a complimentary approach that looks at the most unstable $q = 0$ component of the pairing amplitude. Consider the time dependent wavefunction:

$$|\Psi(t)\rangle = \prod_k \left(u_k(t) + v_k(t) c^\dagger_{k\uparrow} c^\dagger_{-k\downarrow}\right) |0\rangle. \tag{9.14}$$

Here, $\{u_k(t)^2, v_k(t)^2\} = \frac{1}{2}\left(1 \pm \frac{\epsilon_k}{\sqrt{\epsilon_k^2+\Delta(t)^2}}\right)$ are functions of time because the pairing gap $\Delta(t)$ is itself a function of time. The statement that the energy of the collective mode in the pairing channel is ω_q is equivalent to the statement that for small values of $\Delta(t)$, $\Delta(t)$ obeys the equation of motion

$$\frac{\mathrm{d}\Delta(t)}{\mathrm{d}t} = -\mathrm{i}\omega_{q=0}\Delta(t). \tag{9.15}$$

When $\mathrm{Im}\,\omega_q > 0$, we get exponential growth of the pairing amplitude. It turns out that when one focuses on the $q = 0$ mode only, it is possible to derive dynamics even beyond a linearized approximation for Δ (Barankov *et al.* 2004; Yuzbashyan *et al.* 2006). In contrast, the approach discussed here considers modes at all q but is limited to the linearized approximation.

In addition to instantaneous quenches, we can also consider ramps that occur over a finite amount of time. In this case, we can estimate the point in time at which the freezing-in of topological defects occurs via a scaling argument similar to the one of (Zurek 1985). Suppose that the phase transition occurs at the time $t = 0$. Following this time, the susceptibility computed via Eq. (9.13) will have a line of poles with positive imaginary frequencies. As t increases, the system goes deeper into the ordered phase, and the line of poles moves up to larger and larger imaginary frequencies. At each point in time, we can identify the most unstable mode and corresponding wavelength $\Delta_{\mathrm{max}}(t)$ and $q_{\mathrm{max}}(t)$. At approximately the time that the inequality $t \lesssim 1/\Delta_{\mathrm{max}}(t)$ is first satisfied, the fastest growing mode begins to saturate and the defects freeze in. The time dependence of $\Delta_{\mathrm{max}}(t)$ can be estimated from the scaling properties of the phase transition $\Delta_{\mathrm{max}}(t) \sim u^{z\nu}(t)$, where $u(t) = (V_\mathrm{c} - V(t))/V_\mathrm{c}$ is the time-dependent distance to the phase transition, ν and z are the correlation length and the dynamic critical exponents.

At this point, we are seemingly ready to study the pairing instability. However, there is a technical difficulty that one can immediately see with the bare pairing susceptibility. The integral in Eq. (9.12) has a UV divergence: for $k \to \infty$ the integrand becomes $-\frac{1}{2\epsilon_k}$ and thus the integral diverges in two or more dimensions. Mathematically, the divergence originates from using a δ-function (in real space) interatomic potential, which itself is unphysical. In the next section, we will implement a pseudo-potential with a finite effective range to describe the interactions between atoms. We will find that scattering at low energies can be described by a universal scattering matrix that is independent of almost all of the details of the interatomic potential. Indeed, the scattering matrix can be described by only two measurable scattering parameters: namely the scattering length and the effective range. We will use this knowledge to rewrite the pairing susceptibility in terms of these parameters, and thus obtain a universal description of the pairing instability.

9.3.3 Dynamics of magnetization based on time-dependent Hartree approximation

Before proceeding to study the nature of the interatomic interactions, we attack the competing Stoner instability, which at first sight does not seem to suffer from a similar UV divergence. For the Stoner case, we are looking for a response to magnetization M_q, and therefore use the external perturbation

$$H_{\text{ext}} = M_q^{\text{ext}}(\omega)\mathrm{e}^{\mathrm{i}\omega t}\sum_k c_{k,\uparrow}^\dagger c_{k+q,\uparrow} - c_{k,\downarrow}^\dagger c_{k+q,\downarrow}. \tag{9.16}$$

Within the RPA approximation, we find the susceptibility

$$\chi_{\text{FM}}^{\text{RPA}}(\omega, q) = \frac{\chi_{\text{FM}}^{(0)}(\omega, q)}{1 - V\chi_{\text{FM}}^{(0)}(\omega, q)}, \tag{9.17}$$

where

$$\chi_{\text{FM}}^{(0)}(\omega, q) = \int đk\, \frac{n_\uparrow^{\text{F}}(q/2+k) - n_\downarrow^{\text{F}}(q/2-k)}{\omega - (\epsilon_{q/2+k,\uparrow} - \epsilon_{q/2-k,\downarrow})}, \tag{9.18}$$

$$= \frac{N_0}{2}\left(1 + \frac{m^2}{2k_{\text{F}}q^3}\left[(\epsilon_q+\omega)^2 - 4\epsilon_{\text{F}}\epsilon_q\right]\log\frac{\epsilon_q - v_f q + \omega}{\epsilon_q + v_{\text{F}} q + \omega}\right.$$
$$\left. + \frac{m^2}{2k_{\text{F}}q^3}\left[(\epsilon_q-\omega)^2 - 4\epsilon_{\text{F}}\epsilon_q\right]\log\frac{\epsilon_q - v_f q - \omega}{\epsilon_q + v_{\text{F}} q - \omega}\right), \tag{9.19}$$

and $N_0 = \frac{k_{\text{F}}m}{2\pi^2}$ is the density of states at the Fermi surface, and the explicit expression for the susceptibility after carrying out the $đk$ integral (in 3D) is called the Lindhard function.

The susceptibility $\chi_{\text{FM}}^{\text{RPA}}(\omega, q)$ acquires purely imaginary poles $\omega_q = \mathrm{i}\Delta_q$ for $VN_0 > 1$ and has no poles for $VN_0 < 1$. As $\Delta_q > 0$, these poles correspond to unstable collective modes of the system, and therefore $VN_0 = 1$ corresponds to the phase-transition point from Fermi liquid to Stoner ferromagnet (at the RPA level). The growth rate of the instability Δ_q is traced out as a function of q for several values of V in Figure 9.1. In all cases, for small momenta, Δ_q is linear in q. This is a reflection of the fact that magnetization is a conserved-order parameter (the operator for the total spin of the system commutes with the Hamiltonian), therefore making large domains (small q)

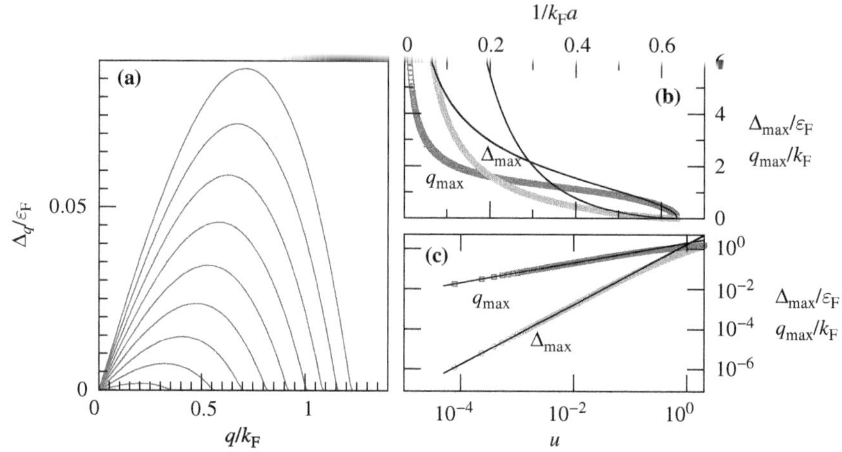

Figure 9.1 Properties of the unstable collective modes associated with the Stoner instability, computed using δ-function interactions. (a) Growth rate Δ_q as a function of wavevector q for $T = 0$ and $1/k_F a = 0.53$ (top line), 0.54, 0.55, ..., 0.63 (bottom line). (b) The most unstable wavevector q_{max} and the corresponding growth rate Δ_{max} versus $1/k_F a$. A fit to the mean-field critical theory ($\nu = 1/2$, $z = 3$) is shown with solid black lines. (c) Details of the critical behaviour of q_{max} and Δ_{max} as a function of distance from the transition point $u = (1/k_F a)_c - (1/k_F a)$, $(1/k_F a)_c = 2/\pi$ on a log–log scale.

requires moving spins by large distances, which is a slow process. As q increases, we are bending the emerging Ferromagnetic texture at shorter and shorter length scales. Eventually, this becomes energetically unfavourable and Δ_q bends over and decreases, becoming zero at $q = q_{cut}$, at which point the imaginary part of the pole of $\chi^{RPA}_{FM}(\omega, q)$ disappears.

After a quench, the fastest-growing modes, i.e., those with the largest Δ_q, will dominate and thus determine the size of the typical domains. In particular, we find that for a given $u = (V - V_c)/V_c$, where V_c is the interaction corresponding to the phase transition, the fastest-growing modes have a $q_{max} \simeq 2k_F u^{1/2}$ and a growth rate of $\Delta_{max} \simeq (16/3\pi)\epsilon_F u^{3/2}$. It is tempting to relate the interaction strength to the scattering length a via $V = 4\pi a/m$. Succombing to this temptation, we plot the growth rate as a function of inverse scattering length, as shown in Figure 9.1. Here, we start to see a problem: as the scattering length becomes larger, so does the growth rate of the ferromagnetic instability. This is clearly a deficiency of the theory, as at unitarity (where the scattering length diverges) the only available scale is the Fermi energy scale, so the distance to the transition u should not appear. Physically, the deficiency lies in using the bare scattering length, which is only meaningful for low-energy collisions, to describe the Stoner instability that involves all energy scales up to the Fermi energy. In the following sections, we shall develop the formalism to describe collisions at all energy scales within a Fermi liquid. Afterwards, we shall come back to the case of the Stoner instability and, using a more realistic interaction potential, fix the divergence at unitarity.

9.4 Feshbach resonance via pseudo-potentials

In this section, our goal is to describe atom scattering in the vicinity of a Feshbach resonance. Our strategy is to first describe the scattering between a pair of atoms in vacuum; in the following section we shall extend this description to include Pauli-blocking, and thus obtain a description of scattering in the many-body system.

The key to controllable interatomic interactions in ultracold atom systems is the so-called Feshbach resonance. Interatomic interactions in ultracold atom experiments are typically of the van der Waals type and therefore intrinsically attractive (the main exceptions to this rule are experiments with dipolar atoms and molecules). However, at low energies, interactions can appear to be repulsive. This is possible via scattering off a shallow bound state, as can be qualitatively appreciated from second-order perturbation theory. A Feshbach resonance corresponds to the appearance of such a bound state. The utility of a Feshbach resonance comes from the fact that the binding energy and, therefore, the interaction strength can be tuned. The tuning is via a magnetic field, as the bound state (or more appropriately the closed channel) has a slightly different magnetic moment from that of the open channel (see (Pethick and Smith 2002; Ketterle and Zwierlein 2008) for details).

Instead of delving into the details of the atomic physics of Feshbach resonances, for pedagogical insight we present a simple model for a so-called geometric resonance. In a geometric resonance, it is assumed that the interatomic interaction can be described by a potential that only depends on the interatomic distance and can be tuned directly, thus avoiding the complication of a true two-channel model.

Consider a pair of atoms with masses m_1 and m_2 interacting via the interatomic potential $V(r_1 - r_2)$. The scattering problem is described by the Hamiltonian

$$H = \frac{1}{2m_1}\nabla_1^2 + \frac{1}{2m_2}\nabla_2^2 + V(r_1 - r_2). \tag{9.20}$$

It turns out that at low energies, the details of V are not important. Our main tool for describing atom scattering will be the T matrix, which is related to the scattering amplitude in the centre of mass frame via $T(E = k^2/2\mu; k, k') = -\frac{2\pi}{\mu} f(k, k')$, where k and k' are the relative momenta of the scattering atoms before and after the collision, E is the total kinetic energy, and $\mu^{-1} = m_1^{-1} + m_2^{-1}$ is the reduced mass of the scattering atoms. In fact, the scattering problem is universal in the sense that many interatomic potentials $V(r_1 - r_2)$ will lead to the same form of the T matrix for low-energy scattering. Therefore, to model scattering, we can come up with any suitable potential that produces the T matrix we want. This type of model potential is usually called a pseudo-potential.

The poles of the T matrix correspond to two-atom bound states. Near a Feshbach resonance, there must be a bound state having a spatial extent of the scattering length a, which corresponds to a binding energy of $E_\mathrm{b} \approx 1/2\mu a^2$. Therefore the T matrix must have the form

$$T(E) = -\frac{2\pi}{\mu}\left(-\frac{1}{a} - \mathrm{i}\sqrt{2\mu E} + r_\mathrm{e}\mu E + O(E^2)\right)^{-1}, \tag{9.21}$$

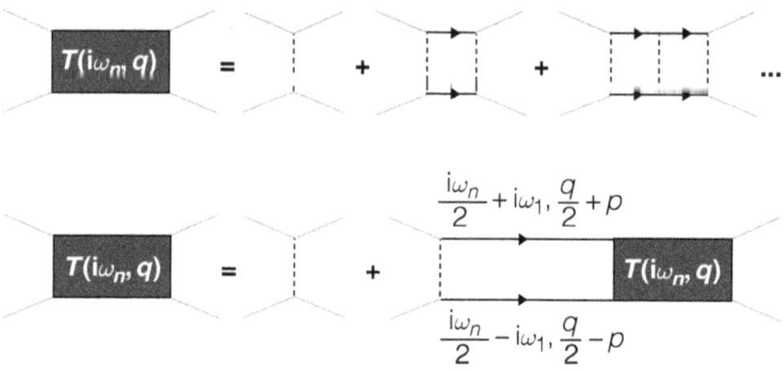

Figure 9.2 Graphic representation of the T matrix as a summation of ladder diagrams (top) and the corresponding Lippmann–Schwinger equation (bottom). Shaded squares represent the T matrix, solid lines fermionic atom propagators, dotted lines interatom interactions, and solid grey lines external fermionic atom legs. The external legs are shown for clarity, and indicate how one would go about connecting the T matrix to fermionic lines.

where a is the scattering length and E is the kinetic energy of the two scattering atoms in the centre-of-mass frame. The effective range r_e is the first correction of the binding energy due to the shape of the interatomic potential and roughly corresponds to its spatial extent. Here, we have specialized to the case of s-wave scattering and thus the T matrix has no angular dependence.

The T matrix may be obtained directly by solving the Schrödinger equation, Eq. (9.20). An alternative, and more instructive approach, is to obtain the T matrix via a resummation of particle–particle ladder diagrams (see Figure 9.2), to yield the Lippmann–Schwinger equation

$$T(\mathrm{i}\omega_n, q; k, k') = V(k - k') + \sum_{\mathrm{i}\omega_1} \int đk_1\, V(k - k_1)\, G_\uparrow\left(\frac{\mathrm{i}\omega_n}{2} + \mathrm{i}\omega_1, \frac{q}{2} + k_1\right)$$
$$G_\downarrow\left(\frac{\mathrm{i}\omega_n}{2} - \mathrm{i}\omega_1, \frac{q}{2} - k_1\right) T\Big(\mathrm{i}\omega_n, q; k_1, k'\Big), \quad (9.22)$$

where $G_\sigma(\mathrm{i}\omega_n, k) = (\mathrm{i}\omega_n - k^2/2m)^{-1}$ is the free-fermion Green function, and we have added the centre of mass momentum q to the labels of the T matrix.

Many different pseudo-potentials will result (at small k) in the same T matrix of Eq. (9.21). In fact, many different pseudo-potentials are in common use, e.g., box potentials, Gaussian potentials, hard sphere potentials, and regularized δ-function potentials, see e.g., (Pethick and Smith 2002; Pilati *et al.* 2010; Chang *et al.* 2011). Here, following (Phillips *et al.* 1998), we shall use a slightly less common form of the pseudo-potential

$$\lambda(k_1, k_2) = \sum_{i,j=0}^{1} \lambda_{ij} k_1^{2i} k_2^{2j}, \quad (9.23)$$

where the matrix λ_{ij}

$$\lambda_{ij} = \begin{pmatrix} C & C_2 \\ C_2 & 0 \end{pmatrix}. \tag{9.24}$$

This pseudo-potential has two tuning parameters C and C_2, which are needed to match both the scattering length and effective range that appear in the T matrix, allowing for the flexibility to describe both wide and narrow resonances. For the special case of a wide resonance, where $r_e \approx 0$, we can drop C_2, thus eliminating the matrix structure and reducing the description to that of (Pethick and Smith 2002). The main advantage of this form of the pseudo-potential is that it is separable, thus considerably simplifying the Lippmann–Schwinger Eq. (9.22). We can now write the T matrix in the same form as the pseudo-potential

$$T(E, q = 0; k_1, k_2) = \sum_{i,j=0}^{1} \tau_{ij}(E) k_1^{2i} k_2^{2j}, \tag{9.25}$$

where the $\tau(E)$ matrix is to be determined. Owing to the translational invariance of the problem, we shall work in the centre-of-mass frame. The T matrix in a moving frame is related to the T matrix in the centre of mass frame via $T(E, q) = T(E - q^2/2(m_1 + m_2), 0)$. From here on, we shall take the step of setting $m_1 = m_2 = m = 2\mu$ to simplify the notation. In matrix form, the Lippmann–Schwinger equation becomes

$$\tau(E) = \lambda + \lambda I(E) \tau(E), \tag{9.26}$$

where

$$I(E) = \begin{pmatrix} \displaystyle\int \frac{\mathrm{d}^3 k}{(2\pi)^3} \frac{1}{E^+ - k^2/2\mu} & \displaystyle\int \frac{\mathrm{d}^3 k}{(2\pi)^3} \frac{k^2}{E^+ - k^2/2\mu} \\ \displaystyle\int \frac{\mathrm{d}^3 k}{(2\pi)^3} \frac{k^2}{E^+ - k^2/2\mu} & \displaystyle\int \frac{\mathrm{d}^3 k}{(2\pi)^3} \frac{k^4}{E^+ - k^2/2\mu} \end{pmatrix}. \tag{9.27}$$

The integrals in Eq. (9.27) come with an upper cut-off Λ, and E^+ stands for $E + \mathrm{i}\delta$. We see that the divergence that appears in the bare pairing susceptibility, Eq. (9.12), is of exactly the same type as the 1,1 component of the $I(E)$ matrix, suggesting that the two are related.

Having obtained a relation between the pseudo-potential and the physically observable T matrix, we can plug the pseudo-potential into the many-body problem to obtain an effective interaction parameters C and C_2 in terms of the scattering length, effective range, and cut-off. Although the parameters C and C_2 depend on the cut-off; as we shall demonstrate, the pairing susceptibility is be independent of it.

9.5 Application to pairing susceptibility

The pairing susceptibility, which we originally obtained using the equation-of-motion approach, can also be obtained diagrammatically. Explicitly, the susceptibility is related to the two-particle propagator, i.e., the Cooperon, which is the many-body version of the T matrix. At the RPA level, the Cooperon $C(\mathrm{i}\omega_n, q)$ corresponds to the solution of the Lippmann–Schwinger equation, Eq. (9.22), with free Green functions replaced with Green functions of fermions in a Fermi sea $G_\sigma(\mathrm{i}\omega_n, q) = (\mathrm{i}\omega_n - k^2/2m + \epsilon_\mathrm{F})^{-1}$, where ϵ_F is the Fermi energy. Thus, at the RPA level, the difference between the Cooperon and the T matrix is that the Cooperon takes into account Fermi blocking so that scattering only occurs on top of the Fermi sea.

The relationship between $C(\mathrm{i}\omega_n, q)$ and $\chi(\mathrm{i}\omega_n, q)$ is shown schematically in Figure 9.4. Explicitly, the relationship is

$$\begin{aligned}\chi(\mathrm{i}\omega_n, q) &= \chi^{(0)}(\mathrm{i}\omega_n, q)\\ &+ \sum_{\mathrm{i}\omega_1,\mathrm{i}\omega_2} \int đk_1\, đk_2\, G_\uparrow\left(\frac{\mathrm{i}\omega_n}{2} + \mathrm{i}\omega_1, \frac{q}{2} + k_1\right) G_\downarrow\left(\frac{\mathrm{i}\omega_n}{2} - \mathrm{i}\omega_1, \frac{q}{2} - k_1\right)\\ &\times C(\mathrm{i}\omega_n, q; k_1, k_2)\, G_\uparrow\left(\frac{\mathrm{i}\omega_n}{2} + \mathrm{i}\omega_1, \frac{q}{2} + k_1\right) G_\downarrow\left(\frac{\mathrm{i}\omega_n}{2} - \mathrm{i}\omega_1, \frac{q}{2} - k_1\right).\end{aligned} \tag{9.28}$$

Using this relation, it can be shown that we can recover Eq. (9.13) for the susceptibility. At this point, we remark that the poles of the Cooperon and the pairing susceptibility match. Therefore, to find the unstable collective modes it is sufficient to look at the poles of the Cooperon.

In matrix form, the analogue of the Lippmann–Schwinger equation, Eq. (9.26), for the T matrix is the Lippmann–Schwinger equation for the Cooperon (using RPA)

$$\mathcal{C}(E, q) = \lambda + \lambda\tilde{I}(E, q)\mathcal{C}(E, q), \tag{9.29}$$

where

$$\tilde{I}(E) = \begin{pmatrix} \displaystyle\int \frac{\mathrm{d}^3k}{(2\pi)^3} \frac{1 - 2n_\mathrm{F}(k)}{E^+ - k^2/2\mu + 2\epsilon_\mathrm{F} - q^2/4m} & \displaystyle\int \frac{\mathrm{d}^3k}{(2\pi)^3} \frac{k^2(1 - 2n_\mathrm{F}(k))}{E^+ - k^2/2\mu + 2\epsilon_\mathrm{F} - q^2/4m} \\ \displaystyle\int \frac{\mathrm{d}^3k}{(2\pi)^3} \frac{k^2(1 - 2n_\mathrm{F}(k))}{E^+ - k^2/2\mu + 2\epsilon_\mathrm{F} - q^2/4m} & \displaystyle\int \frac{\mathrm{d}^3k}{(2\pi)^3} \frac{k^4(1 - 2n_\mathrm{F}(k))}{E^+ - k^2/2\mu + 2\epsilon_\mathrm{F} - q^2/4m} \end{pmatrix}. \tag{9.30}$$

At this point, we can use the relationship between the interaction matrix λ and the physical parameters a and r_e to compute the Cooperon and find its poles. The resulting plot of the poles is shown in Figure 9.3. To understand the pole structure, we first comment on the effect of the cut-off. As the momentum cut-off Λ is increased, the parameters C and C_2 that appear in the λ matrix change. However, as can be clearly

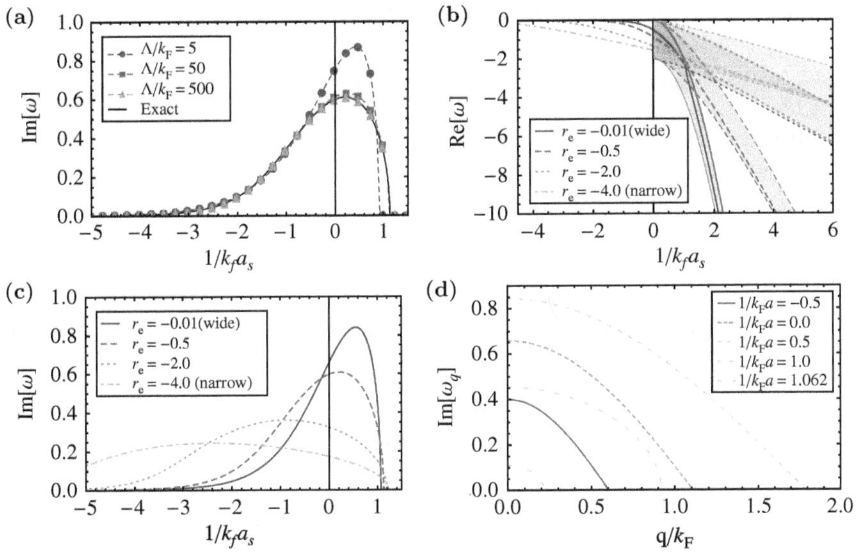

Figure 9.3 (a) Imaginary part of the Cooperon pole trajectory as a function of inverse scattering length for various values of the cut-off Λ and $r_e = -0.5$. As Λ increases, the pole trajectory approaches its asymptotic value. (b) Real part of the Cooperon pole trajectory as a function of inverse scattering length for various values of the effective range r_e. The trajectory of the corresponding vacuum binding energies (T matrix poles) and vacuum binding energies shifted down by the twice the Fermi energy (T matrix poles $-2\epsilon_F$) are indicated by the tops and bottoms of the shaded corridors. In the strong interaction regime (small positive scattering length) the Cooperon pole approaches the bottom of the corridor, i.e., the T matrix pole shifted down by twice the Fermi energy. (c) Imaginary part of the Cooperon pole trajectory (pairing rate) as a function inverse scattering length for various values of the effective range r_e. (d) Imaginary part of the Cooperon pole trajectory (pairing rate) as a function of momentum for various scattering lengths, and a wide resonance ($r_e = -0.01$). The plot indicates that the most unstable pairing mode is always at $q = 0$.

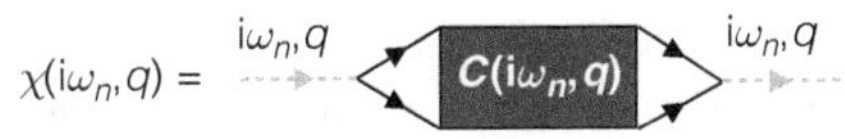

Figure 9.4 Diagrammatic representation of the relation between the Cooperon (many-body T matrix) and the pairing susceptibility. Solid black lines represent fermionic atom propagators, shaded square represents the Cooperon, and dotted grey lines represent external source of the pairing field.

seen in Figure 9.3a, the Cooperon poles converge to their asymptotic values once Λ exceeds the Fermi momentum by a factor of $\sim$100.

Instead of dealing with the cut-off, we can directly cancel the divergences in the Cooperon by comparing it with the T matrix. Comparing Eq. (9.27) and Eq. (9.30), we find that the denominators of $I(E + 2\epsilon_F - q^2/4m)$ and $\tilde{I}(E)$ match, thus the divergences of $C(E, q)$ and $T(E + 2\epsilon_F - q^2/4m, q = 0)$ cancel. Using this fact, we add and subtract the T matrix to the Cooperon Lippmann–Schwinger equation

$$\mathcal{C}^{-1}(E, q) = \lambda^{-1} - \tilde{I}(E) + \tau^{-1}(E + 2\epsilon_F - q^2/4m) - \tau^{-1}(E + 2\epsilon_F - q^2/4m). \quad (9.31)$$

Carefully inverting this equation shows that only the 11 component is non-zero, and we obtain the expression

$$C(E, q) = \left[\frac{\mu}{2\pi}\left(\frac{1}{a} + \mathrm{i}\sqrt{2\mu\left(E + 2\epsilon_F - \frac{q^2}{4m}\right)} - r_e\mu\left(E + 2\epsilon_F - \frac{q^2}{4m}\right)\right)\right.$$
$$\left. + \int đk \frac{n_F(k + q/2) + n_F(k - q/2)}{E + 2\epsilon_F - k^2/m - q^2/4m}\right]^{-1}. \quad (9.32)$$

In Figure 9.3a, we compare the pole structure obtained by gradually increasing the cut-off Λ with the asymptotic pole structure obtained from this expression. In doing so, we verify that the asymptotic expression is indeed correct and corresponds to $\Lambda \to \infty$.

Having understood how to regularize the divergence in the Lippmann–Schwinger equation, we come back to the question of understanding the pole structure. We begin our analysis with the T matrix in vacuum. For each value of a, $T(E, q)$ has a line of poles on the BEC side located at $E = \omega_q + \mathrm{i}\Delta_q = -1/ma^2 + mq^2/4$, corresponding to the binding energy of a Feshbach molecule with centre-of-mass momentum q. As a consequence of energy and momentum conservation the pole frequency is real, indicating that a two-body process in vacuum cannot produce a Feshbach molecule.

The Cooperon is a natural extension of the two-particle scattering amplitude, i.e., the T matrix, to a system with a finite density of atoms. The presence of the Fermi sea shifts the poles of the Cooperon relative to the T matrix in two ways: (1) in the range $-\infty < 1/k_F a \lesssim 1.1$, the Cooperon pole acquires a positive imaginary part Δ_q that corresponds to the growth rate of the pairing instability (see Figure 9.3a); and (2) the real part of the pole Ω_q, which would correspond to the binding energy of a pair in the absence of an imaginary part, uniformly shifts down (see Figure 9.3b).

The shift of the pole into the complex plane is quite surprising, especially on the BEC side of the Feshbach resonance where the interactions are repulsive, and such a shift corresponds to a finite rate of molecule formation. Typically one assumes that a two-particle collision cannot lead to the formation of a molecule as energy and momentum conservation laws cannot be satisfied simultaneously, and hence earlier analysis focused on considering at least a three-body collision (Fedichev *et al.* 1996). Hence, one would think that by analysing poles of the Cooperon, which seems to describe two-particle collisions, we cannot get a pole with a finite imaginary part. An important difference of our system is that we are considering a many-body system.

So even though the Cooperon can be understood as effectively a two-particle scattering amplitude, it describes a scattering event taking place in the presence of a filled Fermi sea. The Pauli exclusion principle plays the role of the 'third body'. That is, the energy–momentum restrictions on molecule (or more precisely Cooper pair) formation are lifted, as the pair forms above the Fermi sea, the excess energy can be absorbed by the two holes that are left behind under the Fermi sea. This process is schematically represented in the inset of Figure 9.7. Within our approximation, there is a sharp cut-off of the molecule formation rate on the BEC side when the distance between particles becomes much larger than the scattering length and the Pauli exclusion principle becomes ineffective. In reality, we expect that the molecule formation rate does not go to zero completely but becomes determined by much slower three-body processes (Esry *et al.* 1999; Nielsen and Macek 1999; Bedaque *et al.* 2000; Petrov 2003).

The uniform shift down of the real part of the pole Ω_q is likewise a result of Pauli blocking (Abrikosov *et al.* 1975), and indicates an appearance of a paired state on the BCS side as well as stronger binding of the pairs on the BEC side. Deep on the BEC side, the Feshbach molecule becomes deeply bound and therefore very small in real space. As a result, the molecule becomes extended far beyond $k_{\rm F}$ in momentum space and Pauli blocking becomes less relevant. Consequently, see Figure 9.3b, the Cooperon pole (solid lines) approaches the T matrix pole (shifted by $2\epsilon_{\rm F}$, dotted lines) deep on the BEC side.

Coming back to the imaginary part of the pole, as depicted in Figure 9.3c, $\Delta_{q=0}$ increases exponentially as one approaches the Feshbach resonance from the BCS side, i.e., the growth rate of the BCS pairing in a wide resonance is equal to the BCS gap at equilibrium $\Delta_{q=0} \approx 8\epsilon_{\rm F} e^{\pi/2k_{\rm F}a-2}$ (Abrikosov *et al.* 1975). For a wide resonance, the growth rate continues to increase on the BEC side, reaching a maximum at $k_{\rm F}a \approx 2$, and finally decreasing to zero at $k_{\rm F}a \approx 1.1$, at which point the Fermi sea can no longer absorb the energy of the Feshbach molecule in a two-body process. Deeper in the BEC regime, pairing takes place via the more conventional three-body process and would round the pairing instability curve near $k_{\rm F}a \approx 1.1$ in Figure 9.3c. As we go from a wide resonance $r_{\rm e} \ll 1/k_{\rm F}$, to a more narrow resonance $r_{\rm e} \approx 1/k_{\rm F}$, the maximum in the paring rate decreases and shifts to the BEC side (see Figure 9.3c).

We comment that pairing at finite q is always slower than at $q = 0$, with Δ_q monotonically decreasing to zero at $q = q_{\rm cut}$ (see Figure 9.3d). Throughout a wide resonance the approximation $q_{\rm cut} \approx (\sqrt{3/2})(\Delta_{q=0}/\epsilon_{\rm F})k_{\rm F}$ works reasonably well except in the vicinity of $k_f a \sim 2$ where $q_{\rm cut}$ reaches the maximal value for a two-body process of $2k_f$.

9.6 More on Stoner instability

In Section 9.3, we left the story of the ferromagnetic instability at the unphysical divergence of its growth rate in the unitary regime. In this section, we use the knowledge gained in the previous two sections to fix this divergence. The reason for the divergence lies in the description of interparticle interactions near unitarity. Even without a Fermi sea, from the form of the T matrix, Eq. (9.21), we see that the interactions are

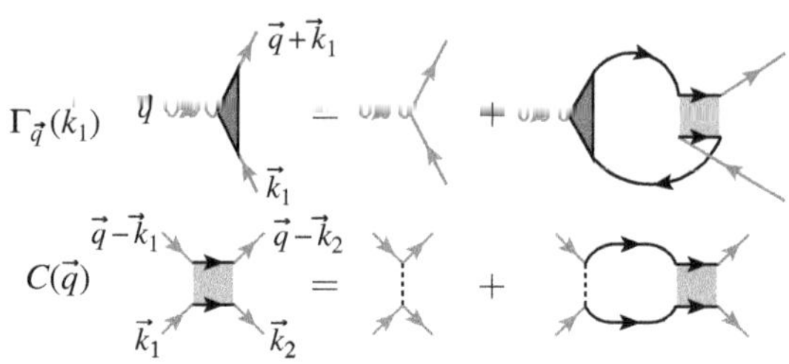

Figure 9.5 Diagrammatic representation of the vertex function $\Gamma_{\vec{q}}(\vec{k}_1)$ that appears in the ferromagnetic susceptibility with the bare two-particle interaction V replaced by the Cooperon $C(\vec{q}_1)$, which is a momentum- and frequency-dependent interaction. Curly lines represent external sources of spin flips, solid lines—fermions, dashed lines—interactions, grey lines—external legs. Figure reprinted from (Pekker *et al.* 2011).

strongly frequency dependent $T(\omega) \approx 4\sqrt{2}\pi/im\sqrt{\omega}$. Only at very low energies does the expression $T(\omega) \approx 4\pi a/m$, which we have used for the interaction strength in the calculation of the Stoner instability, makes sense.

To proceed, we replace the interatomic interaction by the Cooperon. In fact, this programme has been implemented before in the context of the fermionic Hubbard model, see, e.g., (Chen *et al.* 1991). Instead of computing the ferromagnetic susceptibility directly, it is advantageous to compute the vertex function, as indicated in Figure 9.5. The susceptibility is related to the vertex function via

$$\chi_{\mathrm{FM}}(\vec{q}) = \int \mathrm{d}\vec{k}_1 \; G(\vec{q} + \vec{k}_1)\, G(\vec{k}_1)\, \Gamma_{\vec{q}}(\vec{k}_1), \tag{9.33}$$

where we have switched to the notation $\vec{q} = (\omega, q)$ to save space. The poles of the susceptibility arise from the vertex function and not the two Green functions in the previous expression. Therefore, to find the poles of the susceptibility it is sufficient to find the poles of the vertex function. However, this is a rather complicated task, and necessarily involves some approximations. Instead of going through the details, which are presented elsewhere (Pekker *et al.* 2011), here we comment on the physics of the results and the differences between using the Cooperon and the δ-function interactions.

The most significant difference is the disappearance of the divergence of the instability rate near the resonance. Moreover, not only does the rate of the Stoner instability become finite everywhere; the instability persists on the BCS (attractive) side. The singularity of the Stoner instability rate was related to the singularity of the scattering amplitude of two particles in vacuum at zero energy. When analysing a many-body system we need to integrate over energies of the order of the Fermi energy. Since for any finite energy, there is no singularity in the scattering amplitude, this leads to a suppression of the Stoner instability rate. Furthermore, the Pauli blocking by the Fermi sea no longer allows us to make 'ideal' Feshbach molecules. Instead, these molecules are restricted to occupying states outside of the Fermi momentum, which in fact enhances their binding energy. The strongest Stoner instability corresponds to the $k_{\mathrm{F}}a$ at which the scattering at typical energies is strongest. The typical energy scale is the Fermi energy ϵ_F, and scattering is strongest when the bound-state energy corresponds to the typical energy scale. Thus, the Stoner instability is strongest not when the bound state disappears in vacuum but in the vicinity of the point where the bound-state energy plotted in Figure 9.3b 'crosses' ϵ_F. This crossing occurs on the BEC side of the resonance. As a result, the Feshbach resonance in vacuum is not reflected in any singular structure in the presence of a Fermi sea. Instead, the $\Delta_{\max}$ has a maximum on the BEC side, and smoothly decreases to zero on the BCS side.

Another minor difference, for the Stoner instability, between the δ-function and the Cooperon interactions, is the location of the phase transition point, which shifts from $(1/k_F a)_c = 2/\pi$ for δ-function interactions to $(1/k_F a)_c \approx 0.94$ for Cooperon interactions. The shift of the phase transition point is again associated with the form of the interactions at higher energies. However, the behaviour of Δ_q in the vicinity of the phase transition is very similar for both cases, as can be seen by comparing Figures 9.1 and 9.6. Indeed, the power laws describing the behaviour of $q_{\max}$ and $\Delta_{\max}$ remain the same.

Finally, we comment on the nature of approximations that we make in computing the ferromagnetic susceptibility. In the resumming diagrams indicated in Figure 9.5, we did not resum all interaction terms. We resummed only the most divergent contributions as the scattering length went to infinity. In particular, the diagrams

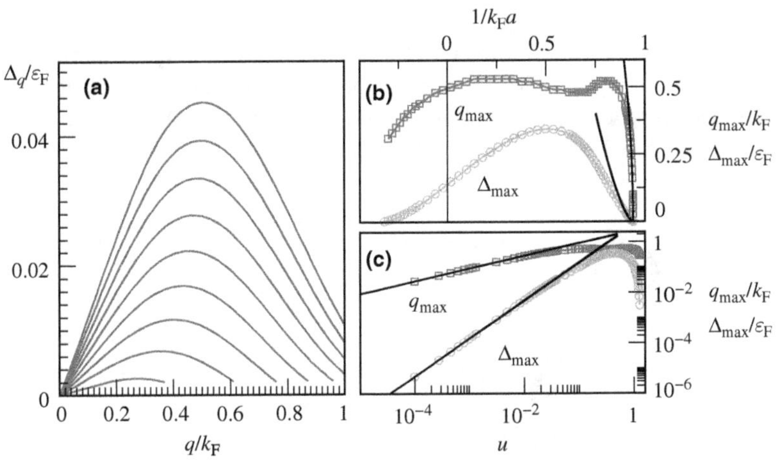

Figure 9.6 Properties of the unstable collective modes associated with the Stoner instability computed using Cooperon interactions, as opposed to the δ-function interactions of Figure 9.1. (a) Near the transition, Δ_q curves look similar to the δ-function case, except that the critical point has changed, so correspondingly the curves were computed for $1/k_F a = 0.85$ (top line), 0.86, 0.87, ..., 0.93 (bottom line). (b) $q_{\max}$ and $\Delta_{\max}$ versus $1/k_F a$. A fit to the mean-field critical theory ($\nu = 1/2$, $z = 3$) is shown with solid black lines. Note that the unphysical divergence at the Feshbach resonance, which was present for δ-function interactions, has disappeared. Further, the instability continues to the attractive side of the resonance. (c) Details of the critical behaviour of $q_{\max}$ and $\Delta_{\max}$ as a function of distance from the transition point $u = (1/k_F a)_c - (1/k_F a)$, $(1/k_F a)_c \approx 0.94$. Figure reprinted from (Pekker *et al.* 2011).

we resum correspond to a very simple time-dependent Hartree-type analysis, in which we neglected changes in the Fermi occupation numbers and screening of interactions. Our justification was that we are interested in short time dynamics, when such processes can be, neglected. However, it is possible to have a hierarchy of timescales. The dynamics of order parameters may be slow, owing to the usual critical slowing down near the quantum phase transition. On the other hand, there may be fast processes on the timescale of ϵ_F^{-1}, which we may not be able to take into account. Although, in the vicinity of the Feshbach resonance and away from the phase transition, the order parameter dynamics also occur on timescales of ϵ_F^{-1}, indicating that there is no timescale separation. A more careful Keldysh-type analysis of non-equilibrium dynamics may be required.

9.7 Discussion

We summarize the results obtained thus far in Figure 9.7. We find that after a quench from the weakly interacting regime to the vicinity of the Feshbach resonance the pairing and the Stoner instabilities compete with each other on both sides of the resonance. The growth rate associated with the pairing instability is always larger than that of the Stoner instability, indicating that a paired state is thus the favoured outcome. In treating both instabilities, we found that it is important to describe the interactions carefully.

Finally, we comment on the interpretation of the MIT experimental observations of (Jo *et al.* 2009) in light of the competition between pairing and ferromagnetism. Following the ramp to the strongly interacting regime, the MIT group let the system evolve for some time before performing their measurements. The most striking results

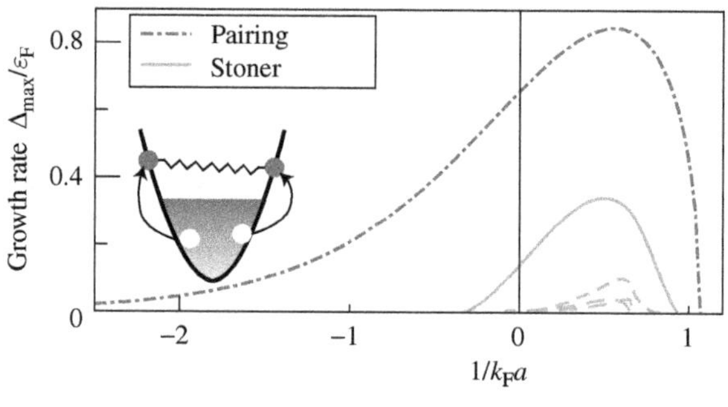

Figure 9.7 Comparison of the growth rates of the pairing instability (dashed-dotted line) and the Stoner instability (solid line) across the Feshbach resonance as a function of $1/k_F a$. The Stoner instability can occur in several 'angular momentum' channels, the subdominant channels are traced by dashed lines. Finally, we also plot the growth rate for the Stoner instability obtained with RPA and δ-function interactions that shows unphysical divergence near the resonance. **Inset**: schematic diagram of the pair creation process showing the binding energy (spring) being absorbed by the Fermi sea (arrows). The figure is reprinted from (Pekker *et al.* 2011).

were obtained for the atom loss measurement, which showed that as the scattering length is increased the atom loss first increases, and then suddenly starts to decreases for $k_F a \gtrsim 2$. The loss rate was measured by rapidly ramping down the magnetic field at the end of the experiment and thus projecting weakly bound 'Feshbach molecules' onto strongly bound molecules away from the resonance. Thus the rapid decrease of the atom loss can have two interpretations: (1) the formation of ferromagnetic domains prevented atomic collisions (which occur only between fermions of different species) and thus resulted in a decrease of atom loss rate, or (2) the atom losses are caused by pair formation and the maximum of the pairing rate near $k_F a = 2$ corresponds to the maximum in the atom loss rate. Interpretation (1) has been studied in a series of theoretical papers (Conduit and Simons 2009b; LeBlanc *et al.* 2009; Pilati *et al.* 2010; and Chang *et al.* 2011) and has been shown to be reasonably consistent with experimental observations. Since we find that the pairing instability always dominates over the Stoner instability, we are forced to conclude that scenario (2), which is also consistent with experimental observations, is more likely.

In addition to the atom loss rate, the MIT group, likewise, measured, changes of the cloud size and the average kinetic energy. Indeed, mean field theory calculations for the Stoner transition in a trap (LeBlanc *et al.* 2009) show similar trends to those found by the MIT experiments. However, the pairing transition is quite similar to the Stoner transition, in the sense that the atoms gain potential energy at the cost of kinetic energy. Thus, we expect that the trends for cloud size and kinetic energy would be similar for the two transitions.

9.8 Concluding remarks

The notion of collective modes is important in understanding not only equilibrium physics but also dynamics. As an example, we have investigated the role of the pairing and the ferromagnetic modes in quenches across the Feshbach resonance where these unstable modes directly compete with each other in real time.

Acknowledgements

We acknowledge our collaborators in the original work (Pekker *et al.* 2011), which formed the basis of this chapter: M. Babadi, L. Pollet, R. Sensarma, N. Zinner, and M. Zwierlein. We also acknowledge stimulating discussions with A. Georges, W. Ketterle, D. Huse, and G. Shlyapnikov. This work was supported by the Army Research Office with funding from the DARPA OLE program, Harvard-MIT CUA, NSF Grant No DMR-07-05472, AFOSR Quantum Simulation MURI, AFOSR MURI on Ultracold Molecules, the ARO-MURI on Atomtronics.

References

Abrikosov, AA, Gor'kov, LP, and Dzyaloshinski, IE (1975). *Methods of Quantum Field Theory in Statistical Physics*. Dover.

Barankov, RA, Levitov, LS, and Spivak, BZ (2004). *Phys. Rev. Lett.*, **93**, 160401.

Barth, M and Zwerger, W (2011). *Ann. Phys.*, **326**, 2544.
Bedaque, PF, Braaten, E, and Hammer, H-W (2000). *Phys. Rev. Lett.*, **85**, 908.
Bray, AJ (2002). *Adv. Phys.*, **51**, 481.
Chaikin, PM and Lubensky, TC (2000). *Principles of Condensed Matter Physics*. Cambridge University Press.
Chang, SY, Randeria, M, and Trivedi, N (2011). *Proc. Nat. Acad. Sci.*, USA **108**, 51.
Chen, L, Bourbonnais, C, Li, T, and Tremblay, A-MS (1991). *Phys. Rev. Lett.*, **66**, 369.
Cherng, RW, Gritsev, V, Stamper-Kurn, DM, and Demler, E (2008). *Phys. Rev. Lett.*, **100**, 180404.
Chuang, I, Turok, N, and Yurke, B (1991). *Phys. Rev. Lett.*, **66**, 2472.
Conduit, GJ and Simons, BD (2009a). *Phys. Rev. A*, **79**, 053606.
Conduit, GJ and Simons, BD (2009b). *Phys. Rev. Lett.*, **103**, 200403.
Conduit, GJ and Altman, E (2010). *Phys. Rev. A*, **82**, 043603.
Conduit, GJ, Green, AG, and Simons, BD (2009). *Phys. Rev. Lett.*, **103**, 207201.
Duine, RA and MacDonald, AH (2005). *Phys. Rev. Lett.*, **95**, 230403.
Esry, BD, Greene, CH, and Burke, JP (1999). *Phys. Rev. Lett.*, **83**, 1751.
Fedichev, PO, Reynolds, MW, and Shlyapnikov, GV (1996). *Phys. Rev. Lett.*, **77**, 2921.
Houbiers, M, Ferwerda, R, Stoof, HTC, McAlexander, WI, Sackett, CA, and Hulet, RG (1997). *Phys. Rev. A*, **56**, 4864.
Jo, GB, Lee, YR, Choi, JH, *et al.* (2009). *Science*, **325**, 1521.
Kanamori, J (1963). *Prog. Theor. Phys.*, **30**, 275.
Ketterle, W and Zwierlein, MW (2008). Ultracold Fermi gases, *Proceedings of the International School of Physics Enrico Fermi*, Course CLXIV, Varenna, 20–30 June 2006, edited by M Inguscio, W Ketterle, and C Salomon.
Kibble, TWB (1976). *J. Phys. A*, **9**, 1387.
Lamacraft, A (2007). *Phys. Rev. Lett.*, **98**, 160404.
Landau, LD (1932). *Physics of the Soviet Union*, **2**, 46.
LeBlanc, LJ, Thywissen, JH, Burkov, AA, and Paramekanti, A (2009). *Phys. Rev. A*, **80**, 013607.
Lieb, E and Mattis, D (1962). *Phys. Rev.*, **125**, 164.
Nielsen, E and Macek, JH (1999). *Phys. Rev. Lett.*, **83**, 1566.
Pekker, D, Babadi, M, Sensarma, R, *et al.* (2011). *Phys. Rev. Lett.*, **106**, 50402.
Pethick, C and Smith, H (2002). *Bose–Einstein Condensation in Dilute Gases*. Cambridge University Press.
Petrov, DS (2003). *Phys. Rev. A*, **67**, 010703.
Phillips, DR, Beane, SR, and Cohen, TD (1998). *Ann. Phys.*, **263**, 255.
Pilati, S, Bertaina, G, Giorgini, S, and Troyer, M (2010). *Phys. Rev. Lett.*, **105**, 30405.
Pines, D and Nozières, P (1989). *The Theory of Quantum Liquids*. Addison-Wesley.
Sadler, LE, Higbie, JM, Leslie, SR, Vengalattore, M, and Stamper-Kurn, DM (2006). *Nature*, **443**, 312.
Stamper-Kurn, D and Ketterle, W (2001). Course 2. Spinor condensates and light scattering from Bose–Einstein condensates. In Kaiser, R, Westbrook, C, and David, F. *Coherent Atomic Matter Waves*, pp 139–217. EDP Sciences and Springer-Verlag.
Stoner, EC (1933). *Phil. Mag*, **15**, 1018.
Tanaka, A and Tasaki, H (2007). *Phys. Rev. Lett.*, **98**, 116402.

Yuzbashyan, EA, Tsyplyatyev, O, and Altshuler, BL (2006). *Phys. Rev. Lett.*, **96**, 97005.
Zener, C (1932). *P. R. Soc. London A Mat.*, **137**, 696.
Zhai, H (2009). *Phys. Rev. A*, **80**, 051605(R).
Zhang, Y and Sarma, SD (2005). *Phys. Rev. B*, **72**, 115317.
Zhang, S and Ho, TL (2011). *arXiv:1102.5687*.
Zhang, S, Hung, H, and Wu, C (2010). *Phys. Rev. A*, **82**, 053618.
Zurek, WH (1985). *Nature*, **317**, 505.

10
Anderson localization of ultracold atoms in a laser speckle

Alain Aspect

Institut d'Optique, Palaiseau, France

Chapter Contents

10.1 Anderson localization for the beginner

Fifty years ago, PW Anderson published a paper presenting a revolutionary suggestion for interpreting the sudden transition of certain materials from conductor to insulator, under a slight change of the amount of disorder in the material [1]. At that time, every physicist would understand (for instance, on the basis of the Drude model of a metal) that increasing the amount of disorder in a conducting material increases the hindrance to the displacement of electrons and thus decreases the conductivity, but no one had predicted a total cancellation of the conductivity, i.e., a transition from a conductor to an insulator, past a certain amount of disorder. The core of the prediction of this phase transition was a fundamental quantum phenomenon, interference between the many quantum amplitudes associated with various trajectories of the electron in a crystal with a certain disorder. Actually, the result was essentially based on a mathematical argument, and it took quite some time for a simple picture to emerge. It is interesting to quote what PW Anderson wrote about 'Anderson Localization' (AL), when he received the Nobel Prize in physics in 1977, almost 20 years after his seminal paper:

... very few believed it at the time, and even fewer saw its importance; among those who failed to fully understand it at first was certainly its author. It has yet to receive adequate mathematical treatment, and one has to resort to the indignity of numerical simulations to settle even the simplest questions about it. Only now, and through primarily Sir Nevill Mott's efforts, is it beginning to gain general acceptance. [2]

More than 30 years after this strong statement, and 50 years after the initial paper, it is surprising that there are still many open questions, even among the simplest. And it is remarkable that physicists working with ultracold atoms have been able to translate the condensed matter situation considered by Anderson into AMO situations, which can be investigated experimentally [3; 4; 5].

Anderson's initial model starts from the so called 'tight binding model' of an electron in a crystal [6], where one considers a periodic lattice of trapping sites for the electron (Figure 10.1a). Actually, the electron can hop from one site to the nearest sites by quantum tunnelling. For large enough tunnelling amplitudes, the stationary solution of the quantum problem is a so called 'extended' wavefunction, describing an electron that can freely propagate over the crystal. This free propagation of an electron in a perfect crystal can alternatively be described by a Bloch wave in a conduction band [6].

If now one adds disorder to the tight binding model, by assuming random values of the energies of the bottoms of the various trapping sites (Figure 10.1b), with a distribution of width W, propagation is hindered, and this corresponds to Ohm conduction with a conductivity that decreases when the amplitude W of the disorder increases. Anderson's conjecture, based on a general mathematical argument, is the existence of a threshold in the amount of disorder, beyond which the conductivity is more than reduced; it is totally cancelled, and the material becomes a perfect insulator. This breakdown of conduction is associated with a sudden change in the electronic wavefunction, which ceases to be extended, and becomes spatially localized, with exponentially decreasing tails: there is no way for the electron to get out from its localized state and propagate in the crystal. The surprise is that this

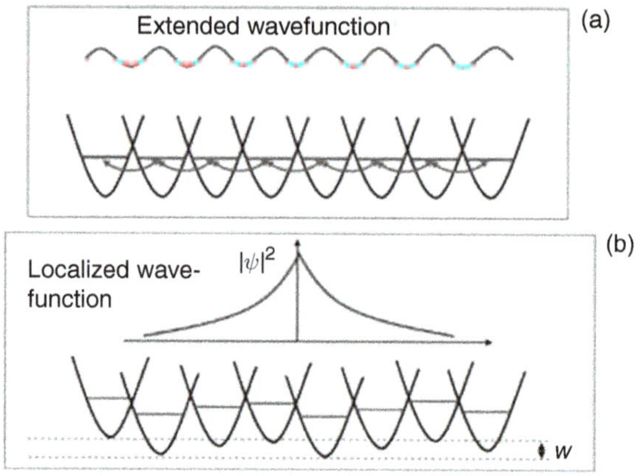

Figure 10.1 Anderson localization in the tight binding model. Starting from bound wavefunctions pinned on each potential well of a regular lattice, and adding tunnelling between neighbour sites, one obtains a solution of the Schrödinger equation that is an extended state, i.e., a freely propagating wave. If one adds disorder to the lattice by changing the depth of the various wells randomly, the solution may become localized, i.e., a wavefunction with an exponentially decaying profile in all three dimensions of space (the figure is drawn in 1D for clarity). This happens when the amplitude W of the disorder is large enough.

happens even if the electron has still the possibility of tunnelling between neighbour sites of the lattice: when one considers the various possible paths to go from one site to another one, the various quantum amplitudes associated with all these paths cancel when they are added. It must be emphasized that this is a quantum interference effect involving *a single quantum particle*. The various amplitudes associated with the various possibilities of going from one place to another belong to the same electron.

After this initial conjecture based on the tight binding model of hopping electrons, an equivalent model based on wave physics emerged, to understand the existence of a conductor–insulator transition for a certain level of disorder.[1] The starting point of this alternative approach is the equivalence between the running Bloch wave describing a particle freely propagating in a conduction band of a perfect crystal, and a matter wave freely propagating in a homogeneous non-absorbing medium. Adding disorder to the latter model can be done by introducing impurities, on which the matter waves scatter (Figure 10.2a). This leads to a diffusive propagation, an alternative model of Ohm conduction, based on the assumption that diffusion is incoherent and that one can add the intensities of the wavelets scattered on impurities. In contrast, when one adds the amplitudes of the scattered waves, as should be done for coherent waves, one can find Anderson localization, resulting from a destructive interference between wavelets multiply scattered onto impurities and the initial wave in the forward direction (Figure 10.2a) or towards any other direction. For this to happen,

[1] It is not easy to point out when exactly this equivalence emerged. It is implicitly present in Mott's remarkable review paper [7].

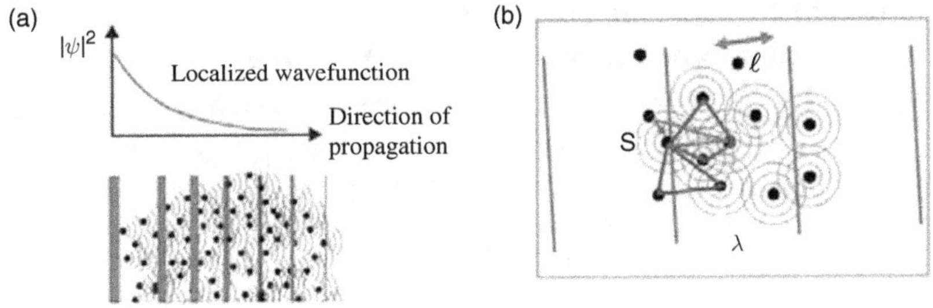

Figure 10.2 Exponential localization of waves in a medium with random scatterers. (a) When a plane wave tries to propagate in a medium with many scatterers, the various rescattered wavelets interfere in the forward direction and add coherently to the 'initial' wave. This may result in an exponential decay of the wavefunction, so that one has a localized state. As suggested in (b), localization happens when many paths of multiple scattering, starting from a scatterer S, return to S in phase. In 3D, where the probability to return at the origin is weak, this happens if the mean free path ℓ between scatterers is less than the wavelength (the figure is drawn in 2D for clarity).

an intuitive condition is that there is a constructive interference for amplitudes associated with multiple scattering paths returning to a scatterer S. This is realized when the mean free path ℓ between two scattering events is smaller than the wavelength λ of the wave, viz. when the dephasing between two scattering events remains small (Figure 10.2b). Introducing the wavevector $k = 2\pi/\lambda$, this condition can be written as $k\ell \leq$ constant ~ 1, which is known as the 'Ioffe–Regel criterion'. This condition, introduced by Mott [7] in the context of 3D Anderson localization, stresses the existence of a threshold—a so called 'mobility edge'—at which the transition from 'localized' to 'extended' happens. Actually, in 2D and 1D, the situation is different (localization happens more easily, see later), because the probability of returning to the origin in a 1D random walk is much larger than in 2D or 3D.

More than 30 years after Anderson's Nobel lecture, and in spite of its wide acceptance, the exact status of the Ioffe–Regel criterion is still not fully clear. Firstly, rigorous derivations of its existence are based on the scaling theory developed by the famous 'gang of four' [8], but they do not yield the value of the constant of the Ioffe–Regel condition. Secondly, the situations in dimensions lower than three have many intriguing features. For instance there is a general theorem that 'in 1D all states are localized', so that there seems to be, a priori, no mobility edge to separate a regime of localized states from a regime of extended states, but as we will see the situation may be more subtle. Dimension two is a marginal case, and as such very interesting from a theoretical point of view [8], and with several open issues. To shed light on these issues, experimental observation of Anderson localization has been looked for in various domains of physics, with electrons in condensed matter physics, naturally, but also in wave physics [9] and in particular with light waves [10; 11; 12; 13], microwaves [14; 15], acoustic waves [16]. Here, we show that ultracold atoms placed in an optical disordered potential form an excellent system to study Anderson localization experimentally.

10.2 Ultracold atoms in optical speckle: a good candidate for the observation of Anderson localization

Anderson was referring to electrons in his original paper, but a direct observation of Anderson localization for electrons is very difficult. Firstly, it is hindered by a number of phenomena that can mask the single-particle quantum effects genuinely induced by disorder: electrons interact with each other by Coulomb electric repulsion; they are affected by the vibrations of the lattice, which is a kind of a disorder that is not static ('quenched' in the usual jargon) and therefore cannot produce AL. Secondly, direct observation of electron wavefunctions in solids is very difficult, and most of the evidence is indirect and stems from conductivity measurements.

In contrast, ultracold atoms allow one to address the core of the phenomenon that Anderson had discovered, since they are genuine quantum particles described as matter waves, and interactions can be reduced to a negligible level so that one can study single particle behaviour. Moreover, with ultracold atoms, single-atom matter waves can be directly visualized by absorption or fluorescence imaging of the atomic density of a dilute, non-interacting, Bose–Einstein condensate. Indeed, such an ideal quantum gas is nothing other than many atoms in the same one-atom wavefunction, which enables the direct observation of that wavefunction density.[2] It then becomes possible to realize experimentally, and investigate, an AL situation, provided that one can generate the adequate disordered potential in which to place the ultracold atoms [19; 20]. This is made possible by using a disordered light pattern, in which atoms are submitted to a dipolar potential directly proportional to the light intensity. If the detuning between the light and the atomic resonance is large enough, there is no other force, and the atoms are submitted to a pure potential and experience no dissipation phenomenon [19]. This is in contrast to experiments with light propagating in media with a high level of scattering, where it is difficult to avoid absorption or ordinary (incoherent) scattering [21], and sophisticated tests must be performed to ascertain that an observed exponential decay is due to genuine AL [15].

We use a disordered potential created by laser speckle, which has remarkable properties. The optical speckle field is produced by passing a laser beam through a diffusing plate that imposes a phase depending randomly on position. As a result, the complex electric field describing the light at any point in the far field can be considered a sum of many independent random variables, and it is thus a Gaussian random process [22]. Note, however, that the light intensity, which is the squared modulus of the complex electric field, is not a Gaussian process. For instance, its probability distribution is a decaying exponential, i.e., a non-symmetric distribution. Nevertheless, it inherits many properties of the underlying Gaussian process, and in particular all the intensity–intensity correlation functions, like all correlation functions of the complex electric field of order larger than two, can be expressed as the second-order correlation function of the electric field, which in turn is nothing other than the diffraction pattern associated with the aperture at the location of the diffusive plate. For instance, in the

[2] Observation of the momentum distribution of ultracold atoms—i.e., of the density of the p-space wavefunction—in a 'kicked rotor' system, has allowed observation and study of the related phenomenon of dynamical localization [17; 18], that is localization in p-space.

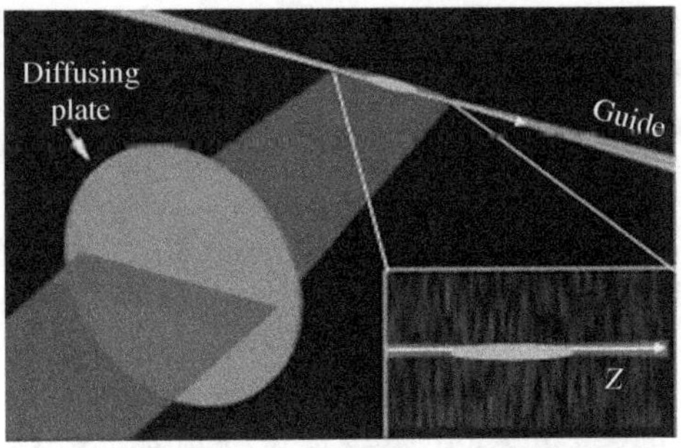

Figure 10.3 Ultracold atoms in a one-dimensional speckle potential. The atoms are held by a matter wave guide, which confines them transversely to the z-axis, but lets them move freely along z. A laser beam passing through an aperture in a diffusing plate elongated along z, and narrow along x, creates a laser speckle, i.e., a disordered intensity pattern, varying quickly along z, and smoothly perpendicular to z. This realizes a situation where particles are submitted to a 1D disordered potential along z.

configuration of Figure 10.3 that we have used to study 1D AL, the scattering aperture is elongated along the z-axis, and narrow along the x-axis, and this results in a speckle pattern, whose 3D spatial autocorrelation function is narrow along z and wide along x (the direction of the narrow side of the aperture) and y (the direction of propagation of the speckle laser). In the experiment of [3], the autocorrelation function of the disordered potential has a half-width of 97 µm along x, 10 µm along y, and 0.82 µm along z. More precisely, the disordered potential autocorrelation function along x is $V_R^2[\sin(\Delta z/\sigma_R)/(\Delta z/\sigma_R]^2$, where the standard correlation length $\sigma_R = 0.26 \pm 0.03$ µm corresponds to a half width $\pi\sigma_R = 0.82$ µm. Such a rapidly varying speckle is obtained with large numerical aperture optics. The amplitude V_R of the disorder is directly proportional to the laser intensity, and the calibration factor is calculated knowing the geometry of the optical system and the properties of ^{87}Rb atoms [19].

Actually, the speckle autocorrelation function has a remarkable property that plays a crucial role. Since the speckle is produced by diffraction, its autocorrelation function has a Fourier transform with a finite support, the size of which is related to the numerical aperture of the optics. In other words, the random process describing the disordered potential has a power spectrum limited by a maximum spatial frequency. In the case of the rectangular aperture discussed here, the disordered potential has no component with a spatial frequency larger than $2/\sigma_R$.

The properties of the disordered potential created by laser speckle have been carefully checked using the atoms themselves as a local probe of the potential [19].

10.3 One-dimensional Anderson localization?

In the experiment sketched in Figure 10.3, the atoms are guided along z by a strong and narrow laser beam, acting as a matter wave guide with a typical diameter of only a few micrometres. They are transversely confined, and cold enough to be in the lowest state of the transverse trapping potential. On the other hand, they can freely move along z. When the anisotropic laser speckle described above is applied, the guided atoms thus experience a potential rapidly varying randomly along z, but transversely invariant (at the scale of their confinement). We are then in a situation allowing us to study Anderson localization in one dimension.

But what is the point of studying 1D Anderson localization? According to the scaling theory of localization [8], there is always localization in 1D. This means that whatever the (non-trivial) potential, for each energy E of a particle of mass M (parameterized by a quantity with dimension of a wavevector, $k = (2ME)^{1/2}/\hbar$) there is a solution of the Schrödinger equation that is a localized wavefunction. At first sight, there is thus no point in addressing the question of 1D Anderson localization. However, when we considered the question of ultracold atoms placed in a 1D optical speckle disorder, we found a yet unexplored approach to it, which allowed us to point out several fundamental features of AL in such a situation [23], and to explore them experimentally [3], as I explain now.

Many theoretical studies of AL of waves are based on a model of point-like scattering impurities described as randomly positioned Dirac peaks, i.e., infinitely high and infinitely narrow potential peaks. Such a disorder is called an *uncorrelated disorder*, i.e., a disorder with an autocorrelation function that is a Dirac peak. In contrast, a laser speckle potential, as drawn in Figure 10.4a,b, has randomly positioned peaks of finite height and finite width, a situation (*correlated disorder*) different in many aspects. Firstly, the peak height distribution is a decaying exponential, which means that it is exponentially improbable to find peaks with a height many times the average value V_{R} of the potential, and in a finite size sample there is a maximum value $V_{\max}$ of the potential. Secondly, the random potential has no spatial variation more rapid than the typical size speckle of a speckle grain (or more precisely no Fourier component with a spatial frequency larger than $2/\sigma_{\mathrm{R}}$). These two properties lead to dramatic consequences.

Let us consider the first property, that in a finite size sample L there will be no peak with a height larger than a maximum value $V_{\max}$ of the potential. We can then distinguish two regimes. If we take a particle with an energy E significantly smaller than $V_{\max}$, it is very likely that we will find two peaks larger than the particle energy in the finite-size sample, and there is then a trivially localized state, corresponding to the classical trapping of a particle bouncing between two potential barriers (Figure 10.4a). But what happens if the particle has an energy E larger than $V_{\max}$, so that a classical particle would propagate from one end of the sample to the other, without any blocking? A numerical solution of the Schrödinger equation in such a weak disorder shows that one can have an exponentially localized wavefunction (Figure 10.4b). One can interpret this localization as resulting from the interference between the many wavelets scattered at the various peaks of the speckle potential, as sketched in Figure 10.4b. In contrast to the trivial case of a particle trapped between two peaks, it is a pure quantum effect of the kind considered by Anderson.

To better understand that quantum localization effect, we can think of the disordered potential as being composed of many standing-wave Fourier components, each characterized by a spatial frequency k, on which a particle with momentum $p = hk/2$

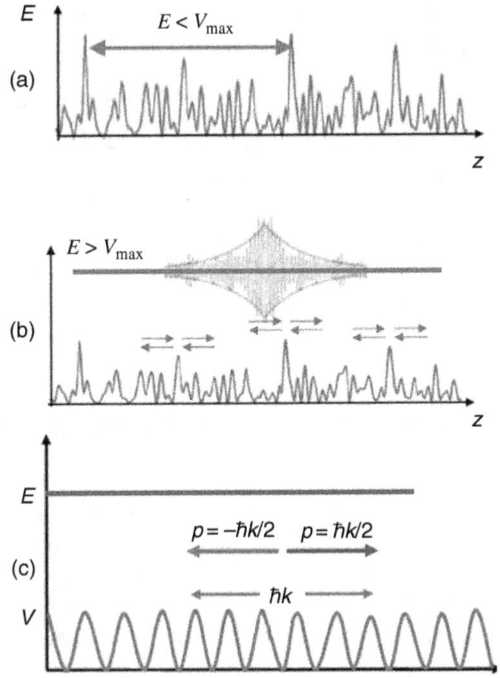

Figure 10.4 Localization of a particle in a 1D laser speckle disordered potential. (a) For an energy smaller than the highest peaks of the potential, localization results from bouncing between two peaks, a trivial classical effect. (b) If the energy is larger than the highest peak, one finds a solution of the Schrödinger equation that is an exponentially localized wavefunction, resulting from the addition of many wavelets due to scattering on the peaks. This is a fully quantum localization phenomenon, equivalent to Anderson localization. (c) If one considers a weak periodic potential of spatial frequency k, a particle with momentum $p = \hbar k/2$ cannot propagate, and is exponentially localized, because it is Bragg-reflected, even if the potential is much weaker than the kinetic energy $\hbar^2 k^2/8M$ of the particle (where M is its mass). This phenomenon can be interpreted as resulting from an interference effect of the many wavelets scattered from the periodic structure. One can generalize the reasoning and consider the localization of Figure (b) as being due to Bragg diffraction of each momentum component on the corresponding periodic component of the disordered potential.

is backscattered by Bragg reflection on the standing wave. This can be interpreted as being due to the rescattering of a photon, from the counterpropagating running-wave component of the standing wave, towards the co-propagating component. The particle then has no possibility of propagating (Figure 10.4c). Again, this is a fully quantum effect, resulting from the interference of the many wavelets scattered from the structure. But there is more. When we take into account the second property of a speckle-disordered potential (no Fourier component with a spatial frequency larger than $2/\sigma_{\mathrm{R}}$), we conclude that there is a cut-off value $k_{\mathrm{co}} = 1/\sigma_{\mathrm{R}}$, such that a matter wave with a momentum larger than k_{co} will not localize. We thus have an *effective mobility edge*, separating the regime of Anderson localization from the regime of free propagation.

To conclude this section, in a 1D disordered potential produced by a laser speckle, we can predict a localization phenomenon that has the characteristic features of AL: (i) it happens in a regime where there is no classical localization, and it is due to interference between the many wavelets scattered on the disordered potential; (ii) there is an effective mobility edge separating a regime of localization (localized states) from a regime where the atom can propagate freely over the whole sample of finite size (extended states) [23]. It is then tempting to investigate experimentally such a situation.

10.4 Direct observation of Anderson localized 1D wavefunctions

Our experiment (sketched in Figures 10.3 and 10.5), starts with a small elongated BEC (1.7×10^4 atoms of ^{87}Rb), trapped in an anisotropic optomagnetic hybrid trap, with trapping frequencies $\omega_z/2\pi = 5.4\,\text{Hz}$ and $\omega_{x,y}/2\pi = 70\,\text{Hz}$. The transverse and longitudinal Thomas Fermi radii are $3\,\mu\text{m}$ and $35\,\mu\text{m}$, respectively, and the chemical potential is $\mu_{\text{ini}}/h = 219\,\text{Hz}$, where h is the Planck constant). The transverse trapping is provided by a far-detuned laser beam (wavelength $1.06\,\mu\text{m}$, to be compared with the resonant wavelength of $0.78\,\mu\text{m}$) in rubidium. It creates an optical waveguide along the horizontal z-axis. A shallow magnetic trap confines the BEC in the longitudinal direction. It is suddenly switched off at $t = 0$, and the BEC begins expanding along z in the waveguide, under the effect of the initial repulsive interaction energy associated with the chemical potential μ_{ini}. A weakly expelling magnetic field compensates the residual longitudinal trapping of the optical waveguide, so that the atoms can freely expand along z over several millimetres. The expanding BEC can be imaged at any chosen time t after release by suddenly switching off the optical guide and irradiating the atoms with a resonant probe of duration $50\,\mu\text{s}$. An ultrasensitive EMCCD camera allows us to make an image of the fluorescing atoms with a resolution of $15\,\mu\text{m}$ and a 1D atomic density sensitivity close to $1\,\text{atom}/\mu\text{m}$. We can then follow the ballistic expansion of the atoms, and check, by looking at the propagation of the forefront of the successive density profiles, that the largest k vector in the expanding atoms corresponds to the initial chemical potential, i.e., $(2M\mu_{\text{ini}})^{1/2}/\hbar$, with M the atomic mass.

When we switch off the longitudinal trapping in the presence of weak disorder, the BEC starts expanding, but the expansion rapidly stops, in stark contrast with the free expansion case. A plot of the density profile, in linear and semilogarithmic coordinates (Figure 10.5c,d), then shows clear exponential wings, a signature of Anderson localization. This observation has been made in a regime allowing AL. Firstly, the disorder is weak enough ($V_{\text{R}}/\mu_{\text{ini}} = 0.12$) that the initial interaction energy per atom is rapidly converted into a kinetic energy of the order of μ_{ini} for atoms in the wings, a value much larger than the amplitude of the disordered potential, so that there is no possibility of a classical reflection on a potential barrier, as in Figure 10.4a. Secondly, the atomic density in the wings of the profiles is small enough (two orders of magnitude

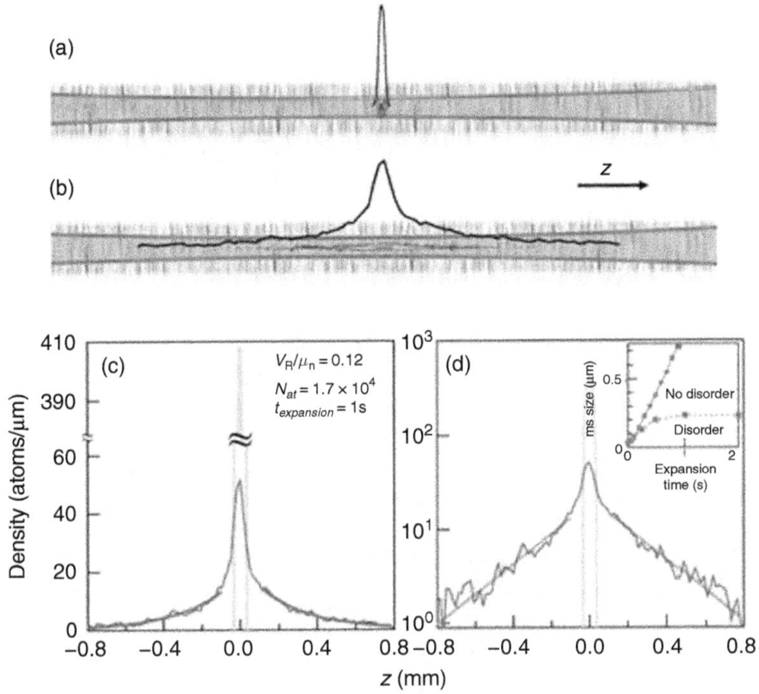

Figure 10.5 Observation of exponential localization. (a) A small BEC (1.7×10^4 atoms) is formed in a hybrid trap, which is the combination of a horizontal optical waveguide, ensuring a strong transverse confinement, and a loose magnetic longitudinal trap. A weak disordered optical potential (disorder amplitude V_R small compared with the chemical potential μ_{ini} of the atoms in the initial BEC), transversely invariant over the atomic cloud, is superimposed (see Figure 10.3). (b) When the longitudinal trap is switched off, the BEC starts expanding and then localizes, as observed by direct imaging of the fluorescence of the atoms irradiated by a resonant probe. On (a) and (b), images and sketched profiles are for illustration, they are not to scale. (c,d) Observed density profile of the localized BEC, 1 s after release, in linear or semilogarithmic coordinates. The inset of (d) (rms width of the profile versus time, with or without disordered potential) shows that the stationary regime is reached after 0.5 s. Solid lines in (c) are exponential fits to the wings, corresponding to the straight lines of (d). The narrow profile drawn at the centre represents the trapped condensate before release.

less than in the initial BEC) that the interaction energy is negligible compared with the atom kinetic energy, and we can consider that we have many non-interacting atoms described by the same wavefunction. Thirdly, the atomic matter-wave k-vector distribution is bounded, with a maximum value k_{max} smaller than half the cut-off in the spectrum of the speckle disordered potential used here, i.e., $k_{max} \cdot \sigma_R < 1$. Indeed, for the observations of Figures 10.5 and 10.6, we have $k_{max} \cdot \sigma_R = 0.65 \pm 0.09$, and all the the k-vector components are below the effective mobility edge.

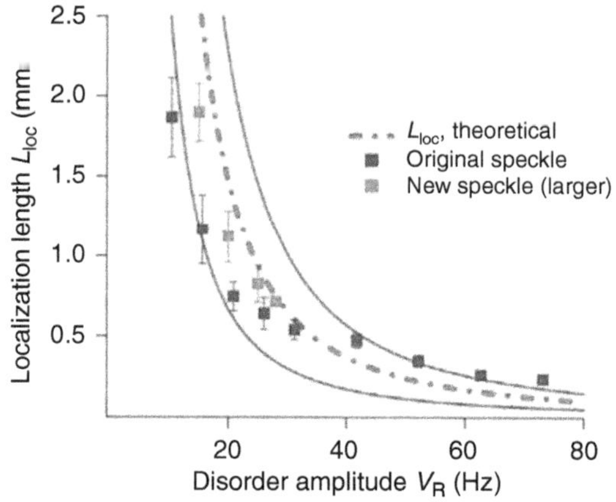

Figure 10.6 Localization length versus amplitude of the disordered potential. The measured localization length L_{loc} is obtained by an exponential fit to the wings of the stationary localized density profiles, as shown in Figure 10.5c,d. The error bars correspond to a confidence level of 95% for the fit. The number of atoms is 1.7×10^4 atoms ($\mu_{ini}/h = 219\,\text{Hz}$). The dash-dotted line represents the theoretical prediction as shown in the displayed equation, where k_{max} is determined from the observed free expansion of the condensate. The two solid lines are the limits of the uncertainty in applying the equation, associated with the uncertainty in k_{max} and V_R. The two sets of data correspond to two different diffusive plates that produce speckle patterns with different homogeneities. When the homogeneous region is broader, the measured values are closer to the dash-dotted line in the case of a long localization length.

An exponential fit to the wings of the density profiles yields the localization length L_{loc}, which we can compare with the theoretical value [23]

$$L_{loc} = \frac{2\hbar^4 k_{max}^2}{\pi M^2 V_R^2 \sigma_R (1 - k_{max} \cdot \sigma_R)}.$$

This expression, based on first-order Born approximation, is valid for $k_{max} \cdot \sigma_R < 1$ and $V_R/\mu_{ini} \ll 1$. In Figure 10.6, we plot the variation of L_{loc} with the amplitude of the disorder, V_R, for the conditions indicated previously. The dash-dotted line is a plot of the displayed equation for the values of k_{max} and σ_R determined as explained already. It shows a good agreement between our measurements and the theoretical predictions. The two solid lines reflect the maximum variations of the dash-dotted line when we take into account the uncertainties on V_R and k_{max}.

An intriguing result of [23] is the prediction of density profiles with algebraic wings when $k_{max} \cdot \sigma_R > 1$, i.e., when the initial interaction energy is large enough that a fraction of the atoms have a k-vector larger than the effective mobility edge $1/\sigma_R$. This predicted behaviour results from the sum of the many exponentially localized wavefunctions with different localization lengths, up to the effective mobility edge where the

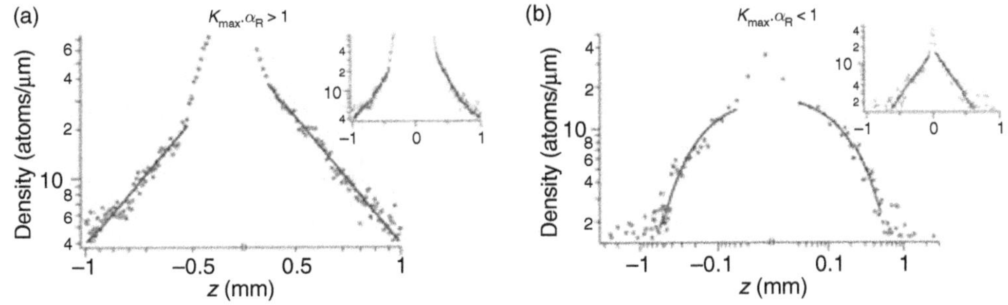

Figure 10.7 Log-log and semilogarithmic plots of the stationary atom density profiles showing the difference between the algebraic ($k_{\max} \cdot \sigma_R > 1$) and the exponential ($k_{\max} \cdot \sigma_R < 1$) regimes. (a) Density profile for $V_R/\mu_{\text{ini}} = 0.15$ and $k_{\max} \cdot \sigma_R = 1.16 \pm 0.14$. The momentum distribution of the released BEC has components beyond the effective mobility edge $1/\sigma_R$. The fit to the wings with a power law decay $1/z^\beta$ yields $\beta = 1.92 \pm 0.06$ for the left wing and $\beta = 2.01 \pm 0.03$ for the right wing. The inset shows the same data in semilogarithmic plot, and confirms the non-exponential decay. (b) For comparison, similar set of plots (log-log and semilogarithmic) in the exponential regime ($V_R/\mu_{\text{ini}} = 0.15$, $k_{\max} \cdot \sigma_R = 0.65 \pm 0.09$) show a clear exponential, non-algebraic, decay.

localization length is infinite. Atoms associated with a k-vector beyond $1/\sigma_R < 1$ eventually escape. We have investigated that regime by repeating the experiment with a BEC containing a larger number of atoms (1.7×10^5 atoms), i.e., a larger initial interaction energy ($\mu_{\text{ini}}/h = 519\,\text{Hz}$), so that a fraction of the atoms have a momentum larger than the effective mobility edge ($k_{\max} \cdot \sigma_R = 1.16 \pm 0.14$). Note that we are still in the weak disorder regime ($V_R/\mu_{\text{ini}} = 0.15$). Figure 10.7a shows a log-log plot of the observed density profile in such a situation, suggesting a power-law decrease in the wings, with an exponent of 1.95 ± 0.10, in agreement with the theoretical prediction of wings decreasing as $1/z^2$. The semilogarithmic plot in the inset confirms that an exponential fit would not work as well. To allow comparison, Figure 10.7b shows a log-log plot and a semilogarithmic plot for the case $k_{\max} \cdot \sigma_R = 0.65 \pm 0.09$ ($V_R/\mu_{\text{ini}} = 0.15$), where we unambiguously conclude in favour of exponential rather than algebraic tails. These data support the existence of a *crossover* from exponential to algebraic regime for a BEC in a speckle potential.

10.5 What happens beyond the 1D effective mobility edge?

The analysis leading to the existence of an effective mobility edge is based on the (lowest-order) Born approximation, and one may wonder what a more exact approach predicts beyond the effective mobility edge. In the case of a speckle disorder, a perturbative expansion to higher orders is possible, since high-order correlators of the potential can easily be calculated. Indeed, for the Gaussian random process describing the light electric field whose squared modulus yields the potential, one can express

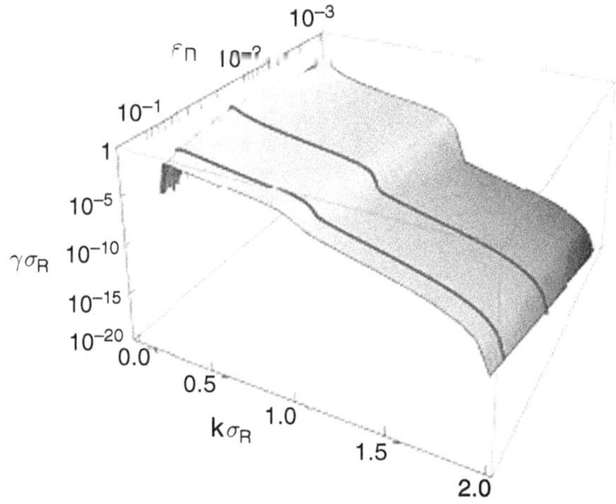

Figure 10.8 Inverse localization length L_{loc} (Lyapunov exponent) calculated two orders beyond the Born approximation for particles in 1D speckle potentials. The speckle (with a correlation length σ_R) is created with a square diffusive plate. The particle momentum $\hbar k$ and the strength of disorder are indicated in reduced units (respectively $k\sigma_R$ and $\varepsilon_R = 2M\sigma_R^2 V_R/\hbar^2$, where V_R is the amplitude of the disorder). The solid lines correspond to $\varepsilon_R = 0.1$ and $\varepsilon_R = 0.02$, and result from numerical calculations.

any higher-order moment as a function of the second-order moment, which is easily calculated by the diffraction theory. A calculation up to order four, as well as a diagrammatic approach, and a numerical calculation, yield interesting results [24]. These are shown on Figure 10.8, where one plots the inverse localization length versus the disorder amplitude V_R and the k-vector associated with the energy of the particle. At the effective mobility edge given by $k \times \sigma_R = 1$, the inverse localization length rapidly drops by several orders of magnitude, so that the localization length becomes long enough that localization is not observable in a system with a finite size smaller than that localization length. From a fundamental point of view, however, there is still localization, in agreement with the general result of the scaling theory that there is always localization in a one-dimensional problem. One observes on the plot that the sharpness of the crossover increases when the amplitude of the disorder decreases, as expected for an expansion where ε_R is the small parameter (see its definition in the figure caption). The figure also shows that there is a second effective mobility edge at $k \times \sigma_R = 2$. This can be interpreted by invoking second-order Bragg diffractions, corresponding to the rescattering, by the moving atom, of two photons from a counter-propagating running component into a co-propagating running component. The result is readily generalized to higher orders, and one predicts the existence of successive effective mobility edges at $k \times \sigma_R = 3, 4$, etc.

Another interesting result of the calculation at the fourth order (in the potential, or more precisely in ε_R) is that, in contrast with the result of the first-order calculation, there is a difference between the localization lengths calculated for a repulsive or attractive speckle potential. This difference results from the third-order term, which is odd in V_R.

10.6 Towards 2D and 3D experimental studies of AL: a quantum simulator with cold atoms

In contrast with the 1D case, where it is possible to solve the problem exactly by standard techniques, the question of Anderson localization in higher dimensions is far from being solved. The scaling theory [8] predicts localization in 3D, for energies smaller than a value associated with Ioffe–Regel criterion, i.e., there is now a true mobility edge, associated with a phase transition. But there is no exact theory giving the exact value of the mobility edge, which is given by the condition $k\ell \leq$ constant ~ 1, where the constant has to be evaluated. Only approximate theories exist, permitting the calculation of that constant, as well as the critical exponents associated with the phase transition. A popular and quite efficient such approximate theory, called the 'self-consistent theory of localization' [25], allows one to calculate these values, and definite results do exist [26; 27]. An experiment allowing one to measure such quantities would then constitute a test of the validity of the self-consistent theory, or of alternative theories yielding precise results.

According to the scaling theory, the phenomenon of Anderson localization dramatically depends on the dimension d of the situation. The 2D case is the marginal situation separating cases where there is always localization ($d < 2$), from cases where a true mobility edge does exist ($d > 2$). Actually, there is always localization in 2D, in principle, but the self-consistent theory predicts that the localization length becomes very large as soon as one crosses the value of energy associated with the Ioffe–Regel criterion. As a consequence, in a finite size system, there is an effective transition between the localized situation (localization length smaller than the size of the system) and the effectively non-localized situation (localization length larger than the size of the system). Here also, a quantitative comparison with the predictions of the theory will constitute a test of the theory.

We have thus embarked into developing apparatuses to study localization of ultracold atoms in a speckle disorder, in 2D and 3D. Figure 10.9 shows our 3D apparatus, which incorporates two important ingredients. Firstly, thanks to a magnetic field with a well-chosen vertical gradient, we can suspend paramagnetic atoms against gravity in a well-chosen magnetic sublevel. The coils have been designed to minimize the inevitable expelling component of the field, and suspended rubidium atoms are submitted to an expelling harmonic potential characterized by an angular frequency of $2\pi\,\mathrm{s}^{-1}$. Secondly, we apply two crossed laser speckle fields, mutually coherent, resulting in an anisotropic 3D speckle pattern, characterized by an autocorrelation function with a half-width, depending on the direction, with an average value of $\sigma_R = 0.4\,\mu\mathrm{m}$.

Figure 10.10 shows our 2D apparatus [28]. Here, the atoms are vertically maintained by a blue detuned sheet of light with a horizontal node, providing a vertical confinement. They can freely move in the horizontal plane. We apply a speckle field

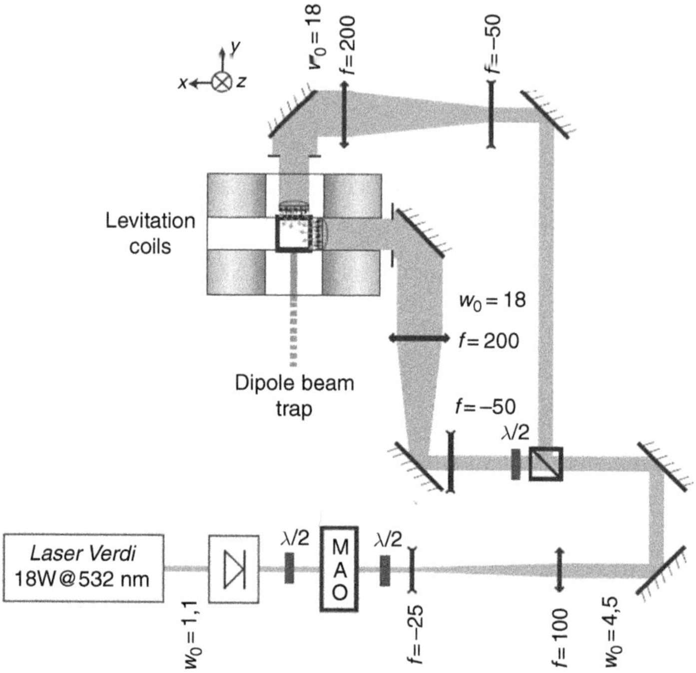

Figure 10.9 Experimental set-up to study 3D localization of ultracold atoms in a 3D laser speckle.

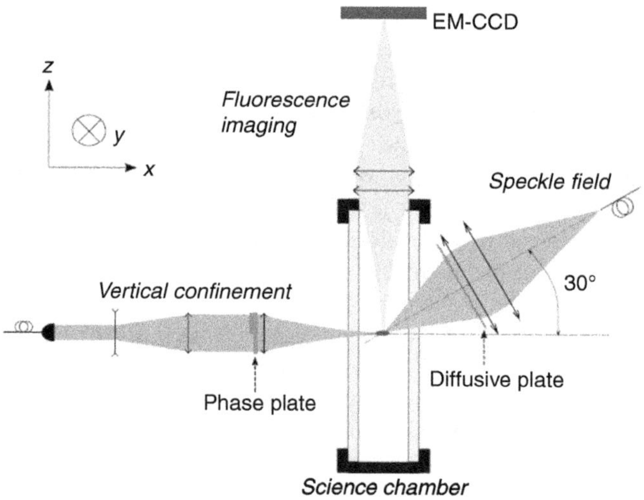

Figure 10.10 Experimental set-up to study 2D localization of ultracold atoms in a 2D laser speckle.

at an angle, so that the atoms are submitted to an anisotropic 2D disorder, with half-widths of $\sigma_y = 0.8\,\mu\text{m}$ and $\sigma_x = 1.6\,\mu\text{m}$. This apparatus has already allowed us to study the (classical) 2D diffusion of cold atoms in an anisotropic disorder [28].

Observation of Anderson localization in 2D and 3D is, a priori, difficult. Indeed, predictions based on the self-consistent theory show that, in contrast to the 1D case, a not-too-large localization length (i.e., small compared with the size of the experiment) demands an average disordered potential V_R of the order of the energy of the atoms to localize. One is then in a situation where, in a classical picture, the atoms do not 'fly' over the disordered potential, but rather follow trajectories around the peaks of the potential. This does not prevent the possibility of a quantum localization due to interferences between the amplitudes associated with different trajectories from one point to another. But it raises two important problems.

Firstly, to observe non-trivial localization, i.e., a localization that cannot be understood as a classical trapping in a potential well, one must use atoms with an energy larger than the classical percolation threshold. In our 2D apparatus, this percolation threshold is a significant fraction of V_R, and one is left with a narrow energy window to perform the experiment. One then has to face the second problem, which is the fact that for a suddenly applied disordered potential, momentum eigenstates have a spread in energy (the width of the so-called spectral function) of the order of V_R, so that only a fraction of the atoms will be localized, and the resulting localized cloud will be the addition of many localized wavefunctions with a broad range of localization lengths. This means that even if localization is observed, comparison to the theory will demand strong theoretical efforts. The situation is analogous in 3D [26]. It is more favourable, however, regarding the percolation threshold, which is negligible (less than 1%) compared to V_R.[3]

When we observe localization of non-interacting ultracold atoms in 2D or 3D, this will be a crucial test of the validity of the theories of Anderson localization. In such a situation, where a full quantum treatment does not exist, and only approximate theories permit us to make predictions, it seems legitimate to consider the experiment as a quantum simulator. This will be even more true when we implement controlled interactions, allowing us to test experimentally how interactions modify Anderson localization, the exact treatment of which is a formidable task for theorists.

Acknowledgements

The work presented here results from a tight collaboration between the theoretical team (L. Sanchez-Palencia, P. Lugan, L. Pezze, and M. Piraud) and experimental teams (P. Bouyer, D. Clément, V. Josse, A. Bernard, J. Billy, P. Cheinet, Z. Zuo, T. Bourdel, B. Allard, J.-P. Brantut, T. Plisson, and M. Robert-de-Saint-Vincent) of the Atom Optics group of the Institut d'Optique at Palaiseau. No experiment would work without our electronics wizards A. Villing and F. Moron.

[3]Note added in proof: we have indeed observed 3D AL with that apparatus: F. Jendrzejewski, A. Bernard, K. Muller, P. Cheinet, V. Josse, M. Piraud, L. Pezze, L. Sanchez-Palencia, A. Aspect, and P. Bouyer. (2012) Three-dimensional localization of ultracold atoms in an optical disordered potential, *Nature Physics*, **8**, 398.

I have learnt a lot from discussions with many physicists, who would be too many to list, but a special mention must be made of P. Chavel, T. Giamarchi, M. Lewenstein, G. Shlyapnikov, and B. Altshuler. I am also happy to acknowledge constructive and friendly exchanges with M. Inguscio and his group.

This research was supported by the French Centre National de la Recherche Scientifique (CNRS), the Ministére de l'Education Nationale, de la Recherche et de la Technologie (MENRT), the Délégation Générale de l'Armement (DGA), the Agence Nationale de la Recherche (ANR), IXSEA, the FINAQS programme of the European Union, the QUDEDIS programme of ESF, and last, but not least, by the Institut Francilien de Recherche sur les Atomes Froids (IFRAF).

References

1. PW Anderson, *Phys. Rev.* **109**, 1492(1958): Absence of diffusion in certain random lattices.
2. PW Anderson, *Rev. Mod. Phys.* **50**, 191 (1978): Local moments and localized states.
3. J Billy, V Josse, Z Zuo, *et al.*, *Nature* **453**, 891 (2008): Direct observation of Anderson localization of matter-waves in a controlled disorder.
4. G Roati, C D'Errico, L Fallani, *et al.*, *Nature* **453**, 895 (2008): Anderson localization of a non-interacting Bose–Einstein condensate.
5. A Aspect and M Inguscio, *Phys. Today* **62**, 30(2008): Anderson localization of ultracold atoms.
6. NW Ashcroft and ND Mermin, *Solid State Physics*, Saunders (1976).
7. NF Mott, *Adv. Phys.* **16**, 49 (1967): Electrons in disordered structures.
8. E Abrahams, PW Anderson, DC Licciardello and TV Ramakrishnan, *Phys. Rev. Lett.* **42**, 673 (1979): Scaling theory of localization: absence of quantum diffusion in two dimensions.
9. BA Van Tiggelen, *in Diffuse Waves in Complex Media*, Nato Advanced Science Institutes Series, Series C, Mathematical and Physical Sciences, Vol. 531 (JP Fouque, ed.), 1999, p. 1: Localization of waves.
10. DS Wiersma, P Bartolini, A Lagendijk, and R Righini, *Nature* **390**, 671 (1997): Localization of light in a disordered medium.
11. M Störzer, P Gross, CM Aegerter, and G Maret, *Phys. Rev. Lett.* **96**, 063904 (2006): Observation of the critical regime near Anderson localization of light.
12. T Schwartz, G Bartal, S Fishman, and M Segev, *Nature* **446**, 52 (2007): Transport and Anderson localization in disordered two-dimensional photonic lattices.
13. Y Lahini, A Avidan, F Pozzi, *et al.*, *Phys. Rev. Lett.* **100**, 013906 (2008): Anderson localization and nonlinearity in one-dimensional disordered photonic lattices.
14. R Dalichaouch, JP Armstrong, S Schultz, PM Platzman, and SL McCall, *Nature* **354**, 53 (1991): Microwave localization by 2-dimensional random scattering.
15. AA Chabanov, M Stoytchev, and AZ Genack, *Nature* **404**, 850 (2000): Statistical signatures of photon localization.
16. H Hu, A Strybulevych, JH Page, SE Skipetrov, and BA van Tiggelen, *Nat. Phys.* **4** 945 (2008): Localization of ultrasound in a three-dimensional elastic network.

17. JL Moore, JC Robinson, C Bharucha, PE Williams, and MG Raizen, *Phys. Rev. Lett.* **73**, 2974 (1994): Observation of dynamical localization in atomic momentum transfer: a new testing ground for quantum chaos.
18. J Chabé, G Lemarié, B Grémaud, D Delande, P Szriftgiser, and J-C Garreau, *Phys. Rev. Lett.* **101**, 255702 (2008): Experimental observation of the Anderson transition with atomic matter waves.
19. D Clément, AF Varon, J Retter, L Sanchez-Palencia, A Aspect, and P Bouyer, *New J. Phys.* **8**, 165 (2006): Experimental study of the transport of coherent interacting matter-waves in a 1D random potential induced by laser speckle.
20. L Fallani, C Fort, M Inguscio, *Adv. At. Mol. Opt. Phys.* **56**, 119 (2008): Bose–Einstein condensates in disordered potentials.
21. F Scheffold, R Lenke, R Tweer, and G Maret, *Nature* **398**, 206 (1999): Localization or classical diffusion of light?
22. JW Goodman, *Speckle Phenomena in Optics*: Roberts Company Publishers (2007).
23. L Sanchez-Palencia, D Clément, P Lugan, P Bouyer, GV Shlyapnikov, and A Aspect, *Phys. Rev. Lett.* **98**, 210401 (2007): Anderson localization of expanding Bose–Einstein condensates in random potentials.
24. P Lugan, A Aspect, L Sanchez-Palencia, *et al.*, *Phys. Rev. A* **80**, 023605 (2009): One-dimensional Anderson localization in certain correlated random potentials.
25. D Vollhardt and P Wolfle, *Phys. Rev. Lett.* **48** 699 (1982): Scaling equations from a self-consistent theory of Anderson localization.
26. SE Skipetrov, A Minguzzi, BA van Tiggelen, and B Shapiro, *Phys. Rev. Lett.* **100**, 165301 (2008): Anderson localization of a Bose–Einstein condensate in a 3D random potential.
27. A Yedjour and BA van Tiggelen, *Eur. Phys. J. D* **59** 249 (2010): Diffusion and localization of cold atoms in 3D optical speckle.
28. M Robert-de-Saint-Vincent, JP Brantut, B Allard, *et al.*, *Phys. Rev. Lett.* **104**, 220602 (2010): Anisotropic 2D diffusive expansion of ultracold atoms in a disordered potential.

The manufacturer's authorised representative in the EU for product safety is Oxford University Press España S.A. of el Parque Empresarial San Fernando de Henares, Avenida de Castilla, 2 – 28830 Madrid (www.oup.es/en or product.safety@oup.com). OUP España S.A. also acts as importer into Spain of products made by the manufacturer.

www.ingramcontent.com/pod-product-compliance
Lightning Source LLC
LaVergne TN
LVHW062112200926
848484LV00007B/1650

* 9 7 8 0 1 9 9 6 6 1 8 8 6 *